AF323295

FLEXOELECTRICITY IN SOLIDS

From Theory to Applications

FLEXOELECTRICITY IN SOLIDS

From Theory to Applications

Editors

Alexander K Tagantsev

Swiss Federal Institute of Technology (EPFL), Switzerland
Ioffe Physical Technical Institute, St. Petersburg, Russia

Petr V Yudin

Swiss Federal Institute of Technology (EPFL), Switzerland
Novosibirsk State University, Novosibirsk, Russia

NEW JERSEY · LONDON · SINGAPORE · BEIJING · SHANGHAI · HONG KONG · TAIPEI · CHENNAI · TOKYO

Published by

World Scientific Publishing Co. Pte. Ltd.

5 Toh Tuck Link, Singapore 596224

USA office: 27 Warren Street, Suite 401-402, Hackensack, NJ 07601

UK office: 57 Shelton Street, Covent Garden, London WC2H 9HE

Library of Congress Cataloging-in-Publication Data
Names: Tagantsev, Alexander K., (Alexander Kirillovich), editor. | Yudin, Petr V., editor.
Title: Flexoelectricity in solids : from theory to applications / Alexander K. Tagantsev
  (Swiss Federal Institute of Technology (EPFL), Switzerland) & Petr V. Yudin
  (Swiss Federal Institute of Technology (EPFL), Switzerland).
Description: New Jersey : World Scientific, 2016. | Includes bibliographical references.
Identifiers: LCCN 2015042588 | ISBN 9789814719315 (hc : alk. paper)
Subjects: LCSH: Solids--Electric properties. | Polarization (Electricity)
Classification: LCC QC176.8.E35 T34 2016 | DDC 530.4/12--dc23
LC record available at http://lccn.loc.gov/2015042588

British Library Cataloguing-in-Publication Data
A catalogue record for this book is available from the British Library.

Desk Editor: Dr. Sree Meenakshi Sajani

Typeset by Stallion Press
Email: enquiries@stallionpress.com

Printed in Singapore by B & Jo Enterprise Pte Ltd

Contents

Chapter 2. First-Principles Theory of Flexoelectricity

31

Massimiliano Stengel and David Vanderbilt

Chapter 3. A Continuum Theory of Flexoelectricity   111

Q. Deng, L. Liu, and P. Sharma

Chapter 4. Mechanical Boundary Conditions for a Case When Thermodynamic Potential Depends on Strain Gradients   **169**

A. S. Yurkov

Chapter 5. Flexoelectric Deformations of Finite-Size Bodies in Framework of the Theory of Continuum   **201**

A. S. Yurkov

Chapter 6.   Flexoelectricity and Phonon Spectra   241

P. V. Yudin, A. Kvasov, and A. K. Tagantsev

Chapter 7.   Impact of Flexoelectric Effect
on Electro-mechanics of Moderate
Conductors   265

A. N. Morozovska, O. V. Varenyk, and S. V. Kalinin

Chapter 8.   Role of Flexoelectricity in Multidomain Ferroelectrics   285

*R. Ahluwalia, A. K. Tagantsev, P. Yudin, N. Setter,
N. Ng, and D. J. Srolovitz*

Chapter 9.   Flexoelectricity Impact on the Domain Wall Structure and Polar Properties   311

A. N. Morozovska, S. V. Kalinin, and E. A. Eliseev

Chapter 10. Bending-Induced Giant Polarization in Ferroelectric MEMS Diaphragm 337

Zhihong Wang and Weiguang Zhu

Chapter 11. Quasi-Amorphous Materials 367

David Ehre, Ellen Wachtel, and Igor Lubomirsky

Preface

During the past decade, flexoelectricity in solids has been attracting appreciable attention from physicists and material scientists. To a great extent, the interest in the flexoelectric effect and related phenomena was triggered by pioneering experimental studies of flexoelectricity in ceramics of perovskite ferroelectrics by Prof. Eric L. Cross and co-authors. Anomalously high values of the flexoelectric coefficients obtained in these materials suggested a possibility for the creation of flexoelectricity-based electromechanical devices. Specifically, flexoelectricity-based piezoelectric meta-materials were fabricated with effective piezoelectric coefficients comparable to those of classical piezoelectrics.

On the theoretical side, the interest was motivated by the experimental work in the field as well as by challenges in the theoretical description of the delicate physics behind the flexoelectric phenomena. All in all, the interest in the flexoelectricity in solids has been steeply growing from year to year as is illustrated by the diagram presented in Fig. 1, which characterizes the publication activity in the field.

Since the flexoelectric effect is allowed by symmetry in non-piezoelectric materials, it suggests a possibility for the creation of electromechanical devices using a much wider variety of materials than just non-centrosymmetric piezoelectrics. The progress in this direction is conditioned by both the experimental studies and theoretical understanding of the obtained results. Here, of importance is to be sure that the theory and experiment are dealing with the

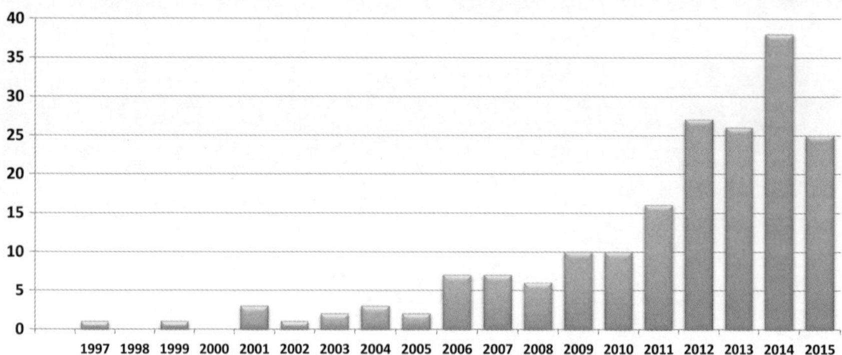

Figure 1.   The number of publications on flexoelectricity in solids per year, up to September 1, 2015. Data from database "webofknowledge.com", search for the keywords "Flexoelectric/flexoelectricity", publications related only to solids are selected.

same phenomenon, which, in view of the general weakness of the flexoelectric effects and the complexity of the physics behind them, might not be a simple task.

The early developments in the field provoked a tendency to oversimplify theoretical explanations of experimentally observed phenomena. When for the first time experimentally and theoretically addressed, the flexoelectric effect was viewed as a simple analogue of its lower-order counterpart, the piezoelectric effect. It was originally termed as "non-local" piezoelectricity, and its theoretical description was believed to boil down to the addition of some extra terms in the bulk constitutive electromechanical equations. However, it was later realized that such a modification of the bulk constitutive equations does not provide an adequate description of the phenomenon and there exist a key difference between the basic properties of piezoelectric and flexoelectric effects. Here, the most remarkable is the impact of the sample surface. The piezoelectric response, as it follows from the theory and has been confirmed by many experimental studies, is, to within a good accuracy (except for extremely thin samples), controlled by the properties of its bulk. On the contrary, the theory of the flexoelectric effect of a finite sample predicts an order-of-magnitude modification of the flexoelectric response of a finite sample by modification of its surface.

Among the other subtleties attesting to the complexity of the flexoelectric phenomena the following two are also remarkable. First, commonly, the flexoelectric effect is interpreted in terms of the so-called static bulk flexoelectric contribution and characterized by flexoelectric or flexocoupling coefficients. However, as was recently shown, the values of the experimentally determined flexocoupling coefficients are incompatible with the stability of the crystalline lattice of the material.

Second, it was believed that in nominally centrosymmetric ceramics, the response of the polarization to an inhomogeneous strain is mainly due to the flexoelectric effect. However, recent studies showed that in these materials a comparable polarization response appears in the case where the applied stain is homogeneous, where the contribution from flexoelectricity is excluded. This is of primary importance because the highest flexoelectric coefficients were reported for nominally centrosymmetric perovskite ceramics.

The big picture of flexoelectricity in solids is available in two recently published review papers: by Zubko *et al.* (2013) [P. Zubko, G. Catalan, and A. K. Tagantsev. *Flexoelectric Effect in Solids.* Volume 43 of *Annual Review of Materials Research*, pp. 387–421. Annual Reviews, Palo Alto (2013)], and by Yudin and Tagantsev (2013) [P. V. Yudin and A. K. Tagantsev. Fundamentals of flexoelectricity in solids. *Nanotechnology.* **24**, p. 432001 (2013)]. At the same time, a systematic in-depth discussion of the flexoelectric phenomena is still missing from the literature. The problem with getting a detailed picture of flexoelectricity in solids is also related to the diversity of methods used by different authors and, moreover, to that of their opinions.

This book is aiming to compile recent advances in both theory and experiment. Except for the first chapter, which gives a simple outlook of the field, the other chapters of the book give in-depth vital aspects of flexoelectricity in solids. The editors did their best to make the chapters of the book compatible with each other. However, the reader will find that the opinions of the authors on the same issue are not always identical. We believe that it is normal for a book on a fast growing field.

Chapter 1

Basic Theoretical Description of Flexoelectricity in Solids

P. V. Yudin

Ceramics Laboratory,
Swiss Federal Institute of Technology EPFL,
CH-1015 Lausanne, Switzerland

Novosibirsk State University, 2 Pirogova street,
Novosibirsk 630090, Russia

A. K. Tagantsev

Ceramics Laboratory,
Swiss Federal Institute of Technology EPFL,
CH-1015 Lausanne, Switzerland

Ferroics Laboratory, Ioffe Physical Technical Institute,
St. Petersburg 194021, Russia

In this chapter, we provide a basic theoretical description for the flexoelectricity in solids, addressing both the phenomenological and microscopic aspects. The term flexoelectricity stands for a number of linear electromechanical phenomena originating from the presence of gradient of polarization and/or gradient of mechanical strain in the material. The most known of these phenomena is the so-called direct bulk flexoelectric effect, which consists of a local linear response of polarization to a strain gradient, in the absence of macroscopic electric field.

1. Introduction

In this chapter, we give an overview of the basic theory of flexoelectricity in solids, using a phenomenological framework and simplest microscopical models. A most simple of the flexoelectric phenomena is the direct flexoelectric effect which is an electromechanical coupling consisting of a linear response of dielectric polarization to a gradient of mechanical strain. This effect may be viewed as a higher-order electromechanical phenomenon with respect to the piezoelectric effect which is a linear response of dielectric polarization to a mechanical strain. At the same time, in contrast to the piezoelectric response, the treatment of the flexoelectric effect in the static (e.g. in a bent plate) and dynamic (in a sound wave) situations generally requires separate treatments[1,2] between piezoelectricity and flexoelectricity is that while the former can be basically described with a single contribution, the description of the latter, in general, requires introduction of a number of them (bulk, surface, static, and dynamic contributions).[3,4] This conditions the structure of this chapter, which is organized as follows. In Section 2, we provide a classical phenomenological description of the static bulk contribution to the flexoelectric effect, which is present in an inhomogeneously deformed sample as well as in an acoustic wave. Then, in Section 3, we will consider flexoelectricity in an acoustic wave both phenomenologically and microscopically. In Section 4, we will describe the total polarization response of a finite sample that is inhomogeneously deformed. Finally, in Section 5 we will discuss some recent advances in the theory of flexoelectricity and comment on limitations of the simple models.

2. Static Bulk Flexoelectric Effect

Following Kogan,[5] we introduce the flexoelectric effect via the constitutive equation for the electric polarization P_i

$$P_i = \chi_{ij} E_j + e_{ijk} u_{jk} + \mu_{klij} \frac{\partial u_{kl}}{\partial x_j}, \tag{1}$$

where E_i, u_{jk}, and $\partial u_{kl}/\partial x_j$ are the macroscopic electric field, the strain tensor, and its spatial gradient, respectively. Hereafter, summation over dummy suffixes is implied. The first two rhs terms of this equation describe the dielectric and piezoelectric responses with the tensor of clamped dielectric susceptibility χ_{ij} and the piezoelectric tensor e_{ijk}, respectively. The last rhs term of equation (1) describes the linear polarization response to a strain gradient — *flexoelectric effect*. The strain tensor is defined as the symmetric part of the tensor $\partial U_l/\partial x_j$, $u_{jk} = 1/2(\partial U_j/\partial x_k + \partial U_k/\partial x_j)$, where U_i is the displacement of point x_j of the medium. The antisymmetric part of the tensor $\partial U_l/\partial x_j$, $\Omega_{jk} = 1/2(\partial U_j/\partial x_k - \partial U_k/\partial x_j)$ does not contribute to the polarization response, because it corresponds to rotations of the sample as a whole. As for the gradients of Ω_{jk}, it can contribute to the polarization response. However, $\partial \Omega_{kl}/\partial x_j$ are not included in this constitutive equation since, as was shown by Indenbom *et al.*,[6] these can always be presented as a sum of the components of tensor $\partial u_{kl}/\partial x_j$. The fourth-rank tensor μ_{klij} controlling the flexoelectric effect in equation (1), *flexoelectric tensor*, is symmetric with respect to the permutation of the first two suffixes. In general, it is not always possible to use for μ_{klij} the 2-suffix Voight tensor notations. However, when possible, e.g. for the crystals of the cubic symmetry, we will use these notations for μ_{klij} as for other fourth rank tensors. The coefficients in the Voight tensor notations used in accordance with the reference text,[7] for μ_{klij} same convention as for the stiffness tensor c_{klij} is applied. The flexoelectric tensor is allowed in materials of any symmetry (including those amorphous) in a sharp contrast to the piezoelectric tensor, e_{ijk}, which is a third rank tensor and allowed only in non-centrosymmetric media. This makes the principle difference between the piezoelectricity and flexoelectricity, as the latter is a general phenomenon having no symmetry restrictions. Since the piezoelectric and flexoelectric tensors describe the properties of a material in the absence of macroscopical electric field, these can also be defined as

$$e_{ijk} = \left(\frac{\partial P_i}{\partial u_{jk}}\right)_{E=0}, \tag{2}$$

$$\mu_{klij} = \left(\frac{\partial P_i}{\partial\left(\frac{\partial u_{kl}}{\partial x_j}\right)}\right)_{E=0}. \tag{3}$$

2.1. *Thermodynamic Phenomenological Description*

A more advanced description for both electromechanical (piezo-electric and flexoelectric) effects is the thermodynamic one which enables the identification of the thermodynamically related converse effects and provides a proper basis for the studies of stability of the system. Such description is given by the following expansion of the thermodynamic potential density in terms of polarization, strain, and their derivatives

$$
\Phi_G = \frac{\chi_{ij}^{-1}}{2} P_i P_j + \frac{c_{ijkl}}{2} u_{ij} u_{kl} + \frac{g_{ijkl}}{2} \frac{\partial P_i}{\partial x_j} \frac{\partial P_k}{\partial x_l} - \vartheta_{ijk} P_i u_{jk}
$$

$$
- f_{ijkl}^{(1)} P_k \frac{\partial u_{ij}}{\partial x_l} - f_{ijkl}^{(2)} u_{ij} \frac{\partial P_k}{\partial x_l} - P_i E_i - u_{ij} \sigma_{ij}. \tag{4}
$$

Its differential is defined as $d\Phi_G = -P_i dE_i - u_{ij} d\sigma_{ij}$. This expansion does not contain anharmonic terms (i.e. the electrostriction term is omitted). When nonlinear effects are of interest, these can readily be incorporated into the framework.

If we set to zero the coefficients for the gradient-containing terms, the bulk equations of state of the material can be found by a simple minimization of potential density (4) with respect to the polarization and strain. Such minimization leads to linear electromechanical equations for a piezoelectric:

$$
E_i = \chi_{ij}^{-1} P_j - \vartheta_{ijk} u_{jk}, \tag{5}
$$

$$
\sigma_{ij} = c_{ijkl} u_{kl} - \vartheta_{ijk} P_i. \tag{6}
$$

It is seen that equation (5) is consistent with the dielectric and piezoelectric responses introduced by (1) with

$$
e_{ijk} = \chi_{il} \vartheta_{ljk}. \tag{7}
$$

In turn, equation (6) describes Hook's law and the converse piezo-electric effect. Thus, the term $\vartheta P u$ of expansion (4) describes the piezoelectric effect in the material. For the clarity of the presentation, we will drop this term in the following discussion. This discussion can be readily generalized to piezoelectrics by taking this term into account.

Thus, we address flexoelectricity using the thermodynamic potential density (4) with $\vartheta = 0$. Here, it is proper to present Φ_G as the sum of two contributions:

$$\Phi_G = \Phi - \frac{f^{(1)}_{ijkl} + f^{(2)}_{ijkl}}{2} \frac{\partial(P_k u_{ij})}{\partial x_l}, \tag{8}$$

$$\Phi = \frac{\chi^{-1}_{ij}}{2} P_i P_j + \frac{c_{ijkl}}{2} u_{ij} u_{kl} + \frac{g_{ijkl}}{2} \frac{\partial P_i}{\partial x_j} \frac{\partial P_k}{\partial x_l}$$
$$- \frac{f_{ijkl}}{2} \left(P_k \frac{\partial u_{ij}}{\partial x_l} - u_{ij} \frac{\partial P_k}{\partial x_l} \right) - P_i E_i - u_{ij} \sigma_{ij}, \tag{9}$$

where $f_{ijkl} = f^{(1)}_{ijkl} - f^{(2)}_{ijkl}$ is called *flexocoupling tensor*. The free energy in the form given by equation (9) was introduced by Indenbom et al.[6] for the description of the static bulk flexoelectricity.

Now that the potential density contains gradient terms, to derive the equations of state, one should minimize the thermodynamic potential of the sample as a whole $\int \Phi_G dV$ (integrating over the volume of the sample), i.e. to apply the Euler equations $\partial \Phi_G / \partial A - \frac{d}{dx}(\partial \Phi_G / \partial(\partial A / \partial x)) = 0$, where A stands for P and u. Such minimization yields the bulk constitutive electromechanical equations in the form proposed by Mindlin[8]

$$E_i = \chi^{-1}_{ij} P_j - f_{klij} \frac{\partial u_{kl}}{\partial x_j} - g_{ijkl} \frac{\partial^2 P_i}{\partial x_j \partial x_l}, \tag{10}$$

$$\sigma_{ij} = c_{ijkl} u_{kl} + f_{ijkl} \frac{\partial P_k}{\partial x_l}. \tag{11}$$

It is seen that, in the case where the strain gradient and the polarization are homogeneous, equation (10) reproduces the flexoelectric effect introduced by (1) with

$$\mu_{klij} = \chi_{is} f_{klsj}. \tag{12}$$

Equation (12) links the flexoelectric and flexocoupling tensors suggesting that the flexoelectric response should be enhanced in materials with high dielectric constants (high-K materials) such as ferroelectrics.[6] It also follows from this equation that via the flexoelectric coupling the strain gradient works as an electric field.

Equation (11) enables us to recognize the thermodynamically conjugated effect to the static bulk flexoelectricity — *converse flexoelectric effect*, which consists of the contribution to the mechanical stress, proportional to the gradient of polarization.[6]

It is worth noting that the last term in equation (8) does not contribute to the bulk constitutive electromechanical equations. This can be concluded directly from the fact that its contribution to the thermodynamic potential of the sample can be transformed to an integral over the surface of the sample: $-\frac{1}{2}(f_1 + f_2)\int uPdS$. Thus, the thermodynamic potential density (9) provides a full phenomenological description of the static bulk flexoelectric effect.

It is instructive to compare the converse effects for the piezoelectric and flexoelectric responses. It can be seen from equations (5), (6), (10), and (11) that in both cases the converse effect is controlled by the same tensor as the direct one.[a] Based on equations (5) and (6), one can speak about the symmetry between the converse and direct piezoelectric effects. Namely, as follows from equation (5), under the short-circuited conditions ($E = 0$), strain induces polarization while, as follows from equation (6), under the mechanically-free conditions ($\sigma = 0$), *polarization induces strain*. However, equations (10) and (11) suggest a certain asymmetry between the direct and converse flexoelectric effects: as clear from equations (10) and (11), at zero electric field, strain gradient induces homogeneous polarization while, for the converse effect, in a mechanically free sample, homogeneous polarization *does not induce* strain gradient. This asymmetry provoked a judgment that a sensor based on the flexoelectric effect will not behave as an actuator.[9] This judgement, however, is not supported by the accurate analysis of the flexoelectric behavior of a finite sample.[10] We will address this issue in Section 4.2.

2.2. *Microscopics of Static Bulk Flexoelectric Effect*

At the microscopic level, the flexoelectric response is controlled by the redistribution of the bound charge of a crystal driven by

---

[a] As it must be for the thermodynamically conjugated effects.

a strain gradient, where ionic and electronic contributions can be distinguished. The theories of this phenomenon provide relationships between the microscopical parameters of the material (e.g. the dynamical matrix which describes the energy of inter-atomic interactions in the crystal) and the flexoelectric tensor introduced phenomenologically. Here, we restrict ourselves with a simple model of point charges,[1] which allows capturing the essence of the flexoelectric effect on the microscopic level. In this model the lattice is assumed to consist of ions with fixed charges which does not depend on their displacements.

An important feature of the flexoelectric effect treated in point-charge approximation is that polarization response is mainly due to parts of atomic displacements, which appear due to discrete nature of the crystal.[11] These parts are known as *internal strains*.[b]

The atomic displacements $w_{n,i}$ can be presented in the form

$$w_{n,i} = \int_{x_j^0}^{R_{n,j}} \frac{\partial U_i}{\partial x_j} dy_j + w_{n,i}^{\text{int}}, \tag{13}$$

where x_j^0 are the coordinates of an immobile reference point, $R_{n,j}$ are the coordinates of each atom (n numerates atoms in the lattice, while j numerates coordinates). The first rhs term in this equation, also known as external strain,[11] represents the contribution of the unsymmetrized strain $\partial U_i/\partial x_j$ taken in the so-called elastic medium approximation, the other contribution is referred to as internal strain. The difference between the external and internal strains can be understood as follows. Consider a crystal and an imaginary continuous medium with elastic constants identical to those of the crystal. Then, let us mark in the medium a mesh corresponding to the positions of the atoms in the crystal (see Fig. 1). If we deform the medium according to the unsymmetrized strain $\partial U_i/\partial x_j$, then the deformation of this mesh indicates the external strains of

---

[b] The terminology where displacements are called strains may be confusing, however we keep it after the classical book by Born and Huang.[11]

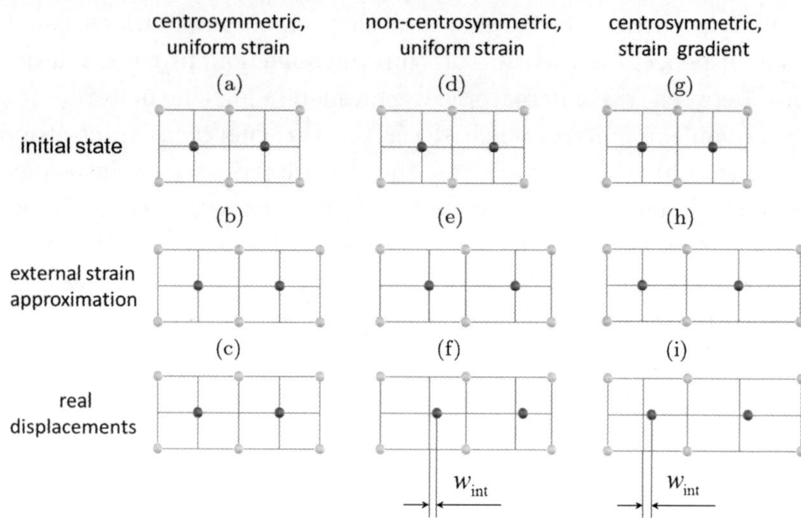

Figure 1. Schematic of atomic displacements in two neighboring unit cells of a crystal caused by application of a macroscopic strain. Cases of uniform macroscopic strain in centrosymmetric (a,b,c) and non-centrosymmetric materials (d,e,f) and of homogeneous strain gradient in centrosymmetric material (g,h,i) are illustrated. Initial state before application of strain is shown (a,d,g); displacements in approximation of external strain (b,e,h) and real displacements, comprising external and internal strains (c,f,i). The mesh is attached to the light-colored atoms. It deforms according to the external strain approximation. After Ref. [4].

the atoms, Figs. 1(b), (e), and (h). A model, where deformation of a crystal is described by external strains only is hereafter referred to as *external strain approximation*. For a material where all atoms are centers of inversion, the external strain fully describes the atomic displacements caused by a homogeneous deformation, Figs. 1(b) and (c). However, in general case in view of the discrete nature of the crystal, it behaves differently than elastic medium and the external strain approximation does not hold. Then there appears a difference between the real displacement of an atom and its external strain.

This difference is known as internal strain[11] and is described by the second rhs term in equation (13). Under a homogeneous deformation, internal strains appear only in material where not all atoms (or none of them) are centers of inversion,[11] Figs. 1(e) and (f).

At the same time, under a deformation gradient, internal strains, in general, appear in materials of any symmetry, Figs. 1(h) and (i). It is also worth mentioning that, typically, the magnitude of external strains is much larger than that of internal strains. For instance, if a sample, the dimensions of which are about L, is under a strain u_{11}, the external strains are about Lu_{11}, while the internal strains are much smaller than the lattice constant of the material. Despite this smallness, in the approximation of fixed ionic charges, it is these displacements (second term in the rhs of equation (13)) that control the bulk flexoelectric effect.

The microscopic theory of the static bulk flexoelectric effect has been developed using two methods: the first is the so-called long-wavelength method, originally introduced into the lattice dynamics theory by Born and Huang.[11] In this method, one considers a sinusoidal wave of elastic deformation and calculates the amplitude of the induced polarization wave, based on lattice mechanics of crystals and the basic definition of polarization. Then the microscopic expressions for the flexoelectric tensor can be found by comparing the results of these calculations with the amplitude of the induced polarization wave calculated within the basic constitutive equation (1).[1,12] Alternatively, one can use a method based on a calculation of the average polarization induced in a finite sample by a homogeneous strain gradient in the absence of macroscopic electric field, in agreement with definition (1–3). Such polarization corresponds to one that can be experimentally measured by integrating the short-circuiting current passing between the plates of a capacitor containing the sample subjected to the strain gradient.[1] Both methods can be readily used for calculating the flexoelectric response in crystals. Below in the sections, we will address the microscopics of the electromechanical response using these methods in the point-charge approximation.

3. Flexoelectric Effect in an Acoustic Wave

Like in case of piezoelectric effect, static bulk flexoelectricity manifests itself in an acoustic wave. However, in contrast to piezoelectricity, where in both static and dynamic situations the effect is described by the same tensor, in case of flexoelectricity there appears an additional dynamic contribution which is absent in quasi-static processes.

3.1. *Dynamic Flexoelectric Effect*

Now we will discuss the so-called *dynamic flexoelectric effect*. While the static bulk flexoelectric effect can be viewed as an analogue of the piezoelectric effect, the phenomenon treated below has no analogue in piezoelectricity. It is conditioned by the fact that, being in accelerated motion, a unit cell of a crystal made of atoms of different masses will be distorted proportionally to the acceleration a. As a result, a polarization wave with the amplitude proportional to a is expected to follow any acoustic wave. Since in an acoustic wave $a \propto \omega^2$ and $\omega \propto q$ (ω and q are the frequency and wave-vector of the wave, respectively), the amplitude of the polarization wave induced this way will be proportional to q^2 i.e. to the strain-gradient amplitude in the wave. Thus, in an acoustic wave, an additional contribution to the flexoelectric response is expected. This contribution is called the dynamic flexoelectric effect. One of the first discussions of the flexoelectric response by Harris[13] actually dealt with the dynamic flexoelectric effect. Microscopic and phenomenological theories of this effect were offered by Tagantsev.[14] It has been shown that microscopically the effect is controlled by the mass difference of the ions making up the crystal so that in a hypothetical case where the masses of all the ions are the same, the dynamic contribution vanishes.

On the phenomenological side, the dynamic flexoelectric effect can be taken into account by adding a mixed term to the density of kinetic energy[2]

$$T_{\mathrm{k}} = \frac{\varrho}{2}\dot{U}_i^2 + \frac{\gamma_{ij}}{2}\dot{P}_i\dot{P}_j + M_{ij}\dot{U}_i\dot{P}_j, \qquad (14)$$

where the effect is controlled by flexodynamic tensor M_{ij}, ϱ is the density and γ_{ij} is a phenomenological tensor describing the dynamics of polarization.

3.2. *Phenomenological Description of the Flexoelectric Effect in an Acoustic Wave*

The dynamic constitutive equations fully incorporating the bulk flexoelectric response can be described by minimizing the action $\int\int (T_{\rm k} - \Phi + u_i\sigma_i)dV\,dt$ (the integral being taken over the volume of the sample and time) with respect to $\overrightarrow{P}$ and $\overrightarrow{U}$. With Φ coming from equation (9) and $T_{\rm k}$ from (14), such minimization yields:

$$E_i = \chi_{ij}^{-1}P_j - f_{klij}\frac{\partial u_{kl}}{\partial x_j} + M_{ij}\ddot{U}_j - g_{ijkl}\frac{\partial^2 P_i}{\partial x_j \partial x_l} + \gamma_{ij}\ddot{P}_j, \qquad (15)$$

$$\varrho\ddot{U}_i = c_{ijkl}\frac{\partial u_{kl}}{\partial x_j} + f_{ijkl}\frac{\partial^2 P_k}{\partial x_l \partial x_j} - M_{ji}\ddot{P}_j. \qquad (16)$$

The last two rhs terms of equation (15) control the spatial and frequency dispersion of the polarization response. However, when we consider macroscopic manifestations of the flexoelectric response (e.g. in a dynamically bent sample or in a macroscopic acoustic wave), where $1/q$ is much larger than the typical microscopic scales and $1/\omega$ is much smaller than the typical optical phonon frequencies, these terms can be neglected. Now, setting $\overrightarrow{E} = 0$ in (15) we see that the $M_{ij}\ddot{U}_j$ term indeed provides a contribution to polarization corresponding to the dynamic flexoelectric effect given by equation

$$P_i = \chi_{is}M_{sj}\ddot{U}_j. \qquad (17)$$

It is instructive to eliminate $\ddot{U}_i$ between equations (15) and (16) to find a relationship controlling the total flexoelectric response in the dynamic case

$$E_i = \chi_{ij}^{-1}P_j - \left(f_{klij} - \frac{1}{\varrho}M_{is}c_{sjkl}\right)\frac{\partial u_{kl}}{\partial x_j}$$
$$- \left(g_{ijkl} - \frac{1}{\varrho}M_{is}f_{sjkl}\right)\frac{\partial^2 P_k}{\partial x_l \partial x_j} + \left(\gamma_{ij} - \frac{1}{\varrho}M_{is}M_{js}\right)\ddot{P}_j. \qquad (18)$$

From this equation we can see that in view of the dynamic flexoelectric effect, the role of flexocoupling tensor f_{klij} is now played by

$$f_{klij}^{\rm tot} = f_{klij} - \frac{1}{\varrho}M_{is}c_{sjkl}. \qquad (19)$$

Thus, in the dynamic case, the flexoelectric response is controlled by the "total" flexoelectric tensor $\mu_{klis}^{tot} = \mu_{klis} + \mu_{klis}^{d}$ where the dynamic contribution is defined as

$$\mu_{klij}^{d} = -\frac{1}{\varrho}\chi_{in}M_{ns}c_{sjkl}.$$

(20)

The phenomenological relationship (20) suggests that, like the static contribution, the dynamic contribution should be enhanced in high-K materials. Order-of-magnitude estimates show that the components of tensors μ_{klij}^{d} and μ_{klij} are expected to be comparable.

A remarkable feature of the bulk flexoelectric effect is that, in an acoustic wave, where a linear relationship holds between the wave-vector q and frequency ω, the relation between the static and dynamic contributions is frequency independent. The dynamic contribution to the flexoelectric effect makes it qualitatively different from the piezoelectric effect. For the latter, the polarization and strain in a moving medium are linked by the same relationships as in the static case, i.e. $P_i = \chi_{ij}E_j + e_{ijk}u_{jk}$.

One should mention that in quasi-static experiments, i.e. where the smallest dimension of the sample is less than the acoustic wavelength corresponding to the frequency of the external perturbation, the contribution of the dynamic effect becomes much weaker (see Ref. [4] for a detailed discussion).

3.3. *Microscopic Description of the Flexoelectric Effect in an Acoustic Wave*

Here, we address the long-wavelength method, which provides an assessment of bulk flexoelectricity in a crystal without dealing with the surface contributions. We will outline the implementation of this method to the calculation of the flexoelectric tensor in the model of point charges, following Ref. [2]. In such model the ions are considered as point charges placed at the points with the coordinates

$$R_{p,j}(\overrightarrow{N}) = N_j + y_{p,j},$$

(21)

where $\overrightarrow{N}$ is the lattice translation vector and $y_{p,j}$ is the vector specifying the position of the pth ion in the elementary unit cell.

Consider an elastic wave characterized by a wave-vector $\vec{q}$ and an angular frequency ω. In general, the ionic displacements in such a wave can be written in the form

$$w_{p,j}(\vec{N},t) = \exp(i\vec{q}\,\vec{R}_p - i\omega t)\widetilde{w}_{p,j}(\vec{q},\omega), \tag{22}$$

where $\widetilde{w}_{p,j}$ are the amplitudes of the ionic displacements and t stands for time.[c] The amplitude, $\widetilde{P}_j$, of the polarization wave

$$P_j(\vec{x},t) = \exp(i\vec{q}\,\vec{x} - i\omega t)\widetilde{P}_j(\vec{q},\omega), \tag{23}$$

can be found using the definition of the polarization

$$\frac{\partial P_j}{\partial x_j} = -\delta\rho, \tag{24}$$

where $\delta\rho$ is the elastic-wave-induced variation of the charge density, which is averaged over a macroscopic scale.

In the point-charge model now considered, the microscopic charge density reads

$$\rho^{\mathrm{mic}}(x_i) = \sum_{p,\vec{N}} Q_p\delta(x_i - R_{p,i}), \tag{25}$$

where Q_p stands for the charge of the pth ion in the unit cell and $\delta(\vec{x})$ denotes the delta-function defined in the three-dimensional space. In this model, the linear response of the charge density to the displacement wave (22) can readily be found in the form

$$\delta\rho^{\mathrm{mic}}(x_i) = -\sum_{p,\vec{N}} Q_p\frac{\partial}{\partial x_j}\delta(x_i - R_{p,i})w_{p,j}, \tag{26}$$

corresponding to the amplitude of the wave of the average charge density[2]

$$\delta\widetilde{\rho}(\omega,\vec{q}) = \frac{i}{v}\sum_p q_j\widetilde{w}_{p,j}Q_p. \tag{27}$$

---

[c] The physical quantities are the real parts of the corresponding complex functions.

Using (27) and the Fourier representation of equation (24), one finds the equation for the amplitude of the polarization wave

$$\widetilde{P}_i q_i = \frac{1}{v} \sum_p q_j \widetilde{w}_{p,j} Q_p. \tag{28}$$

The amplitude of the polarization wave satisfying this equation reads

$$\widetilde{P}_j = \frac{1}{v} \sum_p \widetilde{w}_{p,j} Q_p. \tag{29}$$

Strictly speaking, equation (28) taken in this form defines only longitudinal polarization component. In solution (29) the transverse components of the polarization are introduced consistently with the case of finite-sample, treated below in Section 4, equation (41). In the long-wavelength limit (i.e. at $\vec{q} \to 0$) we are interested in, the amplitude of the atomic displacements $\widetilde{w}_{p,j}$ can be expanded in powers of q, ω, and the amplitude of the acoustic wave $\widetilde{U}_i$ to find[1]

$$\widetilde{w}_{p,j}(\vec{q},\omega) = \widetilde{U}_j + i A^{ik}_{p,j} q_k \widetilde{U}_i - B^{ikl}_{p,j} q_k q_l \widetilde{U}_i - \omega^2 G^i_{p,j} \widetilde{U}_i. \tag{30}$$

Here, the first rhs term is independent of the suffix p. It corresponds to the wave of external strains while the rest of the rhs terms of this equation correspond to the wave of internal strains. The factors A, B, and G, controlling the internal strains, can be expressed in terms of the moments of the dynamic matrix of the crystal[2] (with contribution of the macroscopic electric field being excluded). The factor $A^{ik}_{p,j}$ satisfies the symmetry relationship $A^{ik}_{p,j} = A^{ki}_{p,j}$[11] and obviously $B^{ikl}_{p,j} = B^{ilk}_{p,j}$.

Inserting (30) into (29) one finds the amplitude of the polarization wave, which, using the aforementioned relationships for the factors A and B, can be cast in a form suitable for the comparison with the phenomenological results given above:

$$\widetilde{P}_i = \frac{i}{v} \sum_p Q_p A^{ik}_{p,j} \frac{q_k \widetilde{U}_j + q_j \widetilde{U}_k}{2} - \frac{q_s}{v} \sum_p Q_p \left(B^{ikl}_{p,j} + B^{kil}_{p,j} - B^{lik}_{p,j} \right)$$

$$- \frac{q_k \widetilde{U}_j + q_j \widetilde{U}_k}{2} - \frac{\omega^2}{v} \sum_p G^j_{p,i} Q_p \widetilde{U}_j. \tag{31}$$

In equation (31), $i(q_k \widetilde{U}_j + q_j \widetilde{U}_k)/2$ and $-q_s(q_k \widetilde{U}_j + q_j \widetilde{U}_k)/2$ are nothing but the Fourier components of the strain tensor and its gradient. Thus the expressions for piezoelectric and flexoelectric tensors (e_{ijk} and μ_{ikjl} respectively in equation (1)) can be readily extracted from equation (31) to get

$$e_{ijk} = v^{-1} \sum_p Q_p A^{ik}_{p,j}, \tag{32}$$

$$\mu_{ikjl} = v^{-1} \sum_p Q_p N^{ikl}_{p,j}, \tag{33}$$

$$N^{ikl}_{p,j} = B^{ikl}_{p,j} + B^{kil}_{p,j} - B^{lik}_{p,j}. \tag{34}$$

Note that the first rhs term of (30), corresponding to a wave of external strains, does not contribute to the polarization wave. This follows from the electroneutrality of the elementary unit cell, $\sum_p Q_p = 0$.

In the ω-dependent contribution of equation (31) one readily recognises the dynamic flexoelectric effect. Since in the elastic wave $\omega^2 \propto q^2$, this term will also contribute to the polarization proportionally to the strain gradient in the wave. Comparing equation (31) with equation (17), one derives the relationship for this response:

$$P_i = \chi_{is} M_{sj} \ddot{U}_j = \Lambda_{ij} \ddot{U}_j, \tag{35}$$

with

$$\Lambda_{ij} = \frac{1}{v} \sum_p Q_p G^j_{p,i}. \tag{36}$$

For the case of a crystal with two ions per unit cell, Λ_{ij} can be cast in the form[14]

$$\Lambda_{ij} = \chi_{ij} \frac{m_2 - m_1}{2Q}, \tag{37}$$

where m_1, m_2 are the masses of ions having charges Q and $-Q$, respectively; χ_{ij} is the ionic contribution to the dielectric susceptibility of the crystal. As follows from expression (37), the effect vanishes when the masses of the ions are equal.

Summarizing this subsection we can state that the treatment of the polarization response to a strain gradient in the point-charge approximation given above identifies the bulk flexoelectric response as originating from the internal strains induced by the gradient of the elastic deformation. This is a direct analogy to the piezoelectric response which is controlled by the internal strains that depend on the elastic deformation itself. In addition to the static effect there appears dynamic contribution having no analogue in piezoelectricity. In contrast to the static effect, the dynamic effect explicitly depends on the masses of the ions constituting the crystal.

4. Flexoelectric Effect in a Finite Sample

The above treatment of flexoelectric effect in an acoustic wave was free of surface contributions. In contrast, treatment of flexoelectric response in a finite sample inevitably deals with the surface. Counterintuitively, the impact of surface on the flexoelectric response is not small even for systems with small surface to volume ratio. A theoretical treatment of the flexoelectric effect in a whole sample brings about additional contributions to the effect, namely surface piezoelectricity and surface flexoelectricity. Another remarkable feature of flexoelectricity in a finite sample is a modification of the mechanical and polarization boundary conditions. A comprehensive treatment of flexoelectric phenomena in a finite sample is quite a complicated task to which some other chapters of this book are devoted. In this section, we will present just a simple vision of the important features of the phenomenon.

4.1. *Direct Flexoelectric Effect in a Finite Sample*

Consider the flexoelectric response of a finite crystal which is modeled as consisting of point charges Q_n located at points with coordinates $R_{n,i}$ where n enumerates charges and i is the Cartesian suffix. This can be done by calculating the variation of the average dipole-moment density of the sample

$$\delta P_i = V_{\text{fin}}^{-1} \sum_n Q_n (R_{n,i} + w_{n,i}) - V^{-1} \sum_n Q_n R_{n,i}, \qquad (38)$$

where $w_{n,i}$ is the displacement of the charge (from its original position at $R_{n,i}$) induced by the deformation, V and V_{fin} are the sample volume before and after the deformation; the summation over all the charges of the sample is implied. Here, we address the simple case of vanishing macroscopic electric fields following the original approach by Tagantsev.[1] Strictly speaking, as it was recently understood,[10] even under short-circuit conditions, there may appear inhomogeneous electric fields in a mechanically loaded sample. In this context this treatment under zero macroscopic field, while capturing important features of flexoelectricity in a finite sample, has limited applicability to real systems. We will comment on such limitations in Section 5 where the impact of the electric field inhomogeneities on the total flexoelectric response will be discussed. Thus, here we are interested in the situation where the macroscopic electric field is zero before and during the application of the mechanical perturbation.

The atomic displacements $w_{n,i}$ can be presented in the form given by expression (13):

$$w_{n,i} = \int\limits_{x_j^0}^{R_{n,j}} \frac{\partial U_i}{\partial x_j} dy_j + w_{n,i}^{\text{int}}, \tag{39}$$

where x_j^0 are the coordinates of an immobile reference point. The first rhs term in this equation is the external strain[11] which represents the contribution of the unsymmetrized strain $\partial U_i/\partial x_j$ taken in the so-called elastic medium approximation, the other contribution is the internal strain.

In the lowest, to within the amplitude of the deformation, approximation, the internal strains can be presented as linear functions of the strain tensor and its gradient[1]:

$$w_{n,j}^{\text{int}} = A_{n,j}^{ik} u_{ik} + N_{n,j}^{ikl} \frac{\partial u_{ik}}{\partial x_l}. \tag{40}$$

Where $A_{n,j}^{ik}$ and $N_{n,j}^{ikl}$ are the tensors introduced in the previous section, satisfying symmetry relationships $A_{n,j}^{ik} = A_{n,j}^{ki}$ and $N_{n,j}^{ikl} = N_{n,j}^{kil}$. For the case of an ideal crystalline lattice, $A_{n,j}^{ik}$ and $N_{n,j}^{ikl}$ can be calculated in terms of lattice dynamics theory.[1,11]

Though the lattice dynamics theory behind the calculations of the N and A factors is out of the scope of this chapter, we would like to make an important remark. This theory as used in these calculations deals only with small elastic deformations of the ideal lattice. Thus, only small relative variations of the interatomic distances are considered so that any strong distortions of the lattice (like the dislocation formation or atom hopping in highly unharmonic crystalline latices) are not covered. The flexoelectric effect associated only with such small relative variations of the interatomic distances will be discussed here.

Now inserting (40) and (39) into (38) and keeping the lowest order terms in the amplitude of the deformation we find for the variation of the average polarization of the sample, induced by the mechanical perturbation:

$$\delta P_j = V^{-1} \sum_n Q_n A^{ik}_{n,j} u_{ik} + V^{-1} \sum_n Q_n N^{ikl}_{n,j} \frac{\partial u_{ik}}{\partial x_l} + \delta P^{\text{ext}}_j. \qquad (41)$$

The first and second rhs terms of this equation are conditioned by the internal strains. The first one controls piezoelectricity. For a piezoelectric, the sum $\sum_p Q_p H^{ik}_{p,j}$ taken over the crystalline unit cell is not, in general, zero[11] and the first rhs term of (41) is dominated by the bulk contribution. Then, neglecting the surface contribution to the sum over the sample, one can pass from the summation over the sample to that over the unit cell and by comparing the result with the basic relationship (2) one derives the microscopic expression for the piezoelectric tensor[11]

$$e_{ijk} = v^{-1} \sum_p Q_p A^{ik}_{p,j}, \qquad (42)$$

where the summation is taken over the ions in a unit cell of volume v. A similar treatment leads to the flexoelectric tensor given by

$$\mu_{ikjl} = v^{-1} \sum_p Q_p N^{ikl}_{p,j}, \qquad (43)$$

where the summation is again taken over the ions in a unit cell.

One readily sees that the results of the finite-sample approach, (42) and (43), readily reproduce those for the piezoelectric and static bulk flexoelectric responses obtained in the long-wavelength method (equations (32) and (33)).

Next, we discuss the last rhs term from equation (41). This term is conditioned by the external strains (the integral term in (39)) and the change of the sample volume. As we have shown in the previous section, for an acoustic wave the external strain did not make any contribution to the flexoelectric effect. However, now that the surface of the sample is included into consideration, this term requires delicate treatment. For the case of a response to a homogenous strain in the macroscopic sample of a piezoelectric, this term can be eliminated using the condition of the absence of the macroscopic electric field in the sample.[1] This contribution, however, cannot be fully eliminated in the case of the flexoelectric response bringing about the so-called *surface flexoelectric effect*. It explicitly depends on the termination of the ionic sample, eventual reconstruction of its surface, and the presence of the additional free charges on it.[4]

Equation (41) is also relevant to an additional contribution to the integral flexoelectric response of a finite sample — the so-called contribution of the *surface piezoelectricity* to flexoelectric response. It is related to the fact that the sum $\sum_n Q_n A^{ik}_{n,j}$ taken over the distorted layers adjacent to the faces of the sample should not be, in general, equal to zero (in view of the symmetry-breaking effect of the interface)[1] providing local piezoelectricity. The net dipole moment of the sample equal to the difference between the dipole moments induced in these layers is evidently proportional to the strain gradient and the sample thickness. The resulting polarization (average dipole moment density) appears to be proportional to the strain gradient, contributing to the total flexoelectric response of the sample.[1]

Summarizing this subsection we can state that the microscopic treatment of the polarization response in a finite sample is consistent with the microscopic theory for the polarization response in an acoustic wave. The same expressions are obtained for the static bulk flexoelectric response. The finite sample approach is free of dynamic contribution which was present in an acoustic wave. On balance, it demonstrates the presence of essential additional contributions to the effect, namely surface piezoelectricity and surface flexoelectric effect.

4.2. *Converse Flexoelectric Effect in a Finite Sample*

As was mentioned in Section 2, the electromechanical constitutive equations describing the bulk flexoelectric effect[d]

$$E_i = \chi_{ij}^{-1} P_j - f_{klij} \frac{\partial u_{kl}}{\partial x_j}, \tag{44}$$

$$\sigma_{ij} = c_{ijkl} u_{kl} + f_{ijkl} \frac{\partial P_k}{\partial x_l}, \tag{45}$$

suggest a certain asymmetry between the direct and converse flexo-electric responses. Namely, in the absence of an electric field, a strain gradient induces a homogeneous polarization while a homogeneous polarization has no mechanical yield. A practical situation where such asymmetry might reveal itself is bending experiments with thin electroded plate of a centrosymmetric material (Fig. 2). For instance, a cylindrical bending of such plate about OX_2 axis will bring about non-zero strain gradients $\partial u_{11}/\partial x_3$ and $\partial u_{33}/\partial x_3$. The direct flexoelectric response to such bending can be detected by measuring the induced variation of the charge on the short-circuited electrodes. Such response is described by equation (44), where $E = 0$ (in view of the short-circuit condition) while P_3 is directly linked with the charge. Meanwhile, one might conclude, based on equation (45), that the application of a voltage between the electrodes will not lead

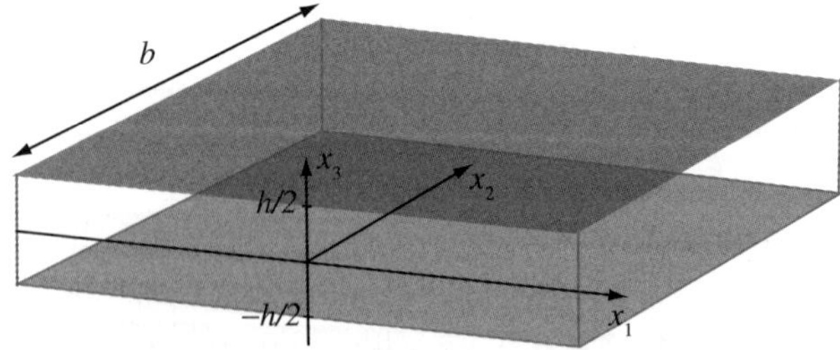

Figure 2.   A thin plate of the material exposed to bending and the reference frame used in calculations. Upper and lower faces of the plate are electroded.

---

[d] In (44) the polarization-gradient term is dropped as being of minor importance for the problem addressed in this subsection.

to any bending of the plate. Indeed, "naturally" assuming that the voltage produces a *homogeneous polarization*, no mechanical yield is seen from equation (45).

Using similar reasoning, it was argued that a flexoelectric-based mechanical sensor, in contrast to piezoelectric based devices, will not behave as an actuator.[9,15] However, there are several reasons to question such statement. First, this statement is in conflict with the results obtained by the group of professor Bursian already in the 60's of the past century. This group reported experimental data on electric-field-induced bending of plates of $BaTiO_3$ crystals.[16] Later, Bursian and co-workers provided a thermodynamic analysis,[17] supporting their experimental findings. Second, the existence of a linear sensor-not-actuator is in a conflict with the general principles of thermodynamics, for instance, based on which one could construct a perpetual motion device.[18]

Thus, there appears to be a contradiction between the straight-forward analysis of the constitutive equations and thermodynamics. Such apparent contradiction was recently identified by Tagantsev and Yurkov[10] and a solution to it was outlined. The reason for this discrepancy is that the polarization-induced bending (*flexoelectric bending*) predicted by Bursian and Trunov[17] is a non-local effect that can only be obtained by considering the thermodynamics of the finite-size sample. Meanwhile, the application of the "local" electromechanical equation (45) to the bulk of the sample does not capture this effect. The resolution for this discrepancy required a comprehensive treatment of the converse flexoelectric response of a finite sample, involving boundary conditions for polarization at the sample surface. Here, we outline the solution for this apparent inconsistency for the case where polarization is set to zero at the surface (so-called blocking boundary condition), the case where an elementary discussion of the problem is available.[10]

We will address the electromechanical bending-mode performance of a thin plate. Consider a plate of a non-piezoelectric material placed in an electric field normal to it (Fig. 2) resulting in the polarization profile schematically shown in Fig. 3. The plate is considered to be macroscopically thick, i.e. its thickness h is much

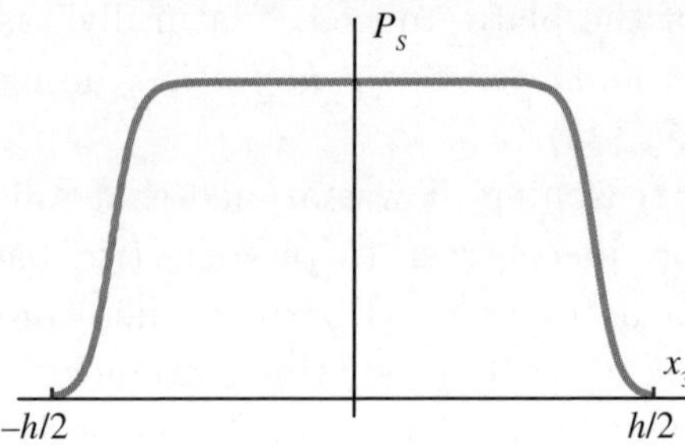

Figure 3.   Schematics for the polarization profile in the plate for the case of blocking boundary conditions.

larger than the spatial scale for the polarization variation at its surface. In the main part of the sample $\frac{\partial P_k}{\partial x_l} = 0$, so that, as clear from (45), the flexoelectricity provides no mechanical input. Meanwhile, at the plate surfaces $\frac{\partial P_3}{\partial x_3} \neq 0$, implying, via equation (45), a certain mechanical yield.[10,19] Let us show that this yield is a plate bending, accounting only for the cylindrical bending about the OX_2 axis. A straightforward way to do this is to consider the equation of balance of the bending moment for the plate. Following the basics of the elasticity theory,[20] to derive such equation, we multiply (45) by x_3 and integrate the result across an X_2X_3 cross-section of the sample (Fig. 2). Finally we get (for simplicity, the Poisson ratio is neglected and only one component of the strain, u_{11}, and stress, σ_{11}, are taken into account):

$$b \int_{-h/2}^{h/2} \sigma_{11} x_3 dx_3 = bf_{13} \int_{-h/2}^{h/2} \frac{\partial P_3}{\partial x_3} x_3 dx_3 + bc_{11} \int_{-h/2}^{h/2} u_{11} x_3 dx_3, \quad (46)$$

where b is the dimension of the sample in the OX_2 direction. Here, the flexoelectric response is described by the first rhs term which can be evaluated using integration by parts:

$$\int_{-h/2}^{h/2} \frac{\partial P_3}{\partial x_3} x_3 dx_3 = - \int_{-h/2}^{h/2} P_3 dx_3 = -h\langle P_3 \rangle, \quad (47)$$

where $\langle P_3 \rangle$ is the averaged polarization induced by the electric field in the plate. Since the spatial scale of the polarization variation at the interface is much smaller than h, with a good accuracy $\langle P_3 \rangle \approx P$, where P — polarization in the bulk. In mechanical equilibrium, in any $X_2 X_3$ cross-section of the sample, the lhs term of (46) is equal to the mechanical moment of the external forces applied to the sample. Without the first rhs term, this equation describes the bending of the sample caused by this moment. In the presence of flexoelectricity, using (47), the equation for the moment balance can be rewritten as

$$M + f_{13} h P = c_{11} \int_{-h/2}^{h/2} u_{11} x_3 dx_3, \tag{48}$$

where M is the bending moment per unit length (in the OX_2 direction) of the plate. It follows from equation (48) that the application of a homogeneous electric field to the plate is equivalent (via the induced polarization and the flexoelectric coupling) to the application of an external bending moment (here we can speak about *flexoelectric bending moment*). Thus, a finite mechanically free ($M = 0$) sample placed in a homogeneous electric field will bend. Note, that though the flexoelectric bending moment is conditioned by a surface effect, it is proportional to the bulk value of the polarization induced by the applied field.

Thus, the above analysis does not support the judgment, stemming from the apparent asymmetry between equations (44) and (45), that a sensor based on the flexoelectric effect will not behave as an actuator.[9] Moreover, it is shown in Ref. [10], that a bending-mode flexoelectric sensor once working as an actuator will, in fact, be characterized by the same effective piezoelectric constant.

Summarizing this subsection we can state that the converse flexoelectric effect in a finite sample manifests itself in a manner quite different from that suggested by a simplified analysis of the constitutive equations for the static bulk flexoelectric effect (44) and (45). Specifically, the mechanical yield requires inhomogeneity of the polarization (or field). In reality, homogeneous electric field readily induced flexoelectricity driven inhomogeneous deformations in a finite sample. Of key importance is that the effect is extremely sensitive to the properties of the sample surface.

5. Recent Advances in the Field

This chapter provides the most simple theoretical vision of flexo-electric phenomena in solid. The attention was mainly paid to the discussion of the bulk (static and dynamic) flexoelectric effects as the most studied ones. At the same time it was clearly demonstrated that, in the case of direct and converse flexoelectric responses of a finite sample (the situation of the most interest for practical applications) the phenomena can be strongly influenced by the impact of the sample surface.

The microscopic treatment of flexoelectricity presented in this chapter was done in a very simple point-charge approximation. Thus, a natural question arises on the limitations of the results presented. In reality, even in ionic crystals the charge is not localized. This can be taken into account by the continuum-charge-density approach offered by Martin.[21] In application to the bulk flexoelectric effects, here, two levels of the generalization of the point-charge theory are possible. First, the simple ionic charges in all above considerations can be replaced by the matrices of transverse Born effective charges that are still localized at the atomic positions. This will not affect the qualitative picture of flexoelectricity pre-sented in the chapter. Specifically, the polarization response will be still fully controlled by internal strains. Such response is typically referred to as ionic contribution. Further comes redistribution of electronic density going beyond the Born charge approximation, which brings about qualitatively new effects. Specifically, now the external stains start contributing to the polarization response, as was shown for the piezoelectric response by Martin.[21] The same holds for the flexoelectric response as was demonstrated by Resta[22] and Hong and Vanderbilt.[12] It is the so-called electronic (frozen-ion) contribution.[12]

5.1. *Electronic Contribution to the Flexoelectric Effect*

For the bulk flexoelectric effects (defined at zero macroscopic electric field), electronic contribution is not enhanced in high K-materials, in contrast to the ionic one. As a result, in such materials, the electronic contribution to bulk flexoelectric effect is expected to be negligible compared to the ionic one. Here one might suggest that the electronic contribution can be neglected for the description of flexoelectric phenomena in high K-materials, which however is not always the case. Let us briefly discuss the situation separately for the case of acoustic wave and static (quasi-static) flexoelectric response of a finite sample. As was done throughout the chapter we will address the case of a non-piezoelectric material.

In the case of a transverse acoustic wave not followed by that of a macroscopic electric field, the polarization response is fully controlled by the static and dynamic bulk flexoelectric effects and, in view of the above said, in high K-materials the electronic contribution can be neglected. However, in the case of a longitudinal acoustic wave which is followed by that of a macroscopic (depolarizing) electric field, the situation changes. Here, typically the depolarizing field suppress the enhancement of the ionic contribution of flexoelectric effect in high K-materials. As a result, the ionic and electronic contributions are expected to be comparable and the latter must be taken into account.

In the case of a finite sample, as was discussed above, the direct flexoelectric response can be essentially controlled by the surface related contributions, e.g. by the surface piezoelectricity. In this case, the response is monitored by macroscopic electric field appeared in the body of the short-circuited sample. The origin of this field is the piezoelectricity of the distorted surface layers, which is, in general, controlled by both electronic and ionic contributions to polarization.[10] Thus, a rigorous description of the direct flexoelectric effect in a finite sample requires taking into account both electronic and ionic contributions. The recent *ab initio* results by Stengel[23] clearly illustrate this issue. The converse flexoelectric effect in a finite sample, in view of its link via thermodynamics to the direct one, must also be sensitive to the electronic contribution to the polarization.

5.2. *Effect of Field Inhomogeneity in a Short-Circuited Sample*

In the sections above the theoretical description of polarization response to a mechanical perturbation was introduced under the condition that the macroscopic electric field is zero throughout the sample. However, as it was recently understood,[10] distorted layers near the electroded surfaces which become oppositely poled under the action of the strain gradient do create a macroscopic electric field in the body of the sample. Taking into account this field[10] leads to a strong enhancements of the effect which now scales as the bulk dielectric constant of the material. Of key importance is that the contribution of the surface piezoelectric effect to the direct flexoelectric response of a finite sample is expected to be of the order of magnitude as that of the bulk static flexoelectric,[10] including the practically important case of high-K materials. Thus, there is no reason to expect that the direct flexoelectric response of a finite sample is mainly controlled by the static bulk flexoelectric effect as it has been generally believed.

5.3. *Details of Converse Flexoelectric Effect in a Finite Sample*

The analysis of converse flexoelectric effect in a finite sample presented in Section 4.2 was done for so-called blocking boundary conditions for polarization, where it was set to zero on the surfaces. Though such analysis give a simple solution to the problem, it does not fully elucidate the matter. The point is that one may imagine such boundary conditions which will allow a homogeneous polarization induced by the applied field. In this case the arguments based on the two inhomogeneously polarized layers near the surfaces will obviously fail. However, even for such a situation the converse flexoelectric effect was found to arise due to a change in mechanical boundary conditions needed in this case.[24] This treatment done by Yurkov showed the appearance of the flexoelectric bending moment, originating from the modified mechanical boundary conditions. However, for configurations identical to those of Section 4.2, the flexoelectric bending moment was found to be exactly two times smaller than in the case of polarization blocked at the sample surfaces. A general case with non-zero polarization at the surface, where its behavior in the surface-adjacent layer required a more involved treatment together with modification of mechanical boundary conditions was considered in Ref. [25].

The surface piezoelectricity also contributes to the total converse flexoelectric effect of a finite sample.[10] In the model discussed for the direct flexoelectric effect, its nature is quite obvious. The piezoelectric coefficients of the two disturbed surface layers of the plate differ in the sign. Thus, in the applied field, signs of the stress applied by these layers to the body plate will also be opposite, resulting in a bending of the plate.

6. Conclusions

The theoretical description of flexoelectricity in solids identifies a number of flexoelectric phenomena which provide their contributions to the total flexoelectric response. These phenomena are static bulk flexoelectric effect, dynamic bulk flexoelectric effect, surface piezoelectricity, and surface flexoelectricity. The magnitudes of the contributions of the different effects are in general case comparable, while their balance is specific to the manifestation of flexoelectricity addressed. Here, we considered two simple cases: acoustic wave and inhomogeneously strained finite sample. The static bulk contribution is present in both cases, and as follows from the microscopic treatment, is controlled by the same flexocoupling tensor. In the point-charge approximation considered the bulk flexoelectric response is only due to internal strains which are allowed by the discrete nature of crystals, and would not appear in the case where the atoms of the crystal obey the law of motion of a continuum elastic media (external strain approximation). This feature of the flexoelectricity is a direct analogy to the piczoelectric response which is also controlled only by the internal strains. In addition to the static effect in acoustic wave there arises dynamic contribution having no analogue in piezoelectricity. In contrast to the static effect, the dynamic effect explicitly depend on the masses of the ions constituting the crystal. In the case of quasi-static flexoelectric response in a finite sample the dynamic contribution is absent. On balance, it demonstrates the presence of essential surface contributions to the effect, namely surface piezoelectricity and surface flexoelectric effect.

Acknowledgments

The authors gratefully acknowledge funding from the Swiss National Science Foundation, Grant No. 200020-144463/1 and from the government of the Russian Federation, Grant No. 2012-220-03-434.

References

1. A. K. Tagantsev. *Phys. Rev. B.* **34**, 5883–5889 (1986).
2. A. K. Tagantsev. *Phase Transit.* **35**(3–4), 119–203 (1991).
3. P. Zubko, G. Catalan, and A. K. Tagantsev. *Flexoelectric Effect in Solids.* Volume 43 of *Annual Review of Materials Research*, pp. 387–421. Annual Reviews, Palo Alto (2013).
4. P. V. Yudin, and A. K. Tagantsev. *Nanotechnology.* **24**, 432001 (2013).
5. S. M. Kogan. *Sov. Phys. Sol. State.* **5**(10), 2069–2070 (1964).
6. V. L. Indenbom, E. B. Loginov, and M. A. Osipov. *Kristalografija.* **26**, 1157 (1981).
7. Landolt–Bornstein: *Numerical Data and Functional-Relationships in Science and Technology, New Series.* Vol. III/16 and III/29a. Springer, Berlin, (1992).
8. R. D. Mindlin. *Int. J. Solids Struct.* **4**(6), 637–642 (1968).
9. L. Cross. *J. Mater. Sci.* **41**, 53–63 (2006).
10. A. K. Tagantsev, and S. A. Yurkov. *J. Appl. Phys.* **112**(4), 044103–044107 (2012).
11. M. Born and K. Huang. *Dynamical Theory of Crystal Lattices.* Oxford University Press, Oxford (1962).
12. J. Hong and D. Vanderbilt. *Phys. Rev. B.* **84**, 180101 (2011).
13. P. Harris. *J. Appl. Phys.* **36**(3), 739–741 (1965).
14. A. K. Tagantsev. *Zhurnal Eksperimentalnoi I Teoreticheskoi Fiziki.* **88**(6), 2108–2122 (1985).
15. B. Chu, W. Zhu, N. Li, and L. Eric Cross. *J. Appl. Phys.* **106**(10), 104109 (2009).
16. E. V. Bursian, and O. I. Zaikovskii. *Sov. Phys. Sol. State.* **10**(5), 1121–1124 (1968).

17. É. V. Bursian, and N. N. Trunov. *Sov. Phys. Sol. State.* **10**, 760–762 (1974).

18. A. S. Yurkov. *Privite communication* (2012).

19. E. A. Eliseev, A. N. Morozovska, M. D. Glinchuk, and R. Blinc. *Phys. Rev. B.* **79**, 165433 (2009).

20. S. Timoshenko, and S. Woinowsky-Kreiger. *Theory of Plates and Shells.* McGraw-Hill Inc. New York (1987).

21. R. M. Martin. *Phys. Rev. B.* **5**(4), 1607 (1972).

22. R. Resta. *Phys. Rev. Lett.* **105**, 127601 (2010).

23. M. Stengel. *Phys. Rev. B.* **90**, 201112(R) (2014).

24. A. S. Yurkov. *Phys. Sol. State.* **57**, 460–466 (2015).

25. A. S. Yurkov. *JETP Lett.* **94**, 455–458 (2011).

Chapter 2

First-Principles Theory of Flexoelectricity

Massimiliano Stengel

*ICREA — Institució Catalana de Recerca i Estudis Avançats,
08010 Barcelona, Spain*

*Institut de Ciència de Materials de Barcelona ICMAB-CSIC,
Campus UAB, 08193 Bellaterra, Spain*

David Vanderbilt

*Department of Physics and Astronomy, Rutgers University,
Piscataway, New Jersey 08854-8019, USA*

In this chapter, we provide an overview of the current first-principles perspective on flexoelectric effects in crystalline solids. We base our theoretical formalism on the long-wave expansion of the electrical response of a crystal to an acoustic phonon perturbation. In particular, we recover the known expression for the piezoelectric tensor from the response at first order in wave-vector $\mathbf{q}$, and then obtain the flexoelectric tensor by extending the formalism to second order in $\mathbf{q}$. We put special emphasis on the issue of surface effects, which we first analyze heuristically, and then treat more carefully by presenting a general theory of the microscopic response to an arbitrary inhomogeneous strain. We demonstrate our approach by presenting a full calculation of the flexoelectric response of a $SrTiO_3$ film, where we point out an unusually strong dependence of the bending-induced open-circuit voltage on the choice of surface termination. Finally, we briefly discuss some remaining open issues concerning the methodology and some promising areas for future research.

1. Introduction

First-principles electronic structure calculations have played an increasingly important role in our understanding of the properties of materials and nanostructures in recent decades. The phrase "first principles" is generally used in the condensed-matter community to convey the notion that the calculations are free of adjustable parameters, taking as input only some list of atoms, their atomic numbers, and some initial guesses at their coordinates in the unit cell. One then solves the Schrödinger equation for the electrons in some approximation, computes the relaxed atomic coordinates, and calculates the desired properties of the crystal. In the condensed-matter community this is typically done in the framework of density-functional theory (DFT),[1] as shall be assumed below, but Hartree–Fock or other quantum-chemical methods can also be used.

While the accuracy and efficiency of DFT methods have improved over the years, of equal importance has been the increasing range of quantities that can be computed. In the context of dielectric properties, the implementation of linear-response theory for phonon and electric-field perturbations in the 1980s and 1990s opened up the calculation of phonon frequencies, dynamical charges, and both electronic and lattice contributions to the dielectric constant.[2] While there was initially some doubt about whether the piezoelectric response was a bulk property at all, a seminal paper of Martin laid this question to rest,[3] and the computation of the piezoelectric tensor is now a standard feature of most DFT codes as well. Strangely, although many of the above properties can be computed as derivatives of the electric polarization **P**, a proper definition of the polarization **P** itself proved more difficult; the physics was clarified, and practical methods for computing it, were developed only in the mid-1990s with the appearance of the "modern theory of polarization."[4–6] Related methods for computing the orbital magnetization of ferromagnets and the properties of crystals in finite electric fields have been developed since the 2000's.[7]

The flexoelectric tensor has been among the few physical properties to have resisted a proper first-principles formulation even until today. The theory of flexoelectricity was pioneered in the 1980s by Tagantsev.[8,9] However, because it encodes a response to a strain gradient, rather than just a strain, and because a strain gradient is inconsistent with ordinary cell-periodic boundary conditions, methods based on Bloch's theorem cannot be straightforwardly applied in the first-principles context. A serious attack on this problem did not begin until 2010, when Hong and collaborators presented the results of calculations on supercell configurations containing strain gradients.[10] Subsequent papers of Resta[11] and Hong and Vanderbilt[12] clarified aspects of the electronic contribution to the flexoelectric response.

More recently, Stengel[13] and Hong and Vanderbilt[14] tackled the problem in a systematic way, and working from slightly different perspectives, arrived consistently at a nearly complete framework for defining, and eventually computing, the flexoelectric tensors fully from first-principles. Some components of the flexoelectric tensor that can be expressed only in terms of bulk current responses, as opposed to charge responses, still require care in their interpretation and await the development of efficient methods for calculating them. However, we can expect these difficulties to be cleared up soon, so we can look forward to a new era in which first-principles calculations of flexoelectric responses can flourish and contribute to a fast-evolving experimental field. The purpose of this chapter is to outline the physical principles underlying these advances in the understanding and computation of flexoelectric responses, and to summarize a few of the preliminary results that have been presented in the literature to date.

2. Theory and Methods

2.1. *Strain, Strain Gradients, and Responses*

We begin by establishing our notation. In continuum mechanics, a deformation can be expressed as a three-dimensional (3D) vector field, $u_\alpha(\mathbf{r})$, describing the displacement of a material point from its reference position at $\mathbf{r}$ to its current location $\mathbf{r}'$,[a]

$$r'_\alpha(\mathbf{r}) = r_\alpha + u_\alpha(\mathbf{r}).$$

The *deformation gradient* is defined as the gradient of u_α taken in the reference configuration,

$$\tilde{\varepsilon}_{\alpha\beta}(\mathbf{r}) = u_{\alpha,\beta}(\mathbf{r}) = \frac{\partial u_\alpha(\mathbf{r})}{\partial r_\beta}. \tag{1}$$

$\tilde{\varepsilon}_{\alpha\beta}(\mathbf{r})$ is often indicated in the literature as "*unsymmetrized* strain tensor", as it generally contains a proper strain plus a rotation. By symmetrizing its indices one can remove the rotational component, thus obtaining the *symmetrized* strain tensor

$$\varepsilon_{\alpha\beta} = \frac{1}{2}(u_{\alpha,\beta} + u_{\beta,\alpha}).$$

This $\varepsilon_{\alpha\beta}$ is a convenient measure of local strain, as it only depends on *relative* displacements of two adjacent material points, and not on their absolute translation or rotation with respect to some reference configuration.

In this work, we shall be primarily concerned with the effects of a spatially inhomogeneous strain. The third-rank *strain gradient* tensor can be defined in two different ways, both important for the derivations that follow. The first (type-I) form consists in the gradient of the *unsymmetrized* strain,

$$\eta_{\alpha,\beta\gamma}(\mathbf{r}) = \frac{\partial \tilde{\varepsilon}_{\alpha\beta}(\mathbf{r})}{\partial r_\gamma} = \frac{\partial^2 u_\alpha(\mathbf{r})}{\partial r_\beta \partial r_\gamma}. \tag{2}$$

Note that $\eta_{\alpha,\beta\gamma}$, manifestly invariant upon $\beta \leftrightarrow \gamma$ exchange, corresponds to the $\nu_{\alpha\beta\gamma}$ tensor of Ref. [12], and to the symbol $\partial \epsilon_{\alpha\beta}/\partial r_\gamma$ of Ref. [8]. Alternatively, the strain gradient tensor can

---

[a] Since we are exclusively interested in linear fexoelectricity here, we shall assume a regime of small deformations henceforth.

be defined (type-II) as the gradient of the *symmetric* strain, $\varepsilon_{\alpha\beta}$,

$$\varepsilon_{\alpha\beta,\gamma}(\mathbf{r}) = \frac{\partial \varepsilon_{\alpha\beta}(\mathbf{r})}{\partial r_\gamma},$$

invariant upon $\alpha \leftrightarrow \beta$ exchange. It is straightforward to verify that the two tensors contain exactly the same number of independent entries, and that a one-to-one relationship can be established to express the former as a function of the latter and vice versa,

$$\eta_{\alpha,\beta\gamma} = \varepsilon_{\alpha\beta,\gamma} + \varepsilon_{\gamma\alpha,\beta} - \varepsilon_{\beta\gamma,\alpha}. \tag{3}$$

The piezoelectric and flexoelectric tensors describe, respectively, the macroscopic polarization response to a uniform strain and to a strain gradient. In type-I form, these are

$$e_{\alpha\beta\gamma} = \frac{dP_\alpha}{d\varepsilon_{\beta\gamma}}, \tag{4}$$

$$\mu^{\mathrm{I}}_{\alpha\beta,\gamma\lambda} = \frac{dP_\alpha}{d\eta_{\beta,\gamma\lambda}}. \tag{5}$$

While the type-I form is more convenient to derive and calculate, the type-II representation is often preferred in applications. The type-II flexoelectric tensor is defined as

$$\mu^{\mathrm{II}}_{\alpha\lambda,\beta\gamma} = \frac{\partial P_\alpha}{\partial \varepsilon_{\beta\gamma,\lambda}}. \tag{6}$$

Note that μ^{I} and μ^{II} are both symmetric under the last two indices, and are related to each other via equation (3) according to

$$\mu^{\mathrm{II}}_{\alpha\lambda,\beta\gamma} = \mu^{\mathrm{I}}_{\alpha\beta,\gamma\lambda} + \mu^{\mathrm{I}}_{\alpha\gamma,\lambda\beta} - \mu^{\mathrm{I}}_{\alpha\lambda,\beta\gamma}, \tag{7}$$

$$\mu^{\mathrm{I}}_{\alpha\beta,\gamma\lambda} = \frac{1}{2} \left(\mu^{\mathrm{II}}_{\alpha\lambda,\beta\gamma} + \mu^{\mathrm{II}}_{\alpha\gamma,\beta\lambda} \right). \tag{8}$$

2.2. *Long-Wave Approach*

A macroscopic strain gradient breaks the translational symmetry of the crystal lattice. For this reason, the response to such a perturbation cannot be straightforwardly represented in periodic boundary conditions. This makes the theoretical study of flexoelectricity more challenging than other forms of electromechanical coupling such as piezoelectricity. To circumvent this difficulty, we shall base our analysis on the study of long-wavelength acoustic phonons. These perturbations, while generally incommensurate with the crystal lattice, can be conveniently described in terms of functions that are lattice-periodic, and therefore are formally and computationally very advantageous.[2]

Consider a crystal lattice spanned by the real-space translation vectors $\mathbf{R}_l$ and by the basis vectors $\boldsymbol{\tau}_\kappa$, in such a way that $\mathbf{R}_{l\kappa} = \mathbf{R}_l + \boldsymbol{\tau}_\kappa$ indicates the location of the atom of sublattice κ and cell l. In full generality, the atomic displacements along the Cartesian direction α associated with a phonon eigenmode of wave-vector $\mathbf{q}$ can be written as

$$u_{\kappa\alpha}(l,t) = u^{\mathbf{q}}_{\kappa\alpha} e^{i\mathbf{q}\cdot\mathbf{R}_{l\kappa} - i\omega t}, \tag{9}$$

where $u^{\mathbf{q}}_{\kappa\alpha}$ (independent of either l or t) is an eigenvector of the dynamical matrix at $\mathbf{q}$, and ω is the frequency.

A convenient description of arbitrary mechanical deformations can be established by choosing an acoustic phonon branch, and by performing a long-wave (small $\mathbf{q}$) expansion of its eigenvector in the vicinity of the Γ point. Provided that the long-range electrostatic fields are adequately screened (see Section 2.3 for a discussion), the aforementioned expansion can be written as

$$u^{\mathbf{q}}_{\kappa\alpha} = U_\alpha \left(\delta_{\alpha\beta} + iq_\gamma \Gamma^{\kappa}_{\alpha\beta\gamma} - q_\gamma q_\lambda N^{\kappa}_{\alpha\beta\gamma\lambda} + \cdots \right), \tag{10}$$

where $\mathbf{U}$ is a Cartesian vector, $\delta_{\alpha\beta}$ is the Kronecker delta, and $\Gamma^{\kappa}_{\alpha\beta\gamma}$ and $N^{\kappa}_{\alpha\beta\gamma\lambda}$ are third- and fourth-order tensors, respectively. (The dots stand for higher-order terms, which are irrelevant in the context of the phenomena described here.) At order zero in $\mathbf{q}$ the phonon eigenmode is a rigid translation of the whole lattice along $\mathbf{U}$ (note the absence of a sublattice index), while the first- and second-order terms describe the internal-strain response of the lattice to a uniform strain or to a macroscopic strain gradient, respectively.

Of course, to obtain the relevant electromechanical coupling coefficients, the sole knowledge of the lattice distortions is not sufficient — one needs to establish a link between atomic displacements and macroscopic polarization. While in a simplified point-charge model such a link is straightforward, in the case of a more realistic quantum-mechanical description of a solid things are significantly more involved, as one needs to understand how the electronic wavefunctions, and not only the nuclei, respond to a macroscopic deformation. If the deformation is sufficiently slow, which is the case of the phenomena described in this chapter, the electronic cloud responds adiabatically to atomic motion by generating a microscopic current density (i.e. the quantum-mechanical probability current). For example, if we displace by hand one atomic sublattice as

$$u_{\kappa\beta}(l,t) = \lambda(t)e^{i\mathbf{q}\cdot\mathbf{R}_{l\kappa}}, \tag{11}$$

the microscopic current density that is linearly induced by such a perturbation can be written as[b]

$$\mathbf{J}(\mathbf{r},t) = \dot{\lambda}(t)\mathbf{P}^{\mathbf{q}}_{\kappa\beta}(\mathbf{r})e^{i\mathbf{q}\cdot\mathbf{r}}. \tag{12}$$

The function $\mathbf{P}^{\mathbf{q}}_{\kappa\beta}(\mathbf{r})$ is the microscopic polarization response; its cell average,

$$\overline{\mathbf{P}}^{\mathbf{q}}_{\kappa\beta} = \frac{1}{\Omega} \int_{\text{cell}} d^3r\, \mathbf{P}^{\mathbf{q}}_{\kappa\beta}(\mathbf{r}), \tag{13}$$

where Ω is the cell volume, describes the contribution of atomic motion to the macroscopic polarization, which is the quantity we are ultimately interested in.

To go from here to the electromechanical tensors we need one more step, i.e. the small-$\mathbf{q}$ expansion of $\overline{\mathbf{P}}^{\mathbf{q}}_{\kappa\beta}$. Again expanding in powers of $\mathbf{q}$ and keeping terms up to second order,[c]

$$\overline{\mathbf{P}}^{\mathbf{q}}_{\kappa\beta} = \overline{\mathbf{P}}^{(0)}_{\kappa\beta} - iq_\gamma \overline{\mathbf{P}}^{(1,\gamma)}_{\kappa\beta} - \frac{q_\gamma q_\lambda}{2}\overline{\mathbf{P}}^{(2,\gamma\lambda)}_{\kappa\beta} + \cdots. \tag{14}$$

---

[b] Recall that, in classical electrostatics, the density of bound currents $\mathbf{J}$ and the microscopic polarization $\mathbf{P}$ are related by $\mathbf{J} = \partial\mathbf{P}/\partial t$.

[c] Note the difference in sign convention between equations (10) and (14). In the former case, the choice of the sign was uniquely determined by the interpretation of $\mathbf{\Gamma}$ and $\mathbf{N}$ as internal-strain response tensors. In the latter case, the adopted convention allows one to identify the $\mathbf{P}^{(n)}$ tensors with the real-space moments of the current-density response.[13,14]

The zeroth order term is the macroscopic polarization response to a rigid translation of the sublattice κ along the direction β. This corresponds precisely to the definition of the Born dynamical charge tensor Z^*,

$$\overline{P}^{(0)}_{\alpha\kappa\beta} = \frac{Z^*_{\kappa,\alpha\beta}}{\Omega}. \tag{15}$$

The remaining **P**-tensors can be regarded as higher-order counterparts of the Born charges. (Physically they are directly related to the moments of the current density induced by the displacement of an isolated atom.[13,14])

Multiplying the lattice-polarization coupling tensors with the phonon eigendisplacements, we can collect terms order-by-order in **q**. The zero-order term (rigid translation) vanishes due to the acoustic sum rule. At first order in **q**, we obtain the explicit expression for the piezoelectric tensor[15]

$$e_{\alpha\beta\gamma} = -\sum_\kappa \overline{P}^{(1,\gamma)}_{\alpha,\kappa\beta} + \frac{Z^*_{\kappa,\alpha\rho}}{\Omega} \Gamma^\kappa_{\rho\beta\lambda}, \tag{16}$$

where the first and second terms are the electronic (frozen-ion) and lattice-mediated terms respectively.[d] Collecting the terms at second order in **q** gives the flexoelectric response, which is again a sum

$$\mu^{\mathrm{I}}_{\alpha\beta,\gamma\lambda} = \bar{\mu}^{\mathrm{I}}_{\alpha\beta,\gamma\lambda} + \mu^{\mathrm{I,mix}}_{\alpha\beta,\gamma\lambda} + \mu^{\mathrm{I,latt}}_{\alpha\beta,\gamma\lambda}, \tag{17}$$

of electronic and lattice terms

$$\bar{\mu}^{\mathrm{I}}_{\alpha\beta,\gamma\lambda} = \frac{1}{2} \sum_\kappa \overline{P}^{(2,\gamma\lambda)}_{\alpha,\kappa\beta}, \tag{18}$$

$$\mu^{\mathrm{I,mix}}_{\alpha\beta,\gamma\lambda} = -\frac{1}{2} \left(\Gamma^\kappa_{\rho\beta\gamma} \overline{P}^{(1,\lambda)}_{\alpha,\kappa\rho} + \Gamma^\kappa_{\rho\beta\lambda} \overline{P}^{(1,\gamma)}_{\alpha,\kappa\rho} \right), \tag{19}$$

$$\mu^{\mathrm{I,latt}}_{\alpha\beta,\gamma\lambda} = \frac{Z^*_{\kappa,\alpha\rho}}{\Omega} N^\kappa_{\rho\beta\gamma\lambda}, \tag{20}$$

---

[d] The unsymmetrized strain is $\tilde{\varepsilon}_{\beta\gamma}(\mathbf{r}) = iU_\beta q_\gamma e^{i\mathbf{q}\cdot\mathbf{r}}$; this can be replaced with the symmetrized strain tensor after observing that both terms on the right-hand side are invariant with respect to $\beta\gamma$ exchange.

where the bar symbol on the first term indicates a purely electronic response and 'mix' and 'latt' refer to "mixed" and "lattice-mediated" contributions, respectively. While the piezoelectric and flexoelectric responses have been developed in parallel until now, we will henceforth concentrate on the latter, referring the reader to Refs. [13, 14] for the detailed treatment of the piezoelectric response. The corresponding type-II flexoelectric responses are

$$\mu^{\mathrm{II}}_{\alpha\lambda,\beta\gamma} = \bar{\mu}^{\mathrm{II}}_{\alpha\lambda,\beta\gamma} + \mu^{\mathrm{II,mix}}_{\alpha\lambda,\beta\gamma} + \mu^{\mathrm{II,latt}}_{\alpha\lambda,\beta\gamma}, \tag{21}$$

where

$$\bar{\mu}^{\mathrm{II}}_{\alpha\lambda,\beta\gamma} = \frac{1}{2} \sum_{\kappa} \left(\overline{P}^{(2,\gamma\lambda)}_{\alpha,\kappa\beta} + \overline{P}^{(2,\lambda\beta)}_{\alpha,\kappa\gamma} - \overline{P}^{(2,\beta\gamma)}_{\alpha,\kappa\lambda} \right), \tag{22}$$

$$\mu^{\mathrm{II,mix}}_{\alpha\lambda,\beta\gamma} = -\Gamma^{\kappa}_{\rho\beta\gamma}\, \overline{P}^{(1,\lambda)}_{\alpha,\kappa\rho}, \tag{23}$$

$$\mu^{\mathrm{II,latt}}_{\alpha\lambda,\beta\gamma} = \frac{Z^{*}_{\kappa,\alpha\rho}}{\Omega} \left(N^{\kappa}_{\rho\beta,\lambda\gamma} + N^{\kappa}_{\rho\gamma,\lambda\beta} - N^{\kappa}_{\rho\lambda,\beta\gamma} \right). \tag{24}$$

For later convenience we rewrite equation (24) as

$$\mu^{\mathrm{II,latt}}_{\alpha\lambda,\beta\gamma} = \frac{Z^{*}_{\kappa,\alpha\rho}}{\Omega} L^{\kappa}_{\rho\lambda,\beta\gamma}, \tag{25}$$

where $L^{\kappa}_{\rho\lambda,\beta\gamma}$ (the type-II counterpart of the type-I internal-strain tensor $\mathbf{N}$) is the quantity in parentheses on the right-hand side of equation (24).

To summarize, according to Eq. (9) a long-wavelength sound wave is comprised of a lattice-periodic distortion pattern $u^{\mathbf{q}}_{\kappa\alpha}$ modulated by a time- and space-dependent complex phase factor. At zero order in $\mathbf{q}$ the deformation can be described as purely "elastic", but at higher orders (i.e. when moving away from the zone center), internal relaxations of the basis atoms in the primitive cell occur, as described by the tensors $\mathbf{\Gamma}$ and $\mathbf{N}$ (or $\mathbf{L}$) at first and second orders in $\mathbf{q}$, respectively. These are related to how the crystal locally responds to a macroscopic strain (first order, "piezo") or strain gradient (second order, "flexo"). The reader is referred to Refs. [13, 14] for the derivation of explicit expressions for these tensors, but we shall highlight the main conceptual issues associated with them in Section 2.4. Each consecutive order in equation (10) gives rise to a corresponding term in the expressions for the flexoelectric tensor in equations (17) and (21).

Regarding the purely electronic term, $\bar{\mu}$, which is associated with the purely elastic part (order-zero in $\mathbf{q}$, also referred to as "frozen ion deformation"), we defer its detailed discussion to Section 2.5. It can be shown that the "mixed" $\mu_{\alpha\lambda,\beta\gamma}^{\mathrm{II,mix}}$ term involving $\Gamma_{\rho\beta\gamma}^{\kappa}$ is active only in crystals that are characterized by Raman-active phonons,[14] which is not the case for simple systems such as cubic rocksalt or perovskite crystals. (Again, we refer the reader to Refs. [13, 14] for the explicit discussion of this term.) By contrast, the $\mu_{\alpha\lambda,\beta\gamma}^{\mathrm{II,latt}}$ term is present in any insulator with IR-active phonons; as this term is very important in practical applications of the flexoelectric effect, we shall discuss it shortly in Section 2.4. First, however, we shall briefly comment on an important issue that is relevant to the above discussion, concerning the treatment of the macroscopic electric fields in the long-wave phonon analysis.

2.3. *Macroscopic Electric Fields*

Depending on their polarity, long-wave phonons in a crystalline insulator generally produce macroscopic electric fields. These are due to the charge perturbation that is generated by the lattice distortion, and have a non-analytic behavior in the vicinity of the Γ point. For example, for a monochromatic perturbation such as that of equation (11), at the lowest order in $\mathbf{q}$ the macroscopic electric field tends to a direction-dependent constant,

$$\overline{\mathbf{E}}_{\kappa\beta}^{\mathbf{q}\to 0} \sim -\frac{\mathbf{q}}{\epsilon_0\Omega}\frac{(\mathbf{q}\cdot\mathbf{Z}^*)_{\kappa\beta}}{\mathbf{q}\cdot\bar{\epsilon}_{\mathrm{r}}\cdot\mathbf{q}}, \tag{26}$$

where $\overline{\mathbf{E}}_{\kappa\beta}^{\mathbf{q}}$ is defined in analogy with equation (13) and $\bar{\epsilon}_{\mathrm{r}}$ is the purely electronic relative permittivity tensor. The main physical consequence of this is the well-known frequency splitting between longitudinal optical (LO) and transverse optical (TO) phonons in polar crystals. In particular, due to the contribution of equation (26) to the dynamical matrix, the LO dispersion curves behave non-analytically already at zero order in $\mathbf{q}$; that is, the eigenvalue and eigenvector associated with an LO branch generally depends on the direction along which one approaches Γ. Such a non-analiticity propagates directly to the electronic and lattice response functions described in the previous section, and needs to be adequately treated in order to be able to apply the Taylor expansions described in equations (10) and (14).[e]

There are several ways to approach this problem. For example, the theory of Ref. [14] was developed for purely transverse and longitudinal phonons separately, leading to flexoelectric coefficients defined at fixed $\mathbf{E}$ and $\mathbf{D}$ (electric displacement field) respectively. Here, we take the approach of removing the macroscopic $\mathbf{E}$-fields[f] in a physically meaningful way by assuming, following Martin,[3] that a very low density of free carriers is present in the insulating crystal, and that these are allowed to redistribute adiabatically in response to a phonon perturbation. In particular, within the Thomas–Fermi approximation, we write the free-carrier density as

$$\rho^{\text{free}}(\mathbf{r}) = -\epsilon_0 k_{\text{TF}}^2 V(\mathbf{r}), \tag{27}$$

where $V(\mathbf{r})$ is as usual the electrostatic potential, and we suppose that the Fermi wave-vector k_{TF} is much smaller than any reciprocal lattice vector of the crystal. In such a regime, the ground-state charge density and wavefunctions are essentially unaffected by the additional screening provided by the free-electron gas. Conversely, in the long-wave limit, the presence of the free carriers drastically alters the

---

[e] The response to an acoustic phonon in a non-piezoelectric insulator is non-analytic only at second order in $\mathbf{q}$, so the situation appears here, at first sight, less serious than in the case of optical phonons. Recall, however, that the flexoelectric tensor is precisely an $\mathcal{O}(q^2)$ property, and therefore it is directly affected by such issues.

[f] It is desirable to remove the macroscopic fields not only for practical reasons, i.e. to make the aforementioned Taylor expansions possible, but also because electromechanical tensors are traditionally defined in short-circuit electrical boundary conditions.

electrostatics; for example, the field of equation (26) becomes[13]

$$\overline{\mathbf{E}}_{\kappa\beta}^{\mathbf{q}\to 0} \sim -\frac{\mathbf{q}}{\epsilon_0 \Omega} \frac{(\mathbf{q} \cdot \mathbf{Z}^*)_{\kappa\beta}}{k_{\mathrm{TF}}^2 + \mathbf{q} \cdot \bar{\epsilon}_{\mathrm{r}} \cdot \mathbf{q}}. \tag{28}$$

Such a modification has the following effects:

- The macroscopic electric fields, and hence all the response properties of the crystal, become *analytic* functions of $\mathbf{q}$.
- The macroscopic electric fields vanish at zero and first order in $\mathbf{q}$, and also at second order in $\mathbf{q}$ provided that we are considering an acoustic phonon branch.
- Both the piezoelectric and flexoelectric tensors calculated in the presence of the free carrier gas are independent of k_{TF}, and therefore can be unambiguously interpreted as the short-circuit versions of the corresponding electromechanical response functions.

In the first-principles calculations, this is done in practice by simply suppressing the $\mathbf{G} = 0$ contribution to the electrostatic energy when computing the self-consistent linear response; this has the same effect as introducing a low-density electron gas as described above.

Based on the above discussion, it would be tempting to conclude that the flexoelectric tensor, like the piezoelectric tensor, is well defined under short-circuit electrical boundary conditions. In writing down equation (27), however, we assumed a particular type of carriers, namely electrons (not holes), and moreover that the band edge for those carriers, the conduction-band minimum (CBM), tracks with the macroscopic electrostatic potential of the crystal. In general, however, the CBM energy may shift relative to the local macroscopic potential as a result of a strain gradient, via the so-called deformation-potential effect. Thus, we can obtain a different flexoelectric tensor depending on what band feature (CBM or other) we choose as the energy reference. We shall come back to this point in Section 2.5.1.

For a given energy reference, the bulk flexoelectric tensor $\boldsymbol{\mu}$ is well defined in short-circuit (fixed $\mathbf{E}$) boundary conditions. If fixed $\mathbf{D}$ boundary conditions are imposed along a specific direction $\hat{\mathbf{q}}$, the induced electric field (defined as the tilt of the corresponding reference potential) can be then easily calculated as[g]

$$\mathbf{\Delta E}^{\text{bulk}} = -\frac{\mathbf{q}}{\epsilon_0}\frac{q_\alpha \mu^{\text{II}}_{\alpha\lambda,\beta\gamma}\mathcal{E}_{\beta\gamma,\lambda}}{\mathbf{q}\cdot\boldsymbol{\epsilon}_{\text{r}}\cdot\mathbf{q}}. \tag{29}$$

Note that equation (29) cannot be written in tensorial form, except for the simplest case of crystals with cubic symmetry, where the denominator reduces to a direction-independent constant.

2.4. *Lattice Response*

To gain some insight into the nature of the lattice-mediated flexoelectric effect it is necessary to understand, in broad terms, the physics behind the internal-strain response (as described by the tensors $\mathbf{N}$ or $\mathbf{L}$) to a strain gradient deformation. To that end, suppose that we perform a computational experiment where we statically freeze in a lattice distortion that corresponds to an acoustic[h] phonon truncated to first order in $\mathbf{q}$, i.e. to the uniform-strain level,

$$u^l_{\kappa\alpha} = \left(\delta_{\alpha\beta} + iq_\gamma \Gamma^\kappa_{\alpha\beta\gamma}\right) U_\beta e^{i\mathbf{q}\cdot\mathbf{R}_{l\kappa}}. \tag{30}$$

(In the simplest crystal structures, where the $\boldsymbol{\Gamma}$ tensor identically vanishes, this corresponds to a purely elastic wave.) As we have perturbed the crystal from its equilibrium configuration, each atom in the lattice (identified, as usual, by a cell index l and a basis index κ) will experience a restoring force $f^l_{\kappa\alpha}$. If the amplitude of the deformation is small (linear-response regime), such forces can be described, as usual, by a lattice-periodic (i.e. l-independent) function that is modulated by a complex phase with the same wave-vector $\mathbf{q}$ as the perturbation. For small $\mathbf{q}$, it can be shown that the magnitude of the induced forces scales as $\mathcal{O}(q^2)$ (first-order terms cannot be present, as we have assumed that uniform-strain effects are already

---

[g] Strictly speaking, this is the contribution from bulk effects; one cannot exclude surface contributions to the internal field, as we shall see in the later sections.
[h] We assume that the long-range Coulomb fields have been removed; see Section 2.3 for details.

included), and can be written as

$$f^l_{\kappa\alpha} \sim -q_\gamma q_\lambda U_\beta T^\kappa_{\alpha\beta,\gamma\lambda} e^{i\mathbf{q}\cdot\mathbf{R}_{1\kappa}}. \tag{31}$$

Here $T^\kappa_{\alpha\beta,\gamma\lambda}$ is, by construction, the type-I flexoelectric force-response tensor. (The detailed derivation can be found in Refs. [13, 14].)

Now one would be tempted, in close analogy to the piezoelectric case, to define the internal-strain response tensor $\mathbf{N}$ by means of the following linear system of equations,

$$\Phi^{(0)}_{\kappa\alpha\kappa'\rho} N^{\kappa'}_{\rho\beta,\gamma\lambda} \stackrel{?}{=} T^\kappa_{\alpha\beta,\gamma\lambda}, \tag{32}$$

where $\Phi^{(0)}_{\kappa\alpha\kappa'\rho}$ is the zone-center force-constant matrix.[i] Unfortunately, the above system is generally not solvable: the sublattice- (κ-) sum of the $\mathbf{T}$-tensor does not vanish, and the $\Phi^{(0)}$ matrix is singular. (It is always characterized by three null eigenvalues, corresponding to rigid translations of the crystal as a whole.) As negative as it sounds, this is nonetheless an important result: it tells us that the internal-strain response to a *static* strain-gradient deformation is *generally* ill-defined. (We shall see later on that there are notable exceptions to this statement, though.)

To understand what went wrong, let us start all over again, but instead of considering a static (frozen-in) deformation, take a dynamical one, i.e. a phonon mode. By performing a long-wave expansion of the equations of motion one obtains,[13,14] for the second-order eigendisplacements,

$$\Phi^{(0)}_{\kappa\alpha\kappa'\rho} N^{\kappa'}_{\rho\beta,\gamma\lambda} = T^\kappa_{\alpha\beta,\gamma\lambda} - \frac{m_\kappa}{M} \sum_{\kappa'} T^{\kappa'}_{\alpha\beta,\gamma\lambda}, \tag{33}$$

where m_κ are atomic masses and $M = \sum_\kappa m_k$. Equation (33) is in all respect analogous to equation (32), except for the additional term that appears on the right-hand side (rhs) of the latter. It is trivial to check that the sublattice sum of the rhs now correctly vanishes,

---

[i] $\Phi^{(0)}$ is the $\mathbf{q} \to 0$ limit of the matrix $\Phi^{\mathbf{q}}_{\kappa\alpha\kappa'\rho}$, which is essentially a dynamical matrix with the mass prefactors set to unity. As for other quantities, $\Phi^{(0)}$ is defined at vanishing macroscopic electric field, i.e. closed-circuit boundary conditions, appropriate for computing transverse optical phonon frequencies.

providing us with well-defined values (modulo a rigid translation) for the **N**-tensor components. This confirms our earlier suspicions that, unlike piezoelectricity, flexoelectricity is a genuinely *dynamical* effect: only in a sound wave are the internal strains well defined, and these internal strains depend explicitly on atomic masses. In retrospect, this conclusion is not entirely surprising. A uniform strain can always be generated and sustained by applying an appropriate distribution of external loads to the surface of the sample. This is not the case for a strain gradient: in general, a uniform force field applied to each material point of the sample is necessary to generate a given component of $\varepsilon_{\beta\gamma,\lambda}$. Such a uniform force can be, e.g. generated by a gravitational field[14] or, as in the above example of the sound wave, by the acceleration of each material point during its periodic oscillation.[13] In either case, the result directly depends on the atomic masses.

To gain further insight into the physical nature of the mass-dependent term in equation (33), it is useful to write the same equation in type-II form,

$$\Phi^{(0)}_{\kappa\alpha\kappa'\rho} L^{\kappa'}_{\rho\lambda,\beta\gamma} = C^{\kappa}_{\alpha\lambda,\beta\gamma} - \frac{m_{\kappa}}{M} \Omega \mathcal{C}_{\alpha\lambda,\beta\gamma}. \tag{34}$$

Here, C^{κ} is the type-II flexoelectric force-response tensor, linked to **T** via the usual permutation of indices,

$$C^{\kappa}_{\alpha\lambda,\beta\gamma} = T^{\kappa}_{\alpha\beta,\gamma\lambda} + T^{\kappa}_{\alpha\gamma,\lambda\beta} - T^{\kappa}_{\alpha\lambda,\beta\gamma} \tag{35}$$

and $\mathcal{C}_{\alpha\lambda,\beta\gamma}$ is the macroscopic elastic tensor. To write equation (34), we have made use of the result

$$\sum_{\kappa} C^{\kappa}_{\alpha\lambda,\beta\gamma} = \Omega \mathcal{C}_{\alpha\lambda,\beta\gamma}, \tag{36}$$

which directly relates flexoelectricity to elasticity.[13] To justify such a sum rule recall that, in the context of linear elasticity, the stress tensor $\sigma_{\alpha\beta}$ (which we allow to be inhomogeneous in space) is directly related to the elastic and strain tensors via

$$\sigma_{\alpha\beta}(\mathbf{r}) = \mathcal{C}_{\alpha\beta\gamma\lambda}\varepsilon_{\gamma\lambda}(\mathbf{r}). \tag{37}$$

Recall also that the divergence of the stress tensor integrated over a finite region of space yields the net force acting on the corresponding volume element of the material,

$$f_\alpha = \int_\Omega d^3 r \, \nabla_\beta \sigma_{\alpha\beta}(\mathbf{r}). \qquad (38)$$

By assuming that the crystal is homogeneous (i.e. that the elastic tensor is a constant), and by assuming that the deformation varies slowly over the volume of a primitive cell, we have

$$\sum_\kappa f_\alpha^\kappa(\mathbf{r}) = \Omega \mathcal{C}_{\alpha\beta\gamma\lambda} \varepsilon_{\gamma\lambda,\beta}(\mathbf{r}). \qquad (39)$$

Assuming that the force on individual atoms is exclusively produced by strain-gradient effects (which is justified, as the relaxations due to the local strain are already included), we can replace f_α^κ with the definition of the flexoelectric force-response tensor, and easily recover equation (36). Thus, in a hand-waving way, one can say that the type-II flexoelectric force-response tensor is a "sublattice-resolved" version of the macroscopic elastic coefficients.

The dynamical nature of the flexoelectric tensor is worrisome if we are to use this theory to rationalize typical experiments — these are typically performed statically. As we shall see in the following, this is not a real issue. If a material is at static equilibrium there might be non-vanishing stress fields due to the application of external loads; nevertheless, the force acting on a material point must vanish everywhere in space. This leads to the following condition on the *strain-gradient* field,

$$\sum_{\beta\gamma\lambda} \mathcal{C}_{\alpha\lambda,\beta\gamma} \varepsilon_{\beta\gamma,\lambda}(\mathbf{r}) = 0. \qquad (40)$$

This means that two or more strain-gradient components will typically be present in any inhomogeneous strain field, in such a way that their respective net forces mutually cancel. By using equation (40) it is straightforward to see that the mass dependence disappears from the resulting polarization field (as obtained by multiplying the flexoelectric tensor by the local strain-gradient tensor), confirming the internal consistency of the theory.

The important message here is that, at the static level, we can define a number of *effective* flexoelectric coefficients; each of them will correspond to a linearly independent set of strain-gradient components that satisfies equation (40). (An explicit example is provided in Section 3.) It is easy to see that the number of such effective static coefficients is always smaller than the number of independent components of the $\boldsymbol{\mu}$-tensor. This means that the latter contains, in fact, more information than is actually needed to predict the outcome of a static measurement. This also means that, in order to determine the full flexoelectric tensor, one cannot rely on static experiments only; additional dynamical data need to be combined with the static results.[16] The resulting values of the tensor components are always inherently dynamic quantities, even if static data are, in part, used to compute them.

2.5. *Electronic Response*

While the lattice-mediated response has a straightforward physical interpretation (i.e. in terms of a polar distortion of the basis atoms that is induced by the macroscopic strain gradient), the purely electronic response (given by the tensor $\bar{\mu}^{\mathrm{II}}_{\alpha\lambda,\beta\gamma}$) is far less intuitive, and therefore deserves a separate discussion. First, recall that $\bar{\mu}^{\mathrm{II}}_{\alpha\lambda,\beta\gamma}$ is defined in terms of the second-order **P**-tensor, $\overline{P}^{(2,\gamma\lambda)}_{\alpha,\kappa\beta}$. To understand the physical meaning of the latter, consider the microscopic current density $J_\alpha(\mathbf{r})$ that is adiabatically induced when displacing an isolated atom (l,κ) with velocity $\dot{u}^l_{\kappa\beta}$ in the Cartesian direction β,[12,13]

$$\mathcal{P}_{\alpha,\kappa\beta}(\mathbf{r}) = \frac{\partial J_\alpha(\mathbf{r} + \mathbf{R}_{l\kappa})}{\partial \dot{u}^l_{\kappa\beta}}. \tag{41}$$

Provided that the macroscopic electric fields have been appropriately screened,[13] one can introduce[14] the moments of the vector field $\mathcal{P}_{\alpha,\kappa\beta}(\mathbf{r})$ at an arbitrary order n,

$$J^{(n,\gamma_1\ldots\gamma_n)}_{\alpha,\kappa\beta} = \int d^3r \, \mathcal{P}_{\alpha,\kappa\beta}(\mathbf{r}) r_{\gamma_1} \ldots r_{\gamma_n}. \tag{42}$$

Then, one can show[13] that the resulting $\mathbf{J}$-tensors coincide with the $\overline{P}$-tensors of the same order apart from a trivial factor of volume,

$$J^{(n,\gamma_1\ldots\gamma_n)}_{\alpha,\kappa\beta} = \Omega \overline{P}^{(n,\gamma_1\ldots\gamma_n)}_{\alpha,\kappa\beta}. \tag{43}$$

This result tells us that the "frozen-ion" (in the sense specified in Ref. [12]) contributions to the piezoelectric and flexoelectric tensors are given in terms of the first and second moments of the current-density response to atomic displacements, respectively.

Direct calculation of the $\overline{P}$-tensors is technically challenging at the time of writing — the required current-density response functions are presently not available in the existing implementations of DFPT. To avoid this complication altogether, Resta[11] proposed to determine the frozen-ion flexoelectric tensor via the sole knowledge of the *charge-density* response to an acoustic phonon, in close analogy with Martin's classic treatment of the piezoelectric problem.[3] In particular, for an elemental crystal (this result was later generalized to arbitrary crystals by Hong and Vanderbilt[12]) Resta demonstrated that the longitudinal component of the response to a longitudinal strain gradient is given by

$$\mu_{\hat{\mathbf{q}}} = \frac{1}{6\Omega} Q^{(3)}_{\hat{\mathbf{q}}}. \tag{44}$$

Here $\hat{\mathbf{q}}$ indicates the spatial direction of interest, and $Q^{(3)}$ indicates the corresponding *third* moment of the charge-density response to atomic displacement (dynamical octupole).

To derive this result in the context of the formalism of Section 2.2, it is useful to introduce the charge-density response to the monochromatic lattice perturbation of equation (11),

$$\overline{\rho}^{\mathbf{q}}_{\kappa\beta} = -iq_\gamma \overline{\rho}^{(1,\gamma)}_{\kappa\beta} - \frac{q_\gamma q_\lambda}{2}\overline{\rho}^{(2,\gamma\lambda)}_{\kappa\beta} + i\frac{q_\gamma q_\lambda q_\delta}{6}\overline{\rho}^{(3,\gamma\lambda\delta)}_{\kappa\beta} + \cdots, \tag{45}$$

where the overline symbol implies cell averaging as in equation (13), and we have pushed the expansion up to *third* order in $\mathbf{q}$. (The zero-order term vanishes because of the condition of charge conservation.) The $\bar{\rho}$ tensors are trivially related via

$$\bar{\rho}_{\kappa\beta}^{(n,\gamma_1\ldots\gamma_n)} = \frac{1}{\Omega}Q_{\kappa\beta}^{(n,\gamma_1\ldots\gamma_n)}, \tag{46}$$

to the moments

$$Q_{\kappa\beta}^{(n,\gamma_1\ldots\gamma_n)} = \int d^3r\, f_{\kappa\beta}(\mathbf{r})r_{\gamma_1}\ldots r_{\gamma_n} \tag{47}$$

of the charge-density response function $f_{\kappa\beta}(\mathbf{r})$,[j] defined as the change in charge density resulting from a single ionic displacement $\kappa\beta$. One can also show that the $\mathbf{\overline{P}}$-tensors and $\bar{\rho}$-tensors are related by[13]

$$\bar{\rho}_{\kappa\beta}^{(n,\gamma_1\ldots\gamma_N)} = \sum_l \overline{P}_{\gamma_l,\kappa\beta}^{(n-1,\gamma_1\ldots[\gamma_l]\ldots\gamma_n)} \quad (n \geq 1), \tag{48}$$

where the symbol $[\gamma_l]$ indicates the absence of the element l in the list. Then one immediately has, for $n = 3$,

$$J_{\alpha,\kappa\beta}^{(2,\gamma\lambda)} + J_{\lambda,\kappa\beta}^{(2,\alpha\gamma)} + J_{\gamma,\kappa\beta}^{(2,\lambda\alpha)} = Q_{\kappa\beta}^{(3,\alpha\gamma\lambda)}. \tag{49}$$

By applying equations (18) and (49) to the case of a longitudinal strain gradient oriented along $\hat{\mathbf{q}}$, one easily recovers equation (44).

Unfortunately, it is not possible to invert equation (49) and extract all components of the $\mathbf{J}^{(2)}$-tensor from the octupolar response tensor, $\mathbf{Q}^{(3)}$. (The fact that $\mathbf{J}^{(2)}$ contains more information than $\mathbf{Q}^{(3)}$ can be already appreciated by counting the maximum number of independent entries in either tensor: 54 in the former, 30 in the latter.[14]) Therefore, working only with the charge-density response at the bulk level is not a viable route to achieving full information over the frozen-ion (electronic) flexoelectric tensor, $\bar{\boldsymbol{\mu}}$.

Such a limitation can be circumvented, at least in cubic crystals, by considering a more general class of deformations that cannot be straightforwardly described as bulk acoustic phonons. For example, as we shall see in the next section, the open-circuit internal field that

---

[j] The function f is, in all respects, analogous to that introduced by Martin in his seminal work on piezoelectricity.[3]

is linearly induced by bending a free-standing slab is a bulk property of the material. Since the electric field is uniquely determined by the induced charge density this gives us, in principle, access to the *transverse* component of the electronic flexoelectric tensor without the need for calculating the polarization response. A bending deformation can be conveniently simulated (although at the price of a significantly higher computational cost) by adopting a slab geometry, and by performing a long-wave analysis analogous to that described here to the corresponding slab supercell.

A formal derivation clarifying whether such a procedure does indeed yield the same transverse flexoelectric component as the **P**-response theory is still missing, due to subtleties at both the conceptual and technical levels. We shall refer to these issues again in the discussion following equation (101). In the remainder of the chapter we shall disregard such issues, and provisionally assume that this relationship holds, i.e. that the bending-induced open-circuit (OC in equation (50)) **E**-field and the corresponding component of the flexoelectric tensor are related by

$$\frac{\partial E_x^{\mathrm{OC}}}{\partial \varepsilon_{yy,x}} = -\frac{\mu_{xx,yy}^{\mathrm{II}}}{\epsilon_0 \bar{\epsilon}_{\mathrm{r}}}, \tag{50}$$

[for a beam bent as in Fig. 6(b)], where $\bar{\epsilon}_{\mathrm{r}}$ is the (isotropic) relative permittivity of the material. We shall use equation (50) from now on, whenever necessary, to resolve the aforementioned indeterminacy in the transverse components of $\bar{\boldsymbol{\mu}}$.

Note that this issue does not apply to the simpler case of the piezoelectric response. In fact, one can write that

$$J_{\alpha\beta}^{(1,\gamma)} + J_{\gamma\beta}^{(1,\alpha)} = Q_{\beta}^{(2,\alpha\gamma)}. \tag{51}$$

(Recall that the basis sum of the $\mathbf{J}^{(1)}$ tensors essentially coincides with the frozen-ion piezoelectric tensor, and that $\mathbf{Q}^{(2)}$ is the dynamical quadrupole tensor.) The above equation can be readily inverted,

$$J_{\alpha\beta}^{(1,\gamma)} = \frac{1}{2}\left[Q_{\beta}^{(2,\alpha\gamma)} + Q_{\gamma}^{(2,\alpha\beta)} - Q_{\alpha}^{(2,\beta\gamma)}\right], \tag{52}$$

which provides an alternative derivation of Martin's theory[3] of piezoelectricity. For completeness, it is useful to mention that, at order zero, the relationship between **J**- and **Q**-tensors is even more direct,

$$J^{(0)}_{\alpha,\kappa\beta} = Q^{(1,\alpha)}_{\kappa\beta} = Z^*_{\kappa,\alpha\beta},\tag{53}$$

where $\mathbf{Z}^*_\kappa$ is the Born effective charge tensor associated with the κ sublattice. Thus, both $\mathbf{J}^{(n)}$ and $\mathbf{Q}^{(n+1)}$ can be regarded as higher-order generalizations of the dynamical charge concept, although starting from $n=2$ (which is relevant for flexoelectricity) the former quantities generally carry more information than the latter ones.

2.5.1. *Spherical term, pseudopotential dependence, and the non-interacting spherical-atom paradox*

As an illustration of the above derivations, it is useful in this context to work out a simple toy model that can be solved analytically; this will be also useful to point out some unconventional aspects of the flexoelectric response that have no counterpart in earlier theories of electromechanical effects in solids. We consider a rocksalt ionic crystal such as NaCl or MgO, and suppose that a longitudinal strain gradient develops along the (100) direction. Here we shall focus on electronic effects only, so that the atomic x coordinates undergo displacements that are a predetermined quadratic function of x with no further relaxations. For the time being we shall also assume that the crystal is perfectly ionic, i.e. that its electronic charge density can be approximated by a superposition of spherical closed-shell ions whose shape is not altered by changes in bond distances, etc. With the above assumptions in mind, one can perform an average of the electrostatic potential in the yz planes, and express the result as a one-dimensional (1D) function of x. The atomic planes will appear as a periodic arrangement of potential wells (each well corresponding to a single charge-neutral monolayer), whose shape will reflect the radial distribution of electrons in the constituent ionic species. For the present purposes, the fine details of the potential wells are irrelevant;

the only important quantity will be

$$K = -e \int_{-\infty}^{\infty} \overline{V}_{\mathrm{ML}}(x)\,dx, \qquad (54)$$

where $\overline{V}_{\mathrm{ML}}(x)$ is the yz-averaged electric (Hartree) potential $V_{\mathrm{ML}}(\mathbf{r})$ generated by one monolayer, and $-e\overline{V}_{\mathrm{ML}}(x)$ is the corresponding electron potential energy. Thus, for purposes of illustration we can represent the potential-energy wells as non-overlapping rectangular dips of area $|K|$, whose shape is fixed and independent of the surrounding neighbors, as sketched in Fig. 1(a). As the wells are all identical and their separations are uniform in the undistorted crystal, the macroscopic electron potential energy obtained by convoluting the corresponding microscopic function with an appropriate low-pass filter,[17] shown as a dashed line in Fig. 1(a), is constant. After freezing in the strain gradient deformation pattern, as shown in Fig. 1(b), the interlayer distance increases linearly along the chosen axis, leading to a constant slope in the macroscopic electrostatic potential and, hence, to a uniform electric field throughout the bulk crystal. This result points to a non-zero flexoelectric coefficient of purely electronic origin, since we explicitly neglected possible internal strains.

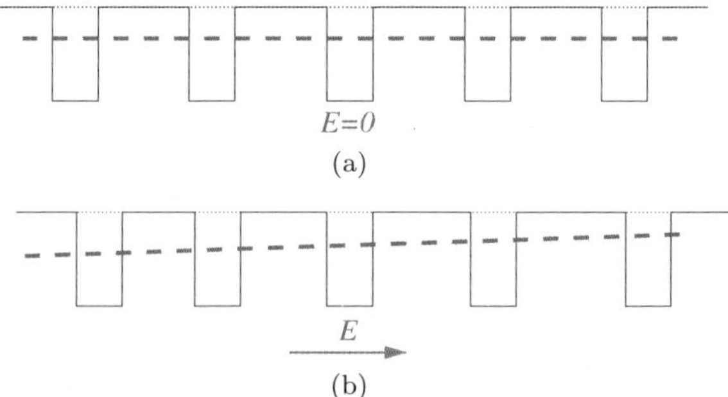

Figure 1.   Simplified sketch of the planar-averaged electron potential energy, $-e\overline{V}(x)$ (black curves) for (a) an undistorted crystal, and (b) a crystal with a uniform longitudinal strain gradient. The dashed lines show the macroscopic averages of the aforementioned functions.

There are several important questions that naturally arise at this stage. The first obvious one is whether (and if yes, how) the outcome of Fig. 1 can be rationalized in the context of the theory developed in this chapter. A second and less obvious issue arises in pseudopotential-based first-principles calculations, where one may wonder whether and how the results depend on choice of pseudopotential. The third question concerns the physical nature of the electric field that we describe in Fig. 1(b). Does it, for example, produce a direct force on charged particles such as electron and hole carriers and ionic cores?

To answer the first question, it suffices to suppose that the electrostatic potential wells are generated by a regular lattice of spherical charge distributions. To make things simple, consider a monatomic lattice, as for a rare-gas solid, that we construct by periodically repeating a spherical charge distribution $\rho_0(\mathbf{r})$. We assume that the volume of the unit cell $\Omega(\mathbf{r})$ depends smoothly on space as a result of an inhomogeneous macroscopic deformation. One can show (see Supplementary Note 1 of Ref. [15]) that the resulting macroscopically averaged (in 3D) electric potential is given by

$$V(\mathbf{r}) = -\frac{1}{6\epsilon_0\Omega(\mathbf{r})} \int d^3s \, s^2 \rho_0(s) = -\frac{1}{6\epsilon_0\Omega(\mathbf{r})} O_{\mathrm{L}}, \qquad (55)$$

where $O_{\mathrm{L}} = 4\pi \int ds \, s^4 \rho_0(s)$ is the isotropic quadrupole moment[k] of the static charge distribution $\rho_0(\mathbf{r})$. Equivalently, it is the longitudinal component $O_{\mathrm{L}} = \sum_\kappa Q_{\kappa x}^{(3,xxx)}$ of the dynamical octupole tensor defined in equation (47), as follows from straightforward algebra.

In the linear regime (small deformations) we have

$$\Omega(\mathbf{r}) \simeq \Omega(1 + \det\left[\varepsilon(\mathbf{r})\right]), \qquad (56)$$

which leads to the variation in the macroscopic electrostatic potential induced by the deformation,

$$\Delta V(\mathbf{r}) = \frac{1}{6\epsilon_0\Omega} \det\left[\varepsilon(\mathbf{r})\right] O_{\mathrm{L}}. \qquad (57)$$

---

[k] That is, the trace of the 3×3 second-moment tensor.

Assuming that the crystal has cubic symmetry and making use of equation (50), this yields, after some algebra, two of the three independent components of the flexoelectric tensor

$$\bar{\mu}^{\mathrm{II}}_{\alpha\alpha,\beta\beta} = \frac{O_{\mathrm{L}}}{6\Omega}. \tag{58}$$

(In the case of a biatomic ionic crystal one simply needs to replace O_{L} with the sublattice sum of the dynamical octupoles of the individual atoms.) By using the relationship between $\mathbf{J}$-tensors and $\mathbf{Q}$-tensors discussed earlier in this section, it is not difficult to deduce that the third component, $\bar{\mu}^{\mathrm{II}}_{\alpha\beta,\alpha\beta}$ with $\alpha \neq \beta$, must be zero. Summarizing the above, the three independent components (longitudinal, transverse and shear) in a rigid-sphere crystal read as

$$\bar{\mu}^{\mathrm{II}}_{xx,xx} = \frac{O_{\mathrm{L}}}{6\Omega}, \quad \bar{\mu}^{\mathrm{II}}_{xx,yy} = \frac{O_{\mathrm{L}}}{6\Omega}, \quad \bar{\mu}^{\mathrm{II}}_{xy,xy} = 0. \tag{59}$$

This demonstrates that the effect illustrated in Fig. 1 is indeed a natural consequence of the theory developed in this chapter.

We now turn to the second question, concerning the use of pseudopotentials in first-principles calculations, as discussed in Ref. [12]. One aspect of the pseudopotential approximation is the replacement of the all-electron charge density $\rho^{\mathrm{AE}}(r)$ by a pseudo charge density $\rho^{\mathrm{PS}}(r)$ in the core region of the atom. Since these charge densities are essentially rigid and spherically symmetric, the above considerations apply to them. As a result, to compensate for the use of the pseudopotential, one should add a "rigid core correction"

$$O^{\mathrm{RCC}}_{\mathrm{L},\kappa} = 4\pi \int ds\, s^4 \left[\rho^{\mathrm{AE}}_{\kappa}(r) - \rho^{\mathrm{PS}}_{\kappa}(r)\right], \tag{60}$$

to the longitudinal dynamic octopole of each atom κ to recover the all-electron result. This propagates into a change $\Delta Q^{(3,xxx)}_{\kappa x} = 3\Delta Q^{(3,xyy)}_{\kappa x} = O^{\mathrm{RCC}}_{\mathrm{L},\kappa}$, and to a change of $\bar{\mu}^{\mathrm{II}}_{xx,xx}$ and $\bar{\mu}^{\mathrm{II}}_{xx,yy}$ (but not $\bar{\mu}^{\mathrm{II}}_{xy,xy}$) by $\sum_{\kappa} O^{\mathrm{RCC}}_{\mathrm{L},\kappa}/6\Omega$.[1]

---

[1] Recall that we work in the framework of equation (50), i.e. we extract the flexoelectric tensor components from the induced electrostatic potential, rather than from the polarization response.

This rigid-core correction is not small, and is not independent of the details of pseudopotential construction. Therefore, two different calculations of the bulk flexoelectric response cannot be directly compared unless this correction has been applied in both cases. Nevertheless, as long as the same pseudopotential is consistently used in the calculation, predictions of physical, experimentally measurable quantities should not be affected by this correction. In particular, we shall see in Section (2.6) than $O_{\mathrm{L},\kappa}^{\mathrm{RCC}}$ makes an equal and opposite contribution to the *surface* contribution. Because of this cancelation, the total (bulk and surface) flexovoltage response [see equation (61)] can be computed without the need for including this correction.

The third question, regarding the physical nature of the resulting electric field, requires taking a closer look at some earlier works on the theory of *absolute deformation potentials*.[18,19] (These can be regarded as the foundation of the modern theory of flexoelectricity, even if they were aimed at addressing a slightly different physical problem.) In a nutshell, if we wish to draw a band diagram of a crystal subjected to a strain-gradient deformation, knowledge of the macroscopic electrostatic field is not sufficient. Indeed, the relative location of the valence-band maximum or conduction-band minimum with respect to the electrostatic reference is itself a function of the local strain (via the so-called *band-structure term*), which implies that each band will "see" a different electric field, see Fig. 2. This means that one band edge may be perfectly flat, and the corresponding carriers feel no force whatsoever, even while the other band edge and/or the mean electrostatic potential can be strongly tilted.[m] In fact, even a metal subjected to a strain gradient will generally have a non-zero internal macroscopic electric field arising from a gradient in the mean electrostatic potential, although no current will flow. Thus, one should be careful not to interpret the macroscopic electric field produced by the flexoelectric effect in a longitudinal acoustic wave as a "real" physical field; it is just the tilt of some arbitrary reference

---

[m] The tilt of the mean electrostatic potential will also depend on choice of pseudopotentials when these are employed, but the tilt of the valence and conduction band edges will not.

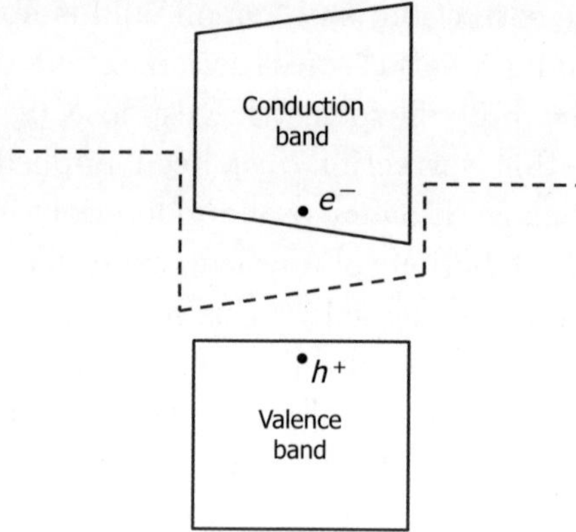

Figure 2.   Sketch of valence bands and conduction bands for a slab with a strain gradient across its width. Dashed line indicates the macroscopic electron potential energy $V(x) = -e\phi(x)$. Hole carriers feel no force because the VB maximum is flat, while electron carriers feel a force to the right, both of which are contrary to naive expectations based on the electric field pointing to the right in the interior of the slab.

energy that may have little to do with the phenomenon of interest in a given specific case. Just as for the notion of a "flexoelectric field", care must be used when speaking of "short-circuit" and "open-circuit" electrical boundary conditions, as these are ambiguous in the non-periodic strain-gradient world.

In light of the above arguments, it is legitimate to wonder whether the bulk flexoelectric effect is experimentally measurable at all. In fact, there are good reasons to believe that the tilt of the mean electrostatic potential does *not* provide a realistic description of the response — at least no more realistic than other reference energies (e.g. the conduction band bottom, or the valence band top, or the Fermi level). First, as we have argued above, the present theory yields a finite open-circuit "flexoelectric field" even in a metal, which is physically inconsistent. Second, if we go back to the example of the non-interacting spherical atoms, there are apparent inconsistencies as well: Since we have assumed that each potential well is independent of its environment, its motion cannot, in principle, be detected by an electrode that is placed at the far-away surface of the sample — and yet, the bulk flexoelectric coefficients do not vanish. We have, therefore, a sort of paradoxical situation, where the presence of a macroscopic electric field inside the material is indisputable, but at the same time there cannot be any open-circuit voltage, because of the hypothesis of rigid potential wells (which excludes long-range effects). To resolve these paradoxes, and place the present theory in the right context regarding experimental measurements, it is necessary to account for surface effects. We shall see how to do this in Section 2.6.

2.6. *Surface Effects*

Knowing whether a given physical property is sensitive to the details of the sample surfaces is a matter of central importance in condensed matter theory. In the majority of cases (e.g. piezoelectricity), surfaces typically start to matter only at small length scales, where they are responsible for deviations in the measured property from the corresponding bulk value. There are situations, however, where such a sensitivity to the crystal termination persists up to the macroscopic scale; flexoelectricity belongs to this category. In the present section, we shall elaborate on this statement from a heuristic point of view, which is anyway sufficient to illustrate the most relevant physical ideas. A more formal discussion, based on a microscopic theory of the response to deformations, will be presented in Sections 2.7 and 2.8.

In order to calculate the flexoelectric response of a finite object such as a slab it is appropriate to consider, rather than the induced macroscopic polarization, the open-circuit voltage ΔV produced by the deformation.[n] We shall only focus, in the following, on contributions that tend to a finite constant in the limit of infinite thickness t, and introduce the *flexovoltage* coefficient,

$$\varphi_{x\lambda,\beta\gamma} = \lim_{t\to\infty} \frac{1}{t} \frac{\partial \Delta V}{\partial \varepsilon_{\beta\gamma,\lambda}}. \tag{61}$$

---

[n] We indicate here by ΔV the total potential step that builds up, as a consequence of the mechanical deformation, between the two vacuum regions located at either side of the slab.

Recall that $\varepsilon_{\beta\gamma,\lambda} = \partial\varepsilon_{\beta\gamma}/\partial r_\lambda$ is the gradient of the symmetric strain tensor along the Cartesian direction r_λ, and x indicates the direction normal to the surface.[o] For simplicity, here we shall also restrict our analysis to strain gradients of the type $\varepsilon_{\alpha\alpha,x}$, i.e. a diagonal (either longitudinal or transverse) component of the symmetric strain tensor that is linearly growing across the slab thickness. (These are sufficient to describe the bending of a free-standing slab; a more general analysis, including the shear component, is deferred to Section 2.8.) We shall write the flexovoltage coefficient as a sum of bulk and surface-specific contributions,

$$\varphi_{xx,\alpha\alpha} = \varphi_{xx,\alpha\alpha}^{\text{bulk}} + \varphi_{xx,\alpha\alpha}^{\text{surf}}, \tag{62}$$

whose explicit forms will be derived in the following paragraphs.

2.6.1. *Electronic surface response*

First let us consider only the purely electronic (frozen-ion) response. Strain gradients of the type $\varepsilon_{\alpha\alpha,x}$ are governed by equation (50); in our present notation this implies that the (open-circuit) uniform electric field that builds up in the interior of the slab as a consequence of the deformation is uniquely given in terms of the bulk flexoelectric coefficient of the material and its macroscopic dielectric constant by

$$\left.\frac{\partial E_x^{\text{slab}}}{\partial\varepsilon_{\alpha\alpha,x}}\right|_{\text{frozen}-\text{ion}} = -\frac{\bar{\mu}_{xx,\alpha\alpha}^{\text{II}}}{\epsilon_0\bar{\epsilon}_{\text{r}}}. \tag{63}$$

Here ϵ_0 and ϵ_{r} are the vacuum and relative permittivities, respectively, while μ^{II} is the type-II flexoelectric tensor; as before, we use the bar symbol to distinguish frozen-ion quantities from fully relaxed ones. Since the electric field is minus the derivative of the potential, the bulk internal field contribution to the overall open-circuit voltage

---

[o] In spite of its notation, $\varphi_{x\lambda,\beta\gamma}$ should not be thought as a tensor. First, the surface contribution depends on the specific details of the crystal termination, and is therefore not a simple function of the surface plane orientation. Second, the bulk contribution is defined in fixed-**D** boundary conditions and therefore it has a non-analytic behavior [see equation (29)] in all materials except those characterized by cubic crystal symmetry.

is then proportional to t, leading to a finite contribution to the overall flexovoltage coefficient that we identify with $\bar{\varphi}^{\text{bulk}}$,

$$\bar{\varphi}^{\text{bulk}}_{xx,\alpha\alpha} = -\left.\frac{\partial E^{\text{slab}}_x}{\partial \varepsilon_{\alpha\alpha,x}}\right|_{\text{frozen-ion}} = \frac{\bar{\mu}^{\text{II}}_{xx,\alpha\alpha}}{\epsilon_0 \bar{\epsilon}_r}. \tag{64}$$

The surface contribution φ^{surf} in equation (62) originates from the fact that a surface can always be characterized by a potential offset ϕ between the macroscopic potential just inside and just outside the surface, and that this offset is different for the two surfaces in the presence of a strain gradient. Consider first the case of a *uniform* strain $\varepsilon_{\alpha\alpha}$ applied to a slab of thickness t, as shown in Figs. 3(a) and (b). The figure shows the derivative of the macroscopic electric field (Panel a) and electron potential energy (Panel b) with respect to the applied uniform strain $\varepsilon_{\alpha\alpha}$, and φ^{surf} is the corresponding derivative of the potential offset ϕ. The variation of ϕ with strain can be regarded as resulting from the fact that the surface, by virtue of its lack of inversion symmetry, is locally piezoelectric.[P] For the slab as a whole, however, a uniform strain does not produce a net voltage, since the induced potential offsets on either side of the slab cancel each other, consistent with the fact that the overall slab is non-piezoelectric.

In the case of a *strain-gradient* deformation, on the other hand, the local strains at the opposite surfaces are opposite in sign, and do not cancel out, as illustrated in Figs. 3(c) and (d). The slab is taken to extend over $-t/2 < x < t/2$ with local strain $\varepsilon_{\alpha\alpha}(x) = x\varepsilon_{\alpha\alpha,x}$, reaching values of $\varepsilon_{\alpha\alpha} = \pm(t/2)\varepsilon_{\alpha\alpha,x}$ at the two surfaces. The figure shows the derivative of the field (Panel c) and potential (Panel d) with respect to $(\varepsilon_{\alpha\alpha,x}t)$. This means that the induced potential offsets at the two opposite surfaces have the same sign and add up in a flexoelectric experiment, leading to a surface contribution of the form

$$\bar{\varphi}^{\text{surf}}_{xx,\alpha\alpha} = \left.\frac{\partial \phi}{\partial \varepsilon_{\alpha\alpha}}\right|_{\text{frozen-ion}}. \tag{65}$$

[P] In another language, we are basically describing a strain dependence of the surface work function, although technically the latter is referenced to the valence band maximum rather than the average potential in the subsurface region.

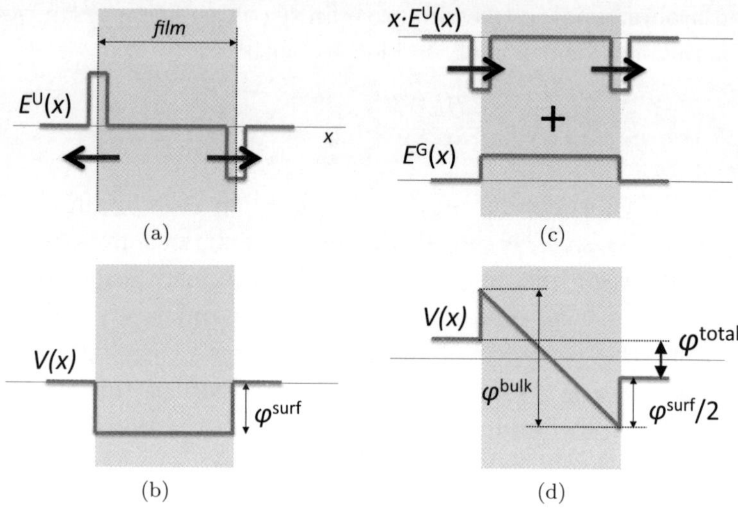

Figure 3.   (a) and (b): Linear response of the electric field (a) and the electric potential (b) to a *uniform* strain applied to a film of thickness t. The function E^U (where U indicates "uniform") can be regarded as representing a surface piezoelectric response (surface dipole layer appearing in response to strain). (c) and (d): Linear response of the field (c) and the potential (d) in the case of a strain *gradient*. The sketch shows the derivative with respect to $(\varepsilon_{\alpha\alpha,x}t)$ for a strain variation $\varepsilon_{\alpha\alpha}(x) = x\varepsilon_{\alpha\alpha,x}$ in a slab extending from $-t/2 < x < t/2$. The field response contains two contributions. The first is given by E^U, appropriately scaled by the linearly varying local strain. Note that the induced surface dipoles, schematically illustrated by arrows, now point in the *same* direction. The other is given by genuine strain-gradient effects contained in E^G, which essentially reflects the bulk flexovoltage response. The resulting potential response in (d) thus consists in a macroscopic internal field plus a surface dipole contribution.

The total flexovoltage coefficient then reads as

$$\bar{\varphi}_{xx,\alpha\alpha} = \frac{\bar{\mu}^{II}_{xx,\alpha\alpha}}{\epsilon_0 \bar{\epsilon}_r} + \bar{\varphi}^{surf}_{xx,\alpha\alpha}. \tag{66}$$

The above derivation allows us to solve the paradoxes that we mentioned at the end of the previous section. First, recall that we encountered some difficulties in giving a physical interpretation to the "internal electric field" that is induced by a strain gradient, as such a field depends on the reference energy (i.e. Bloch electrons in different eigenstates do not experience the same electrical force). This issue is easily solved by observing that the surface potential offset ϕ suffers from the same ambiguity as the bulk flexoelectric field; we defined it relative to the macroscopically averaged electrostatic potential under the surface, but we could have used the valence or conduction band edge instead. It is easy to see that the respective ambiguities contained in the surface and bulk terms exactly cancel, yielding an overall flexovoltage coefficient that is uniquely defined. Next, we have observed that there is an apparent physical inconsistency in the rigid-spherical-atom model, in that there should be no overall voltage response to a strain gradient, and yet the bulk flexoelectric coefficient does not vanish. It is easy to see that, once the surface contribution is taken into account, the total flexovoltage response of a slab made of non-interacting spherical atom is zero as it should be. Indeed, when such a slab is subjected to a uniform strain, its surface potential voltage response is

$$\bar{\varphi}^{\text{surf}}_{xx,xx} = \bar{\varphi}^{\text{surf}}_{xx,yy} = -\frac{O_{\text{L}}}{6\epsilon_0\Omega}, \tag{67}$$

since a positive longitudinal or transverse strain increases the spacing between the atomic spheres and thereby reduces the surface potential offset. But, using equations (58) and (64) (and the fact that $\bar{\epsilon}_{\text{r}} = 1$ for this model), this is exactly $-\bar{\varphi}^{\text{bulk}}_{xx,\alpha\alpha}$, leading to the claimed cancelation in equation (66). This cancelation also explains why the replacement of the all-electron by the pseudo core charge density in the context of pseudopotential calculations has no effect on the total flexovoltage response, so that the rigid-core correction of equation (60) can be neglected, as was claimed in Section 2.5.1.

Note that the spherical atom model, in spite of its simplicity, is crucial to understand how flexoelectricity works in real materials. As we shall see in the results section, there generally tends to be a large cancelation between surface and bulk contributions to the flexoelectric effect. This happens because, even in covalently-bonded materials, the electronic charge distribution that is dragged along by each atom during its motion is largely constituted by a spherical shell, with comparatively smaller aspherical components. Spherical objects do not contribute to the overall flexovoltage coefficient of a slab, hence the aforementioned cancelation.

This gives a measure of the importance of the surface contribution — only when it is correctly taken into account together with the bulk term we obtain a meaningful physical quantity. Therefore, asking whether the surface contribution is "large or small" compared to the bulk effect is a poorly formulated question; the two must always go hand in hand. Instead, a more physically meaningful question is "How strong is the dependence of the surface contribution on its atomic and electronic structure?"

Based on these considerations, one can attempt to give an answer to a long-standing question that has been somewhat controversial in recent years: "Is flexoelectricity a bulk property?" As we said above, if by "flexoelectricity" we refer to the result of a typical flexoelectric experiment (i.e. where the induced current upon bending a short-circuited slab is measured), the answer is *no*. Conversely, if by the same name we call the current flowing through the bulk of the material while well-defined internal electrical boundary conditions are imposed, then the answer is *yes*. The problem is that the internal electrical boundary conditions depend on the externally-applied ones in a way that is surface-dependent, and unlike in the case of most known material properties, such a dependence persists in the limit of a macroscopically thick sample. All in all, in the present context we would rather stay away from the traditional rigid classification into bulk properties and surface properties, as flexoelectricity, strictly speaking, does not belong cleanly to either category.

2.6.2. *Lattice surface response*

We now discuss how the above conclusions need to be modified when full ionic relaxations are incorporated — these are, of course, of the utmost importance for a realistic description of the flexo-electric effect. Essentially, the above conclusion still hold, except for two important details: (i) the frozen-ion quantities (flexoelectric coefficient, dielectric constant, surface potential response) need to be replaced with their relaxed-ion counterparts; (ii) an *effective* deformation, given by an appropriate linear combination of, e.g. a longitudinal and transverse strain gradient, need to be considered in place of the individual tensor components.

To illustrate the implications of (i) and (ii) in a practical situation, it is useful to work out the explicit formulas for the simplest case of an unsupported slab subjected to bending.[q] Linear elasticity dictates that a transverse strain gradient (corresponding to a "frozen-ion" bending deformation) at static equilibrium must be accompanied by a longitudinal strain gradient, which for most materials will have opposite sign compared to the transverse one. In fact, the top layers of the slab ("top" here means furthest from the bending center) are under tensile strain, and this typically induces a longitudinal contraction of such layers, whose magnitude is related to the Poisson's ratio of the material. Conversely, the bottom layers are transversely compressed, and will therefore expand longitudinally by an equal amount. This means that, to calculate the static flexovoltage coefficient of a bent slab, we need to consider the "effective" deformation

$$\varepsilon_{yy,x} = \varepsilon_{\text{eff},x}; \quad \varepsilon_{xx,x} = -\nu\varepsilon_{\text{eff},x}, \tag{68}$$

rather than the individual strain-gradient tensor components, where

$$\nu = \frac{\mathcal{C}_{xx,yy}}{\mathcal{C}_{xx,xx}} \tag{69}$$

is uniquely given by the elastic constants of the bulk material. Consequently, when the ions are relaxed, we shall be concerned with an effective flexovoltage coefficient reflecting the aforementioned mechanical equilibrium condition,

$$\varphi_{xx,\text{eff}} = \frac{\mu^{\text{II}}_{xx,\text{eff}}}{\epsilon_0\epsilon_{\text{r}}} + \frac{\partial\phi}{\partial\varepsilon_{\text{eff}}}, \tag{70}$$

where

$$\varepsilon_{yy} = \varepsilon_{\text{eff}}; \quad \varepsilon_{xx} = -\nu\varepsilon_{\text{eff}} \tag{71}$$

refers to an analogous linear combination of the *uniform* strain components.

---

[q] We shall exclusively focus, for the time being, on the *plate*-bending regime, where any deformation (e.g. anticlastic bending) along the main bending axis is forbidden. More general situations will be considered in the later sections.

The fact that, even at the level of the surface contribution, we have an *effective* response to a combined transverse and longitudinal strain is fully consistent with the behavior of an unsupported slab subjected to uniform in-plane tension. In such a situation, the relaxation will affect not only the surface atoms, but will also extend to the entire slab, leading to a contraction in the 3D proportional to the bulk coefficient ν. Thus, for a free film in a relaxed-ion context, it is only meaningful to consider the response of the surface potential offset ϕ to $\varepsilon_{\mathrm{eff}}$, and not to the individual ε_{yy} or ε_{xx} components; the former is precisely the quantity that enters the total flexovoltage coefficient in equation (70).

Of course, one generally needs to consider more realistic mechanical boundary conditions than that of a free-standing film. In such cases, some of the specifics of the above example are no longer valid (e.g. the absence of surface loads). Still, the points (i) and (ii) are applicable to the most general case.

2.7. *Electronic and Lattice Response Revisited: Curvilinear Coordinates*

In the early sections of this chapter we have described a fundamental theory of the bulk flexoelectric effect, based on a first-principles quantum-mechanical description of the insulating crystal. Later, in Section 2.6 we have argued, based on heuristic arguments, that there are important surface contributions to the flexoelectric response of a finite sample, and that these need to be accounted for when discussing experimental results. Here, we shall put the derivations of Section 2.6 on firmer theoretical grounds by developing an alternative approach. In particular, we shall clarify how to describe the microscopic charge and current responses to an arbitrary inhomogeneous strain field in terms of cell-periodic response functions. Such a formalism is necessary in order to treat, in full generality, the response of a finite (and hence, spatially inhomogeneous) body to a deformation. As we shall see later, this will be useful not only for the formal derivation of the surface contributions to the flexoelectric effect in finite samples, but also for the practical calculation of the transverse bulk components of the flexoelectric tensor. (Recall that such components are presently difficult to access at the bulk level.) Given its rather technical character, and the fact that the most relevant physical results have already been presented in Section 2.6, this section and the following can be skipped on a first reading.

2.7.1. *A simple 1D example*

In order to establish a microscopic theory of deformations, the first issue one needs to address concerns the proper representation of the scalar and vector fields that describe the physical property of interest (e.g. atomic positions, electronic charge density, etc.). To appreciate the nature of the problem, it is useful to analyze the charge-density response of a simple lattice to a macroscopic deformation. Consider a 1D chain of equally spaced atoms, which we represent as a regular array of Gaussian charge distributions as in Fig. 4(a). Its unperturbed charge density is

$$\rho(x) = \sum_n \rho_0(x - R_n), \qquad \rho_0(x) = \frac{1}{\sigma\sqrt{\pi}}e^{-x^2/\sigma^2}, \qquad (72)$$

where $R_n = na$ is an integer multiple of the lattice parameter a. Now, we apply a uniform expansion to the chain by displacing each atom by

$$u_n = \varepsilon R_n, \qquad (73)$$

and we look at how the charge density responds to such a perturbation. In the linear limit (small lambda) we obtain the response function $\partial\rho(x)/\partial\varepsilon$ that is plotted as the black curve in Fig. 4(b). The form of $\partial\rho(x)/\partial\varepsilon$ is manifestly problematic: such a function grows linearly when moving away from the origin, i.e. it is clearly non-periodic, which contrasts with the fact the system remains periodic after the application of the perturbation. Moreover, it introduces an undesirable dependence of the result on the arbitrary location of the coordinate origin. Such issues become even more severe when

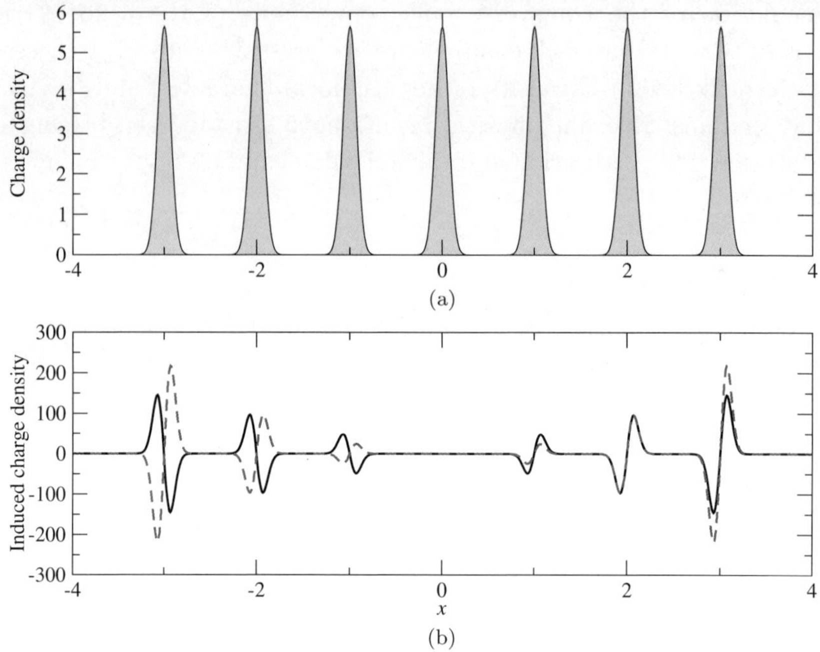

Figure 4. (a) Unperturbed charge density of model 1D crystal composed of Gaussian charge packets. (b) Change of charge density in linear response to a uniform strain (solid line) or strain gradient (dashed line).

considering a strain-gradient perturbation of the type

$$u_n = \frac{\eta}{2} R_n^2. \tag{74}$$

The charge density response, plotted as the dashed curve in Fig. 4(b), now grows *quadratically* with the value of the unperturbed atomic position, and extracting any relevant physical information from such a function appears difficult.

The solution of the above problems comes from the realization that the fixed laboratory frame is a poor choice of coordinate system if we wish to represent the response to a macroscopic elastic deformation. In such a frame, the boundary atoms in a large crystallite have to move very far from their original location even if the applied strain is small; if we naively take the difference in the charge density from the original to the current state we obtain a result that has little physical meaning, and most likely will strongly deviate from the linear regime that we have in mind. A viable alternative is to treat an elastic deformation as a deformation of *space*, rather than an atomic displacement pattern. This implies applying a *coordinate transformation* that exactly reproduces the macroscopic elastic deformation.[r] From this viewpoint, the atoms do not explicitly move from their original location, although they do move with respect to the laboratory frame because the coordinate system itself is changing.

To be explicit, consider a distortion $\mathbf{r}' = \mathbf{r} + \mathbf{u}(\mathbf{r})$ that maps point $\mathbf{r}$ in the original periodic crystal into point $\mathbf{r}'$ of the distorted crystal, and such that a nucleus at $\mathbf{R}_{l\kappa}$ would be carried to $\mathbf{R}'_{l\kappa} = \mathbf{R}_{l\kappa} + \mathbf{u}(\mathbf{R}_{l\kappa})$ if one neglects the additional internal displacements arising from the lattice effects described in Section 2.4. If the initial charge density $\rho_0(\mathbf{r})$ were also carried along by this distortion, the new charge density would be

$$\rho_{\text{ref}}(\mathbf{r}') = \rho_0(\mathbf{r}) \det^{-1}(\mathbf{h}), \tag{75}$$

where the Jacobian factor involving $h_{\alpha\beta} = \partial r'_\alpha / \partial r_\beta = \delta_{\alpha\beta} + \partial u_\alpha / \partial r_\beta$ is needed to reflect the dilution or concentration of charge density. In fact, the actual charge density $\rho(\mathbf{r}')$ has to be computed from the appropriate physical laws (e.g. first-principles DFT calculations), so it will not be equal to $\rho_{\text{ref}}(\mathbf{r}')$. However, we may hope that the difference $\rho(\mathbf{r}') - \rho_{\text{ref}}(\mathbf{r}')$ is small, and we want to express this difference in terms of the *original* spatial variable $\mathbf{r}$. This is conveniently done by defining

$$\hat{\rho}(\mathbf{r}) = \rho(\mathbf{r}') \det(\mathbf{h}), \tag{76}$$

so that our small quantity is $\Delta\hat{\rho}(\mathbf{r}) = \hat{\rho}(\mathbf{r}) - \rho_0(\mathbf{r})$. Note that $\hat{\rho}(\mathbf{r})$ describes the actual charge density after the deformation, but transformed back to the original coordinate system; the hat symbol is

---

[r] Recall that a deformation of a continuum is a 3D–3D mapping of each material point to its perturbed location, i.e. it has the exact same mathematical form as a coordinate transformation.

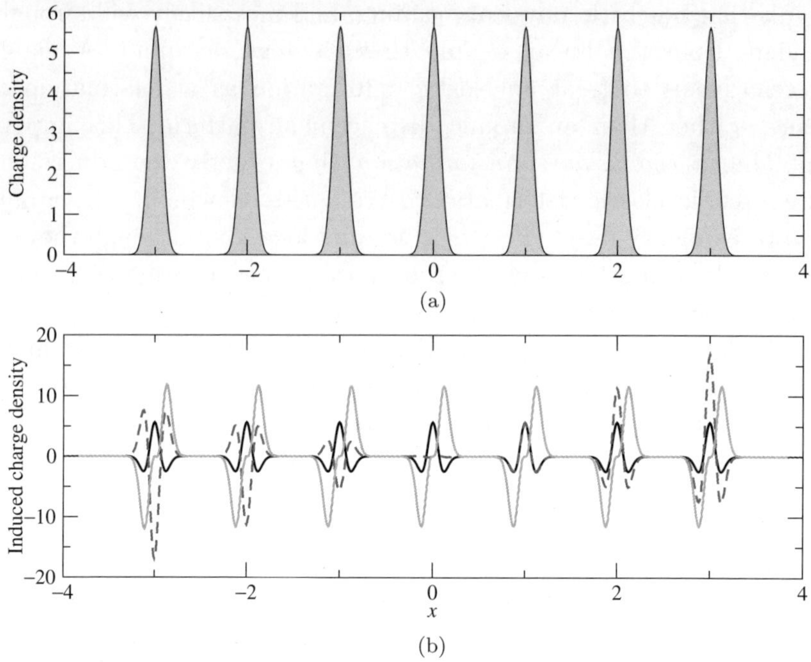

Figure 5.   (a) Same as in Fig. 4(a). (b) Change in charge density, when expressed in transformed coordinates, for a uniform strain (solid black line) or a strain gradient (dashed dark gray line). The latter, while not periodic, can be expressed as a sum of black and light gray contributions (the latter was magnified by a factor of 50), as explained in the text.

used henceforth to highlight quantities that describe the transformed system from the curvilinear-coordinate point of view.

In Fig. 5, we again perform the same analysis as in Fig. 4, illustrating how the use of coordinate transformations effectively solves the problems that we pointed out earlier. Figure 5(a) shows the same charge density at rest. As before, in this model we assume that the actual charge densities shift rigidly with the nuclei. In Fig. 5(b) we plot as the black solid curve the induced density $\partial\hat{\rho}(x)/\partial\varepsilon$ for a uniform strain. The response is now periodic and much smaller in magnitude than before (note the scale change). We shall denote this response function as $\rho^{\mathrm{U}}(x)$, where 'U' indicates a 'uniform' strain. The response to a strain gradient, shown as the dashed curve, is still not periodic, although it now has a milder dependence on the spatial coordinate, growing only linearly rather than quadratically with x. Remarkably, however, we can write this response as a linear combination of two *cell-periodic* functions,

$$\frac{\partial \hat{\rho}(x)}{\partial \eta} = x \rho^{\mathrm{U}}(x) + \rho^{\mathrm{G}}(x), \tag{77}$$

where $\rho^{\mathrm{U}}(x)$ is the same as above (response to uniform strain), and $\rho^{\mathrm{G}}(x)$ is a new quantity, reflecting the genuine strain-gradient effects (shown as a thick light gray curve in the figure, where it has been magnified by a factor of 50 to better illustrate its functional form). Since we are considering a uniform strain gradient above, we have $\varepsilon(x) = x\eta$, so that we can write

$$\Delta\hat{\rho}(x) = \varepsilon(x)\rho^{\mathrm{U}}(x) + \frac{d\varepsilon(x)}{dx}\rho^{\mathrm{G}}(x) + \cdots . \tag{78}$$

In other words, we have achieved a closed expression for the induced charge density $\Delta\hat{\rho}(x)$ that depends only on proper measures of the local deformation, with the only hypothesis that the local strain $\varepsilon(x)$ varies slowly on the scale of the interatomic spacings.

Several questions naturally arise from the above discussion. First, how general is such an analysis? For our illustrative example above we have used a trivially simple system, and a single (longitudinal) strain (or strain gradient) type, so it is legitimate to wonder whether the same procedure is applicable to a full first-principles simulation in 3D. Second, what do we do with $\Delta\hat{\rho}(x)$ once we have calculated it? To make the discussion relevant for flexoelectricity it is necessary to trace a direct link between $\Delta\hat{\rho}(x)$ and measurable electrical quantities, such as the macroscopic polarization in short-circuit, or the induced voltage in open circuit. We shall address both questions in the remainder of this section.

2.7.2. *General formalism in 3D*

Regarding the general applicability of the coordinate transformation method, there are several conceivable ways to proceed. One could, for example, directly incorporate the curvilinear-coordinates formalism at the level of the Kohn–Sham equations (borrowing from the adaptive coordinate scheme of Gygi[20]) and, in a similar spirit as in Ref. [21], directly perform the perturbation expansion with respect to the metric tensor and its gradients. Alternatively — and we shall follow this latter strategy throughout this chapter — one can go back to the phonon analysis that we have introduced in Section 2.2, this time focusing on the microscopic charge-density response functions; the challenge here lies in converting these to the curvilinear representation outlined in this section. We thus consider a deformation

$$r'_\beta(\mathbf{r}) = r_\beta + U_\beta e^{i\mathbf{q}\cdot\mathbf{r}}, \tag{79}$$

which generates a simple frozen phonon

$$u^l_{\kappa\beta} = U_\beta e^{i\mathbf{q}\cdot\mathbf{R}_{l\kappa}}. \tag{80}$$

For the moment we neglect the internal displacements leading to the lattice response of Section 2.4, so that equation (80) is equivalent to equations (9) and (10) with the $\mathbf{q}$-dependent terms neglected, but they will be restored shortly in Section 2.7.4.

In the linear limit, the charge density responds as

$$\rho(\mathbf{r}) = \rho_0(\mathbf{r}) + \rho^{\mathbf{q}}_\beta(\mathbf{r})\, e^{i\mathbf{q}\cdot\mathbf{r}}, \tag{81}$$

where the cell-periodic part $\rho^{\mathbf{q}}_\beta(\mathbf{r})$ gets modulated by the same phase factor as in equation (79). Inserting this in equation (76) gives

$$\hat{\rho}(\mathbf{r}) = \left(\rho_0(\mathbf{r}') + U_\beta\rho^{\mathbf{q}}_\beta(\mathbf{r}')e^{i\mathbf{q}\cdot\mathbf{r}'}\right)\left(1 + iq_\gamma U_\gamma e^{i\mathbf{q}\cdot\mathbf{r}'}\right), \tag{82}$$

where the last term in parentheses is the value of $\det(\mathbf{h})$ resulting from equation (79). We now expand the cell-periodic response function up to second order in $\mathbf{q}$,

$$\rho^{\mathbf{q}}_\beta(\mathbf{r}) = \rho^{(0)}_\beta(\mathbf{r}) - iq_\gamma\rho^{(1,\gamma)}_\beta(\mathbf{r}) - \frac{q_\gamma q_\lambda}{2}\rho^{(2,\gamma\lambda)}_\beta(\mathbf{r}). \tag{83}$$

Since we are only collecting terms to first order in $\mathbf{U}$ in equation (82), we can ignore the distinction between $\mathbf{r}$ and $\mathbf{r}'$ in the cross terms, but for the direct term we have

$$\rho_0(\mathbf{r}') = \rho_0(\mathbf{r} + \mathbf{U}e^{i\mathbf{q}\cdot\mathbf{r}}) = \rho_0(\mathbf{r}) - U_\beta \rho_\beta^{(0)}(\mathbf{r})e^{i\mathbf{q}\cdot\mathbf{r}}, \tag{84}$$

where we have used that $\partial_\beta \rho_0(\mathbf{r}) = -\rho_\beta^{(0)}(\mathbf{r})$. Collecting all the terms linear in $\mathbf{U}$, we obtain

$$\Delta\hat{\rho}(\mathbf{r}) = U_\beta e^{i\mathbf{q}\cdot\mathbf{r}}\left[iq_\beta\rho(\mathbf{r}) - iq_\gamma\rho_\beta^{(1,\gamma)}(\mathbf{r}) - \frac{q_\gamma q_\lambda}{2}\rho_\beta^{(2,\gamma\lambda)}(\mathbf{r})\right]. \tag{85}$$

The $\rho^{(0)}$ terms have now canceled, as expected from the fact that the coordinate transformation has removed the translational part from the response.

After observing that the unsymmetrized strain and strain gradient are related to partial derivatives of the displacement field,

$$\tilde{\varepsilon}_{\beta\gamma}(\mathbf{r}) = iq_\gamma U_\beta e^{i\mathbf{q}\cdot\mathbf{r}}, \tag{86}$$

$$\eta_{\beta,\gamma\lambda}(\mathbf{r}) = -q_\gamma q_\lambda U_\beta e^{i\mathbf{q}\cdot\mathbf{r}}, \tag{87}$$

we can readily write

$$\Delta\hat{\rho}(\mathbf{r}) = \tilde{\varepsilon}_{\beta\gamma}(\mathbf{r})\left[\delta_{\beta\gamma}\rho(\mathbf{r}) - \rho_\beta^{(1,\gamma)}(\mathbf{r})\right] + \frac{\eta_{\beta,\gamma\lambda}(\mathbf{r})}{2}\rho_\beta^{(2,\gamma\lambda)}(\mathbf{r}). \tag{88}$$

Finally, one can replace $\tilde{\varepsilon}_{\beta\gamma}$ with the symmetrized counterpart, $\varepsilon_{\beta\gamma}$ (the quantity in the square brackets is invariant upon $\beta\gamma$ exchange[13]), and replace $\eta_{\beta,\gamma\lambda}$ with the type-II strain gradient tensor $\varepsilon_{\beta\gamma,\lambda}$. This leads to an expression that is in all respects analogous to equation (78),

$$\Delta\hat{\rho}(\mathbf{r}) = \varepsilon_{\beta\gamma}(\mathbf{r})\rho_{\beta\gamma}^{\mathrm{U}}(\mathbf{r}) + \frac{\partial\varepsilon_{\beta\gamma}(\mathbf{r})}{\partial r_\lambda}\rho_{\beta\gamma,\lambda}^{\mathrm{G}}(\mathbf{r}), \tag{89}$$

where the uniform (U) and gradient (G) terms are defined as follows,

$$\rho_{\beta\gamma}^{\mathrm{U}}(\mathbf{r}) = \delta_{\beta\gamma}\rho(\mathbf{r}) - \rho_\beta^{(1,\gamma)}(\mathbf{r}), \tag{90}$$

$$\rho_{\beta\gamma,\lambda}^{\mathrm{G}}(\mathbf{r}) = \frac{1}{2}\left[\rho_\beta^{(2,\gamma\lambda)}(\mathbf{r}) + \rho_\gamma^{(2,\lambda\beta)}(\mathbf{r}) - \rho_\lambda^{(2,\beta\gamma)}(\mathbf{r})\right]. \tag{91}$$

This result formalizes and generalizes the arguments of the first part of this section: it shows that the microscopic charge density response to an arbitrary inhomogeneous deformation can indeed be computed (and rigorously expressed in terms of well-defined response quantities) in a first-principles context, and for an arbitrary combination of the relevant 3D deformation tensor components.

As a side note, one can show that the quantity $\rho^{\mathrm{U}}_{\beta\gamma}(\mathbf{r})$ essentially coincides (apart from a trivial scaling factor) with the first-order charge-density as defined by Hamann *et al.*[21] within their linear-response theory of strain based on the metric tensor. It will be interesting in the near future to draw even closer connections between the two formalisms, which bear several similarities at the conceptual level.

2.7.3. *Microscopic polarization response*

In the above derivations we have focused on the charge-density response of the system to an inhomogeneous deformation, but we could have worked just as well with the microscopic polarization response instead. This quantity is well defined only for infinitesimal transformations, otherwise it depends on the specific path followed by the system during its evolution; this is not an issue here, since we are exclusively interested in the linear-response regime. The microscopic polarization $\mathbf{P}(\mathbf{r})$ is related to the adiabatic current-density response of the system to a time-dependent perturbation. Thus, in order to construct an appropriate definition of this quantity in a generic curvilinear frame, we need first to examine the transformation laws of the current-density field $\mathbf{J}(\mathbf{r})$. To this end, let $\mathbf{r}' = \mathbf{r} + \mathbf{u}(\mathbf{r}, t)$ be a generic time-dependent coordinate transformation, which we suppose to coincide, as usual, with the displacement field associated with the mechanical deformation of the sample. (We suppose now that such a deformation happens slowly over a finite interval of time.) In a curvilinear framework the four-current, defined as $J^{\mu} = (\rho, j_1, j_2, j_3)$, transforms as

$$\bar{J}^{\mu} = \frac{\partial \bar{x}^{\mu}}{\partial x^{\alpha}}\, J^{\alpha}\, \det{}^{-1}\left[\frac{\partial \bar{x}^{\beta}}{\partial x^{\gamma}}\right], \tag{92}$$

where $x^{\mu} = (t, x_1, x_2, x_3)$ is the coordinate four-vector and the barred (unbarred) symbols refer to the deformed (original) frame. We work here in the non-relativistic limit with $\bar{t} = t$ which implies that $\bar{t}$ does not depend on $\mathbf{r}$, so that

$$\bar{\rho} = \rho \det^{-1}(\mathbf{h}), \tag{93}$$

$$\bar{J}_i = \left(\rho \frac{\partial u_i}{\partial t} + h_{ij} J_j \right) \det^{-1}(\mathbf{h}). \tag{94}$$

(Latin indices refer to the 3D Cartesian space.) This leads to the definitions

$$\hat{\rho} = \bar{\rho} \, \det(\mathbf{h}), \tag{95}$$

$$\hat{J}_i = (\mathbf{h}^{-1})_{ij} \left[\bar{J}_j - \bar{\rho} \frac{\partial u_j}{\partial t} \right] \det(\mathbf{h}). \tag{96}$$

The above expression for the curvilinear-frame charge density $\hat{\rho}$ coincides with that postulated earlier in equation (76), showing that this definition is, in fact, dictated by the fundamental transformation laws of a scalar density field. Equation (96), on the other hand, gives the desired expression for the curvilinear-frame current density $\hat{J}_i$, which we will use in the following to derive the microscopic polarization response to an acoustic phonon perturbation.

As in the former case of the charge density response, we consider a monochromatic acoustic phonon as in equation (79), again without internal cell relaxations. This time, however, we allow the amplitude of the displacement to depend on time,

$$\mathbf{r}' = \mathbf{r} + \mathbf{u}(\mathbf{r}, t), \quad u_\beta(\mathbf{r}, t) = U_\beta(t) \, e^{i\mathbf{q}\cdot\mathbf{r}}. \tag{97}$$

(Henceforth we shall go back to using Greek indices for 3D space coordinates, as we did in the previous sections.) In the adiabatic limit, one has

$$\mathbf{J}(\mathbf{r}, t) = \dot{U}_\beta(t) \frac{\partial \mathbf{P}(\mathbf{r})}{\partial U_\beta},$$

and, after dropping all terms that are quadratic in either U_β or $\dot{U}_\beta$ [this implies setting $h_{ij} = \delta_{ij}$ in equation (96)],

$$\hat{\mathbf{J}}(\mathbf{r}, t) = \dot{U}_\beta(t) \frac{\partial \mathbf{P}(\mathbf{r})}{\partial U_\beta} - \dot{\mathbf{U}}(t) \rho_0(\mathbf{r}).$$

Now, the microscopic polarization response can be written, as usual, as a cell-periodic part times a phase, $\mathbf{P}(\mathbf{r}) = U_\beta\, e^{i\mathbf{q}\cdot\mathbf{r}}\, \mathbf{P}^{\mathbf{q}}_\beta(\mathbf{r})$, which immediately leads to

$$\hat{P}_\alpha(\mathbf{r}) = U_\beta e^{i\mathbf{q}\cdot\mathbf{r}}\left[P^{\mathbf{q}}_{\alpha,\beta}(\mathbf{r}) - \delta_{\alpha\beta}\rho_0(\mathbf{r})\right]. \tag{98}$$

By following the same steps as for the case of the charge-density response, we now proceed to expand $P^{\mathbf{q}}_{\alpha,\beta}(\mathbf{r})$ in powers of $\mathbf{q}$,

$$\mathbf{P}^{\mathbf{q}}_\beta(\mathbf{r}) = \mathbf{P}^{(0)}_\beta(\mathbf{r}) - iq_\gamma \mathbf{P}^{(1,\gamma)}_\beta(\mathbf{r}) - \frac{q_\gamma q_\lambda}{2}\mathbf{P}^{(2,\gamma\lambda)}_\beta(\mathbf{r}). \tag{99}$$

From translational invariance, it is then easy to show that the zeroth-order term

$$P^{(0)}_{\alpha,\beta}(\mathbf{r}) = \delta_{\alpha\beta}\rho_0(\mathbf{r}), \tag{100}$$

exactly cancels with the last term involving ρ_0 in equation (98). Eventually, we arrive at a provisional result for the linearly induced polarization currents in the curvilinear frame of the deformed body in the form

$$\Delta\hat{P}_\alpha = \varepsilon_{\beta\gamma}(\mathbf{r})P^{\mathrm{U}}_{\alpha,\beta\gamma}(\mathbf{r}) + \frac{\partial\varepsilon_{\beta\gamma}(\mathbf{r})}{\partial r_\lambda}P^{\mathrm{G}}_{\alpha\lambda,\beta\gamma}(\mathbf{r}), \tag{101}$$

where the cell-periodic vector fields $\mathbf{P}^{\mathrm{U,G}}$ are

$$P^{\mathrm{U}}_{\alpha,\beta\gamma}(\mathbf{r}) = -P^{(1,\gamma)}_{\alpha,\beta}(\mathbf{r}), \tag{102}$$

$$P^{\mathrm{G}}_{\alpha\lambda,\beta\gamma}(\mathbf{r}) = \frac{1}{2}\left[P^{(2,\gamma\lambda)}_{\alpha,\beta}(\mathbf{r}) + P^{(2,\lambda\beta)}_{\alpha,\gamma}(\mathbf{r}) - P^{(2,\beta\gamma)}_{\alpha,\lambda}(\mathbf{r})\right]. \tag{103}$$

To arrive at this equation, however, we have had to assume that the currents generated by a global rotation are the same as those obtained by rigidly rotating a classical charge density that is equal to the true quantum-mechanical one. This assumption, which is implicit in equation (50), was used to conclude that $P^{(1,\beta)}_{\alpha,\gamma}(\mathbf{r}) = P^{(1,\gamma)}_{\alpha,\beta}(\mathbf{r})$, and hence to replace the unsymmetrized ($\tilde{\varepsilon}$) with the symmetrized (ε) strain tensor (see Section V.C of Ref. [13]). While such an assumption was indeed valid in the charge-density case, it is not obvious that it is justifiable in the present case of the microscopic polarization current. Discussing this point in detail would take us far from the scope of the present chapter; nevertheless, the reader is warned that there are still some unresolved formal issues in the theory of the current-density response.

Regardless of such issues, one can show that the functions $\Delta\hat{\rho}$ and $\Delta\hat{\mathbf{P}}$ enjoy the fundamental relationship

$$\hat{\nabla} \cdot \Delta\hat{\mathbf{P}}(\mathbf{r}) = -\Delta\hat{\rho}(\mathbf{r}), \tag{104}$$

where $\hat{\nabla} \cdot \hat{\mathbf{A}}$ indicates the divergence of the vector field $\hat{\mathbf{A}}$ in the curvilinear frame. (The hat is used to emphasize that the differentiation is with respect to $\mathbf{r}$ rather than $\mathbf{r}'$.) This reflects a well-known fact, which is important in the specific context of flexoelectricity: the induced charge density can be readily deduced from the polarization, but not the other way around. As a matter of fact, taking the divergence annihilates the solenoidal part of the $\Delta\hat{\mathbf{P}}$-field, which does contribute to the bulk flexoelectric tensor. (Recall the relationship between dynamical octupoles and second moments of the current-density response: the additional information contained in the latter can indeed be ascribed to divergenceless polarization currents that arise in response to an atomic displacement.)

2.7.4. *Atomic relaxations*

We now return to the inclusion of the internal atomic relaxations, and describe how they can be conveniently incorporated in the aforementioned theory; the practical implications regarding surface effects will be discussed in Section 2.8.

The first important observation is that, given a strain field $\varepsilon_{\beta\gamma}(\mathbf{r}, t)$ that depends slowly on space and time, the internal-strain tensors that were introduced in Section 2.2 can readily be identified with the microscopic lattice response of the crystal in the curvilinear frame of the deformed body,

$$u_{\kappa\alpha}^{l}(t) = \varepsilon_{\beta\gamma}(\mathbf{R}_{l\kappa}, t)\Gamma_{\alpha\beta\gamma}^{\kappa} + \frac{\partial\varepsilon_{\beta\gamma}(\mathbf{R}_{l\kappa}, t)}{\partial r_{\lambda}}L_{\alpha\lambda,\beta\gamma}^{\kappa} + \cdots \tag{105}$$

In particular, for a macroscopic strain gradient, the above relationship reduces to

$$\frac{\partial u_{\kappa\alpha}^{l}}{\partial \varepsilon_{\beta\gamma,\lambda}} = R_{l\kappa\lambda}\,\Gamma_{\alpha\beta\gamma}^{\kappa} + L_{\alpha\lambda,\beta\gamma}^{\kappa}. \tag{106}$$

Next, it is easy to show that equation (101) still holds, provided
that we replace the purely electronic response functions with their
relaxed-ion counterparts,

$$P_{\alpha,\beta\gamma}^{\mathrm{U}}(x) = \bar{P}_{\alpha,\beta\gamma}^{\mathrm{U}}(x) + P_{\alpha,\kappa\rho}^{(0)}(x)\Gamma_{\rho\beta\gamma}^{\kappa}, \tag{107}$$

$$P_{\alpha\lambda,\beta\gamma}^{\mathrm{G}}(x) = \bar{P}_{\alpha\lambda,\beta\gamma}^{\mathrm{G}}(x) - P_{\alpha,\kappa\rho}^{(1,\lambda)}(x)\Gamma_{\rho\beta\gamma}^{\kappa} + P_{\alpha,\kappa\rho}^{(0)}(x)L_{\rho\lambda,\beta\gamma}^{\kappa}. \tag{108}$$

(We have used the bar symbol here to denote the purely electronic
$P^{\mathrm{U,G}}$ response functions and thereby distinguish them from the fully
relaxed quantities.) This formally extends the microscopic linear-
response theory discussed in this section to the relaxed-ion case.

2.7.5. *Electrostatics in a curved space*

Having established a convenient form for the microscopic charge and
polarization response functions, we still need to figure out how to use
them, e.g. how to calculate the voltage response $V(\mathbf{r})$. This requires
some attention, given the fact that we are no longer working in the
Cartesian frame of the laboratory. In particular, one needs to replace
Gauss's law with its "curvilinear" generalization[22,23]

$$\hat{\nabla} \cdot (\hat{\boldsymbol{\epsilon}} \cdot \hat{\mathbf{E}}) = \hat{\rho}, \tag{109}$$

where the vacuum permittivity has been replaced with the tensor

$$\hat{\boldsymbol{\epsilon}} = \epsilon_0 \det(\mathbf{h})\mathbf{g}^{-1}, \tag{110}$$

the hat on $\hat{\nabla}$ is again a reminder that the gradient is in the curvilinear
frame, $\mathbf{g} = \mathbf{h} \cdot \mathbf{h}^{\mathrm{T}}$ is the metric of the deformation, and

$$\hat{E}_{\alpha}(\mathbf{r}) = h_{\alpha\beta}E_{\beta}(\mathbf{r}') \tag{111}$$

is the transformed electric field.[s] As $\hat{\nabla}V(\mathbf{r}) = -\hat{\mathbf{E}}(\mathbf{r})$, the above strategy allows one to compute the induced electrostatic potential from the induced polarization (or charge density).[t]

In the linear limit of small deformations, one has

$$\epsilon_0 \hat{\nabla} \cdot (\Delta\hat{\mathbf{E}} + \Delta\mathbf{E}^{\mathrm{met}}) = \Delta\hat{\rho}(\mathbf{r}), \qquad (112)$$

where $\Delta\hat{\mathbf{E}}(\mathbf{r}) = -\hat{\nabla}[\Delta V(\mathbf{r})]$ is minus the (curvilinear) gradient of the induced electrostatic potential, ΔV, and $\Delta\mathbf{E}^{\mathrm{met}}$, coming from the linearization of $\hat{\epsilon}$, reads as

$$\Delta E_\alpha^{\mathrm{met}}(\mathbf{r}) = \varepsilon_{\beta\gamma}(\mathbf{r}) \left[\delta_{\beta\gamma}E_\alpha(\mathbf{r}) - \delta_{\alpha\beta}E_\gamma(\mathbf{r}) - \delta_{\gamma\alpha}E_\beta(\mathbf{r})\right], \qquad (113)$$

where $E_\gamma(\mathbf{r})$ is the electric field in the unperturbed system. The choice of notation $\Delta\mathbf{E}^{\mathrm{met}}$ is meant to suggest a "metric contribution to the curvilinear-frame electric field", but this is somewhat problematic as it is not an irrotational field. Alternatively, $\epsilon_0\Delta\mathbf{E}^{\mathrm{met}}$ could be regarded as a "metric contribution" to the polarization, since one can rewrite equation (112) in terms of $\Delta\mathbf{P}$ as

$$\epsilon_0 \hat{\nabla} \cdot \Delta\hat{\mathbf{E}} = -\hat{\nabla} \cdot \left(\Delta\mathbf{P} + \epsilon_0\Delta\mathbf{E}^{\mathrm{met}}\right). \qquad (114)$$

However, this is not entirely satisfactory either, as $\Delta\mathbf{E}^{\mathrm{met}}$ does not really originate from the displacement of charged particles. It is probably most appropriate to interpret $\epsilon_0\Delta\mathbf{E}^{\mathrm{met}}$ as a *displacement current* arising from the effective change of permittivity associated with the deformation of the reference frame.

In any case, equation (112) shows that the induced potential $\Delta V(\mathbf{r})$ contains, in addition to contributions from the rearrangement

---

[s] One can arrive at equation (109) by observing that, in a curvilinear frame, Poisson's equation reads as $\sqrt{g}^{-1}\partial_\mu\left(\sqrt{g}g^{\mu\nu}\partial_\nu V\right) = -\rho/\epsilon_0$ where $g = \det(\mathbf{g}) = \det^2(\mathbf{h})$, i.e. the Laplacian must be replaced with the Laplace–Beltrami operator. Then, by defining $\hat{\rho} = \sqrt{g}\rho$, $\hat{\mathcal{E}}_\nu = -\partial_\nu V$, and $\hat{\epsilon}^{\mu\nu} = \epsilon_0\sqrt{g}g^{\mu\nu}$, one immediately recovers Eq. (109).

[t] It is interesting to note the close connection between the curvilinear-frame electrical quantities ($\hat{\mathbf{E}}$ and $\hat{\mathbf{P}}$) described here and the *reduced* electrical variables (respectively $\bar{\varepsilon}_\alpha$ and p_α) of Ref. [24], where a linear mapping was implicitly used to connect lattice and Cartesian coordinates.

of the electron cloud occurring during the deformation (these are contained in $\Delta\hat{\rho}$), also a term that depends on the local variation of the metric at fixed charge density. We shall come back to this point in the discussion of surface contributions to the flexoelectric effect in Section 2.8. Note that, by construction, the microscopic electric-field response to an arbitrary deformation enjoys an analogous representation as the charge density response,

$$\Delta\hat{E}_\alpha(\mathbf{r}) = \varepsilon_{\beta\gamma}(\mathbf{r})E^{\mathrm{U}}_{\alpha,\beta\gamma}(\mathbf{r}) + \frac{\partial\varepsilon_{\beta\gamma}(\mathbf{r})}{\partial r_\lambda}E^{\mathrm{G}}_{\alpha\lambda,\beta\gamma}(\mathbf{r}), \qquad (115)$$

where both E^{U} and E^{G} are lattice-periodic functions whose explicit expressions can be readily derived from equation (112).

2.7.6. *Treatment of the macroscopic electric fields*

Equation (112) specifies $\Delta\hat{\mathbf{E}}(\mathbf{r})$ modulo an $\mathbf{r}$-independent integration constant, $\Delta\bar{\hat{\mathbf{E}}}$ whose value is fixed by the electrical boundary conditions (EBC) of the problem. The electronic response functions are typically defined (and calculated) by assuming $\Delta\bar{\hat{\mathbf{E}}} = 0$, i.e. short-circuit (SC) EBCs, but any other EBC choice can be recovered if the charge-density (and/or polarization) response to a macroscopic electric field is known.[u] For example, in the case of the polarization one can write

$$\Delta\hat{P}_\alpha(\mathbf{r}) = \Delta\hat{P}^{\mathrm{SC}}_\alpha(\mathbf{r}) + \Delta\bar{\hat{E}}_\beta P^{E_\beta}_\alpha(\mathbf{r}), \qquad (116)$$

where $P^{E_\beta}_\alpha(\mathbf{r}) = \partial P_\alpha(\mathbf{r})/\partial E_\beta$ is the microscopic P response to an applied field along β.[25] The contribution of the macroscopic field can be readily incorporated into the strain-gradient term ($\mathbf{P}^{\mathrm{G}}$ in this case), as the macroscopic electric-field response in a non-piezoelectric crystal vanishes at the uniform-strain level. Therefore, equations (89), (101), and (115) remain valid in arbitrary EBC.

---

[u] Note that $\Delta\bar{\hat{\mathbf{E}}}$ is first order in the perturbation, and therefore its contribution to the electronic response functions is only due to the microscopic dielectric properties of the unperturbed system.

2.8. *Surface Effects in Curvilinear Coordinates*

Recall that, in order to quantify the flexoelectric response of a free-standing slab, we have introduced the flexovoltage coefficient $\varphi_{x\lambda,\beta\gamma}$. Here, we shall demonstrate how this quantity can be rigorously derived in the context of the microscopic theory developed in the previous section. To express $\varphi_{x\lambda,\beta\gamma}$ in terms of well-defined response functions of the system, we shall follow the strategy of Section 2.7, now specializing to the case of a supercell geometry. As before, we shall derive the microscopic response of the system (charge density, polarization, and atomic displacements) to a strain-gradient deformation via a long-wave analysis of its acoustic phonons. With the help of a coordinate transformation to the curvilinear frame of the perturbed body, one can express such microscopic response functions in terms of "proper" measures of the local deformation, i.e. in a translationally and rotationally invariant form,

$$\Delta \hat{f}(\mathbf{r}) = \varepsilon_{\beta\gamma}(\mathbf{r}) f^{\mathrm{U}}_{\beta\gamma}(\mathbf{r}) + \frac{\partial \varepsilon_{\beta\gamma}(\mathbf{r})}{\partial r_\lambda} f^{\mathrm{G}}_{\beta\gamma,\lambda}(\mathbf{r}) + \cdots . \qquad (117)$$

Here $\hat{f}$ can stand for the charge density ($\hat{\rho}$), polarization ($\hat{\mathbf{P}}$) or electric field ($\hat{\mathbf{E}}$) expressed in the curvilinear frame. Note that the cell-periodic functions f^{U} and f^{G}, referring respectively to uniform and gradient terms, are characterized by an oscillatory behavior on the scale of an interatomic distance, due to the discreteness of the atomic lattice. As it is customary in the space-resolved analysis of many other physical properties (e.g. dielectric response), we shall assume in the following (unless otherwise specified) that such oscillations have been filtered out by means of an appropriate nanosmoothing[17,26] technique.

2.8.1. *Surface polarization and metric*

To illustrate the above arguments in the present context, consider a symmetrically terminated slab of a cubic material (we assume that the surfaces are parallel to the yz plane), and perturb it with a strain-gradient deformation. We assume for the moment that the ionic coordinates simply follow the deformation; we shall lift

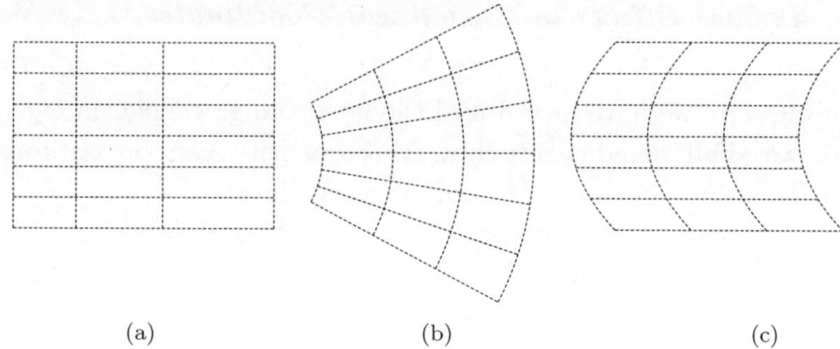

(a)   (b)   (c)

Figure 6. Schematic representation of the three types of strain gradient described in the text. (a) Longitudinal, $\varepsilon_{xx,x}$. (b) Transverse, $\varepsilon_{yy,x}$. (c) Shear, $\varepsilon_{xy,y}$. The x and y axes correspond to the horizontal and vertical directions in the figure respectively. (Adapted from Ref. [27].)

this limitation in the next subsection. In order to calculate the flexovoltage coefficient of the slab, we shall first derive the electric field $\mathbf{E}(\mathbf{r})$ induced by the deformation under open-circuit electrical boundary conditions. Then, by performing a line integral of $\mathbf{E}(\mathbf{r})$ across the slab thickness, one can readily obtain the desired value of $\varphi_{\alpha\lambda,\beta\gamma}$. To calculate $\mathbf{E}(\mathbf{r})$ we need, in turn, two basic ingredients: the microscopic polarization response, $\Delta\mathbf{P}(\mathbf{r})$, and the "metric" contribution to the polarization, $\Delta\mathbf{E}^{\mathrm{met}}$. Regarding the former, after nanosmoothing $\mathbf{P}^{\mathrm{U,G}}$ are functions of x only, and we can write

$$\left.\frac{\partial\hat{P}_\alpha(\mathbf{r})}{\partial\varepsilon_{\beta\gamma,\lambda}}\right|_{\mathrm{frozen-ion}} = r_\lambda\bar{P}^{\mathrm{U}}_{\alpha,\beta\gamma}(x) + \bar{P}^{\mathrm{G}}_{\alpha\lambda,\beta\gamma}(x). \tag{118}$$

The two response functions $P^{\mathrm{U}}_{\alpha,\beta\gamma}(x)$ and $P^{\mathrm{G}}_{\alpha\lambda,\beta\gamma}(x)$ have the physical meaning of a *local* piezoelectric and flexoelectric coefficient, respectively. Note that $P^{\mathrm{U}}_{\alpha,\beta\gamma}(x)$ differs from zero only in the vicinity of the surface, as the bulk material is non-piezoelectric. The metric term, on the other hand, reads as

$$\frac{\partial E^{\mathrm{met}}_\alpha(\mathbf{r})}{\partial\varepsilon_{\beta\gamma,\lambda}} = r_\lambda\left[\delta_{\beta\gamma}E_\alpha(x) - \delta_{\alpha\beta}E_\gamma(x) - \delta_{\gamma\alpha}E_\beta(x)\right]. \tag{119}$$

This quantity, just like $\mathbf{P}^{\mathrm{U}}$, is active only at the surface: the electric field of the undistorted slab is non-zero (and directed perpendicular to the surface plane) only in a small region where the crystal lattice is perturbed by the truncation of the bonding network.

The details of the derivation differ, from now on, depending on the specific type of flexovoltage response that one wishes to calculate. It can be shown that, given the symmetry of the slab, there are only three types of strain-gradient deformation that yield a net open-circuit voltage in a cubic material: a variation of ε_{xx} with x (longitudinal), ε_{yy} with x (transverse), or ε_{xy} with y (shear), which are responsible for the flexovoltages $\varphi_{xx,xx}$, $\varphi_{xx,yy}$, and $\varphi_{xy,xy}$ respectively, in the notation of equation (61).

Longitudinal and transverse cases. In these two cases (which we shall indicate as $xx, \alpha\alpha$), the system remains periodic in-plane, and the problem becomes essentially 1D. We have, in particular,

$$\left. \frac{\partial \hat{P}_x(x)}{\partial \varepsilon_{\alpha\alpha,x}} \right|_{\text{FI}} = x \bar{P}^{\text{U}}_{x,\alpha\alpha}(x) + \bar{P}^{\text{G}}_{xx,\alpha\alpha}(x), \tag{120}$$

$$\frac{\partial E^{\text{met}}_x(x)}{\partial \varepsilon_{\alpha\alpha,x}} = x E_x(x)(1 - 2\delta_{\alpha x}), \tag{121}$$

(where FI is shorthand for 'frozen-ion'). The polarization response functions P^{U} and $x P^{\text{U}} + P^{\text{G}}$ are schematically illustrated for the transverse case in Figs. 7(a) and (c) respectively. Given that both vector fields are irrotational and vanish at infinity, one can safely simplify equation (114) by removing the divergence sign on both sides to get

$$\left. \frac{\partial \hat{E}_x(x)}{\partial \varepsilon_{\alpha\alpha,x}} \right|_{\text{FI}} = -\frac{1}{\epsilon_0} \left[x \bar{P}^{\text{U}}_{x,\alpha\alpha}(x) + \bar{P}^{\text{G}}_{xx,\alpha\alpha}(x) \right] - x E_x(x)(1 - 2\delta_{\alpha x}). \tag{122}$$

The frozen-ion flexovoltage coefficient of equation (61) can then be calculated by writing the open-circuit potential associated with the above field as

$$\Delta V = - \int_{-\infty}^{+\infty} dx \, \left[x E^{\text{U}}_{x,\alpha\alpha}(x) + E^{\text{G}}_{xx,\alpha\alpha}(x) \right], \tag{123}$$

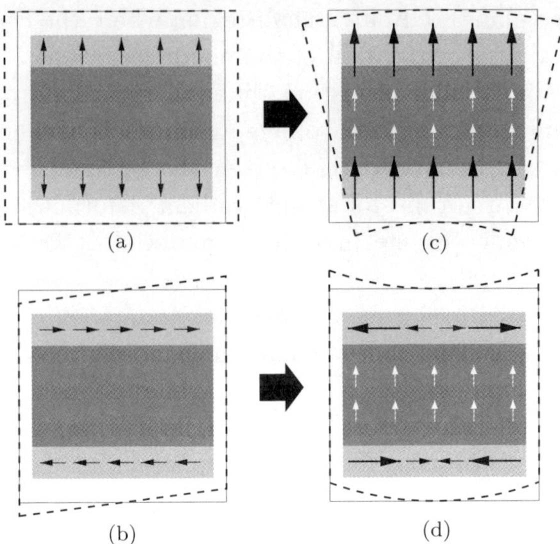

Figure 7.   Schematic representation of the polarization fields induced by different macroscopic deformations of a slab of thickness t, drawn in the undistorted reference frame. The x coordinate is vertical here. Left panels (a–b) illustrate uniform strains; right panels (c–d) refer to uniform strain gradients. Top (a, c) are transverse deformations (the situation is qualitatively identical in the longitudinal case, not shown), bottom (b, d) are shear patterns. The surface region and corresponding polarization field are indicated by light gray shading and black arrows. White arrows on a dark gray background refer to the bulk region. The dashed black frames indicate the type of deformation in each case. (Adapted from Ref. [15].)

where

$$E^{\mathrm{U}}_{x,\alpha\alpha}(x) = -\frac{1}{\epsilon_0}\bar{P}^{\mathrm{U}}_{x,\alpha\alpha}(x) - E_x(x)(1 - 2\delta_{\alpha x}), \qquad (124)$$

$$E^{\mathrm{G}}_{xx,\alpha\alpha}(x) = -\frac{1}{\epsilon_0}\bar{P}^{\mathrm{G}}_{xx,\alpha\alpha}(x). \qquad (125)$$

The above functions enjoy a number of useful properties:

(i) $E^{\mathrm{U}}_{x,\alpha\alpha}(x)$ vanishes everywhere except for a small region near the surface at $x \sim \pm t/2$.

(ii) $E^{\mathrm{U}}_{x,\alpha\alpha}(x - t/2) = -E^{\mathrm{U}}_{x,\alpha\alpha}(-x + t/2)$ is antisymmetric, and *independent* of t for a sufficiently thick slab.

(iii) $E^{\mathrm{G}}_{xx,\alpha\alpha}(x)$ corresponds to minus the bulk flexocoupling coefficient in the slab interior,

$$E^{\mathrm{G}}_{xx,\alpha\alpha}(x \sim 0) = -\bar{\varphi}^{\mathrm{bulk}}_{xx,\alpha\alpha} = -\frac{\mu^{\mathrm{bulk}}_{xx,\alpha\alpha}}{\epsilon_0 \bar{\epsilon}_{\mathrm{r}}},$$

and only deviates from this value in a small region near the surface;

(iv) $E^{\mathrm{G}}_{xx,\alpha\alpha}(x - t/2) = E^{\mathrm{G}}_{xx,\alpha\alpha}(-x + t/2)$ is symmetric, and again independent of t for a sufficiently thick slab.

Based on these observations, in the limit of large slab thickness one can approximate equation (123) as

$$\Delta V \sim -t \int_0^{+\infty} dx\, E^{\mathrm{U}}_{x,\alpha\alpha}(x) + t\bar{\varphi}^{\mathrm{bulk}}_{xx,\alpha\alpha}. \tag{126}$$

(We assume that $x = 0$ is the center of the slab and $x = +\infty$ is deep in the vacuum region.) As the integral in the last equation is independent of t, we can readily write

$$\bar{\varphi}_{xx,\alpha\alpha} = - \int_0^{+\infty} dx\, E^{\mathrm{U}}_{x,\alpha\alpha}(x) + \bar{\varphi}^{\mathrm{bulk}}_{xx,\alpha\alpha}, \tag{127}$$

whence we obtain

$$\bar{\varphi}^{\mathrm{surf}}_{xx,\alpha\alpha} = - \int_0^{+\infty} dx\, E^{\mathrm{U}}_{x,\alpha\alpha}(x). \tag{128}$$

The last equation states that the surface contribution to the flexovoltage response of a slab corresponds to minus the line integral of the induced electric field upon application of a uniform strain. The latter is, of course, the electrostatic potential offset response to uniform strain, $\partial\phi/\partial\varepsilon_{\alpha\alpha}$, that we already discussed in Section 2.6. We have, therefore, rigorously demonstrated that the flexovoltage response of the slab indeed contains both bulk and surface contributions, and that their nature is correctly described by equation (66). Note that the derivations presented here, in addition to corroborating the arguments of Section 2.6, allow us to make one step further and split $\bar{\varphi}^{\text{surf}}_{xx,\alpha\alpha}$ into a polarization current and a metric term,

$$\frac{\partial \phi}{\partial \varepsilon_{\alpha\alpha}} = \frac{1}{\epsilon_0} \int_0^{+\infty} dx\, P^{\text{U}}_{x,\alpha\alpha}(x) + (2\delta_{\alpha x} - 1)\phi_0. \qquad (129)$$

In the last term on the right, $\phi_0 = -\int_0^{+\infty} dx\, E_x(x)$ is the potential offset before the perturbation.

In summary, in the present case (longitudinal or transverse strain gradient) the induced electric field in the interior of the film is a bulk property of the material — it is given by the flexoelectric coefficient divided by the macroscopic dielectric constant. The surface contribution, on the other hand, acts as an induced potential offset that grows linearly with slab thickness, and therefore scales similarly to the bulk contribution, as illustrated in Fig. 3.

Shear case. The case of a shear deformation (xy, xy) is qualitatively different from the former two cases.[v] Here, we have

$$\left. \frac{\partial \hat{P}_\alpha(\mathbf{r})}{\partial \varepsilon_{xy,y}} \right|_{\text{FI}} = \delta_{\alpha y} y P^{\text{U}}_{y,xy}(x) + P^{\text{G}}_{\alpha y,xy}(x), \qquad (130)$$

$$\frac{\partial E^{\text{met}}_\alpha(\mathbf{r})}{\partial \varepsilon_{xy,y}} = -\delta_{\alpha y} y E_x(x). \qquad (131)$$

Figures 7(b) and (d) illustrate the polarization fields that are linearly induced by shear deformations. By taking the divergence of the above vector fields, we can write the curvilinear Poisson's equation, equation (114), as a function of x only,[w]

$$\epsilon_0 \frac{\partial \hat{E}_{xy,xy}(x)}{\partial x} = -\bar{P}^{\text{U}}_{y,xy}(x) - \frac{\partial \bar{P}^{\text{G}}_{xy,xy}(x)}{\partial x} + \epsilon_0 E_x(x). \qquad (132)$$

---

[v] We mention this case for completeness, as it is not relevant for an unsupported slab after full atomic relaxation, provided that we consider a region that lies far (compared to the thickness, t) from the edges and/or the mechanical loading points. We shall come back to this issue in Section 2.8.2.

[w] Note that the partial derivatives along y of both P^{U} and P^{G} vanish identically, as these nanosmoothed functions are periodic in plane, and therefore only depend on x.

[We have used the short-hand notation $\hat{E}_{xy,xy}(x) = \partial\hat{E}_x(x)/\partial\varepsilon_{xy,y}$.] Assuming that the field vanishes inside the slab,[x] we can then calculate the macroscopic electric field in the vacuum region,

$$\epsilon_0 \left. \frac{\partial\hat{E}_x(x=+\infty)}{\partial\varepsilon_{xy,y}} \right|_{\text{FI,SC}} = \sigma^{\text{surf}}_{xy,xy} + \sigma^{\text{bulk}}_{xy,xy} + \sigma^{\text{met}}_{xy,xy}. \tag{133}$$

(SC stands for 'short-circuit.') The three quantities on the right-hand side have the physical dimension of a surface charge density and are given by

$$\sigma^{\text{bulk}}_{xy,xy} = \bar{\mu}^{\text{bulk}}_{xy,xy}, \tag{134}$$

$$\sigma^{\text{surf}}_{xy,xy} = - \int_0^{+\infty} dx\, \bar{P}^{\text{U}}_{y,xy}(x), \tag{135}$$

$$\sigma^{\text{met}}_{xy,xy} = \epsilon_0 \int_0^{+\infty} dx\, E_x(x) = -\epsilon_0\phi_0. \tag{136}$$

In order to derive the flexocoupling coefficient, we need to switch to open-circuit boundary conditions by imposing an external electric field that exactly cancels the above vacuum field. We obtain

$$\bar{\varphi}_{xy,xy} = \frac{1}{\epsilon_0\bar{\epsilon}_{\text{r}}} \left[\bar{\mu}^{\text{II}}_{xx,\alpha\alpha} - \int_0^{+\infty} dx\, P^{\text{U}}_{y,xy}(x) - \epsilon_0\phi_0 \right]. \tag{137}$$

The total surface contribution coming from both the polarization currents and from the metric is thus

$$\bar{\varphi}^{\text{surf}}_{xy,xy} = \frac{1}{\epsilon_0\bar{\epsilon}_{\text{r}}} \left[- \int_0^{+\infty} dx\, P^{\text{U}}_{y,xy}(x) - \epsilon_0\phi_0 \right]. \tag{138}$$

Note that, in contrast with the transverse and longitudinal cases, the internal electric field is no longer a bulk property here; the

---

[x] This is the natural choice for the electrical boundary conditions when considering shear strain gradients — recall that they directly relate to transverse acoustic phonons.

surface terms contained in $\bar{\varphi}^{\text{surf}}_{xy,xy}$ manifest themselves as surface *charge* densities that tend to a constant in the limit of large slab thickness t, rather than dipole densities that grow linearly with t. (In either case, the surface contribution to the flexovoltage response of the slab scales similarly to the bulk term for increasing t.) These, unlike in the previous two cases, need to be divided by the bulk permittivity, as the bulk material dielectrically screens the additional electric field produced by surface effects.

Spherical atom model. The formalism that we have developed in this section allows us to complete the solution of the toy model that we described at the end of Section 2.6, consisting of a finite slab made of a lattice of rigid (and non-interacting) closed-shell atoms. The solutions for all the contributions to the flexovoltage response, now including the shear case and the aforementioned separation of the surface term into polarization charge and metric terms, are schematically illustrated in Fig. 8. (The details of the derivations can be found in the Supplementary Notes of Ref. [15].) As expected, in all cases the net voltage vanishes, consistent with the physical expectations (a rigid displacement of spherical charge distributions cannot lead to a long-range electrical perturbation). By the same token, the pseudopotential rigid-core correction of equation (60) has no effect on the net flexovoltage. Note, however, that the bulk, surface-metric and surface-polarization terms cancel each other in a non-trivial way depending on the strain-gradient component, indicating that a consistent treatment of all three terms is crucially important for having a physically meaningful solution.

2.8.2. *Atomic relaxations*

The contribution of atomic relaxations to the flexovoltage coefficient of a bent slab has been extensively treated in Section 2.6.2. It is easy to show that, by using the formalism presented in Section 2.7.4 we recover equation (70), which describes the total response in terms of bulk- and surface-specific quantities. What remains to be discussed is the shear case. In Section 2.8.1, we postulated that this type of strain-gradient deformation is not relevant for a fully relaxed unsupported

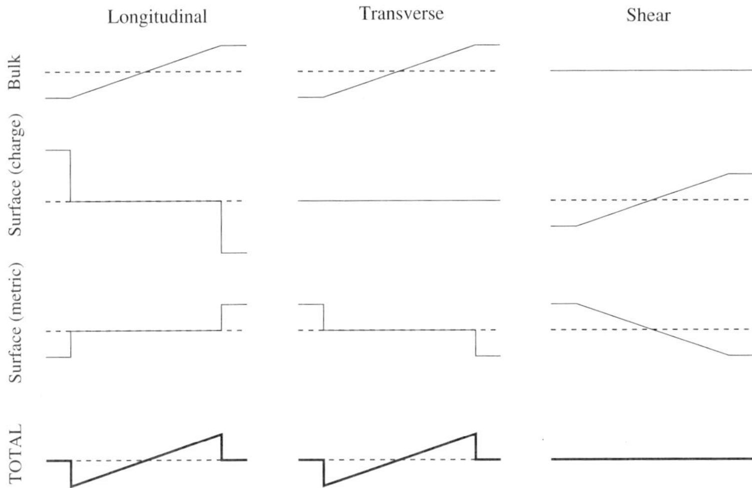

Figure 8. Flexoelectric response of a slab of non-interacting spheres. The total induced potential is decomposed into three contributions, consistent with the formalism developed in the main text. In all panels, the vertical axis corresponds to the potential (dashed line indicates the zero), and the horizontal axis is the spatial coordinate x along the surface normal. The combined effect of the surface (including both induced polarization-charge and metric contributions) and the bulk flexoelectric response yields a vanishing bias potential, regardless of the type of strain gradient. (From Ref. [15].)

slab. Here, we shall substantiate this statement in light of the results presented so far. Recall equation (106), which describes the microscopic atomic relaxation pattern induced by a strain gradient in terms of the internal-strain response tensors $\mathbf{\Gamma}$ and $\mathbf{L}$, and let $X_{l\kappa}$ and $Y_{l\kappa}$ denote the x and y components of $\mathbf{R}_{l\kappa}$. In the case of a shear strain gradient of the type $\varepsilon_{xy,y}$ in Fig. 6(c), equation (106) reads as

$$\frac{\partial u_{\kappa\alpha}^{l}}{\partial \varepsilon_{xy,y}} = Y_{l\kappa}\,\Gamma_{\alpha xy}^{\kappa} + L_{\alpha y,xy}^{\kappa}. \tag{139}$$

Now, regardless of the microscopic details of the slab, rotational invariance dictates that

$$\Gamma_{\alpha xy}^{\kappa} = -X_{0\kappa}\delta_{\alpha y}, \tag{140}$$

i.e. under a uniform shear the slab rigidly rotates to accommodate the deformation of the supercell, without feeling any restoring force because the repeated images of the slab are decoupled. By combining equations (139) and (140) we obtain

$$\frac{\partial u^l_{\kappa\alpha}}{\partial \varepsilon_{xy,y}} = -Y_{l\kappa} X_{l\kappa} \delta_{\alpha y} + L^\kappa_{\alpha y,xy}. \tag{141}$$

Now, recall that a macroscopic strain gradient can be written in terms of the components of the type-I strain-gradient tensor as

$$u^l_{\kappa\beta} = \frac{\eta_{\beta,\gamma\lambda}}{2} (\mathbf{R}_{l\kappa})_\gamma (\mathbf{R}_{l\kappa})_\lambda.$$

Equation (141) states that a shear strain gradient of amplitude $\eta_{x,yy} = \eta$ is always accompanied, in a fully relaxed unsupported film, by a second strain gradient component of the type $\eta_{y,xy} = -2\eta$. The overall effect, in type-II notation, is that of a negative *transverse* strain gradient, $\varepsilon_{yy,x} = -\eta$. This means that, for a free-standing film, the shear case reduces exactly to the transverse one. The basic concept is illustrated in Fig. 9, where we compare the configurations obtained by periodically subjecting a slab to a transverse strain gradient $\varepsilon_{yy,x}$ as in Fig. 6(b), or a (negative) shear strain gradient $\varepsilon_{xy,y}$ of the kind shown in Fig. 6(c). If internal atomic relaxations are allowed while still preserving the overall undulation along y, the two configurations will clearly relax to the exact same geometry.

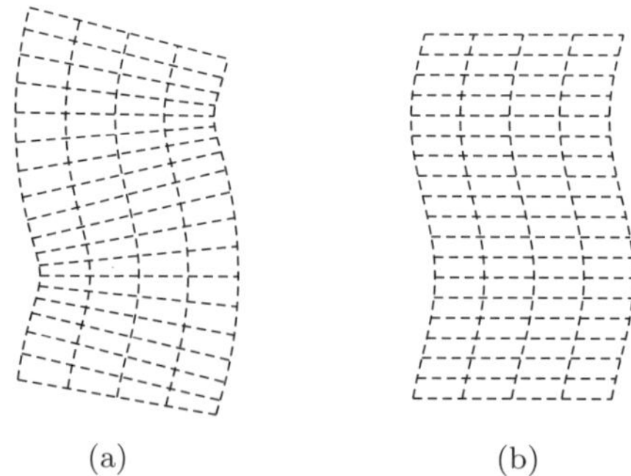

(a)   (b)

Figure 9.   Sketch of slab subjected to a periodic transverse strain of the type shown in Fig. 6(b), or a negative shear strain of the type shown in Fig. 6(c). After internal atomic relaxations, the two configurations become equivalent.

Thus, we have rigorously demonstrated the result that we heuristically presented in Section 2.6: flexoelectric effects in a free-standing film of sufficiently high symmetry (e.g. cubic or in-plane hexagonal) are governed by only one response coefficient $\varphi_{xx,yy}$, as given by equation (70). The induced voltage at a given location is then given by $\varphi_{xx,yy}(t/\xi_y + t/\xi_z)$, where $\xi_y = \varepsilon_{yy,x}^{-1}$ and $\xi_z = \varepsilon_{zz,x}^{-1}$ are the radii of curvature (along the Cartesian axes) of the film at that specific point.[y] This includes the plate-bending and beam-bending limits as special cases.

It is interesting to note that, in contrast with what happens in the bulk, here we have a notable case where the flexoelectric effects induced by a sound wave are identical to those associated with a static deformation. (Equivalently, one can say that the same strain gradient field can be induced either by dynamic or static means.) Indeed, any 2D object such as a slab is characterized by a transverse acoustic phonon branch, usually referred to as ZA, with zero sound velocity, corresponding to a bending mode. A long-wavelength ZA phonon coincides, therefore, with the static bending case described above, and produces the same flexoelectric response.

2.9. *Summary*

In this section we have presented a fundamental theory of flexoelectricity, based on a quantum-mechanical description of the electronic and lattice response to a strain-gradient perturbation. In particular, we have used a long-wave expansion of acoustic phonons to derive, in the linear limit, the relevant electromechanical response functions of a crystalline solid. Our formalism is fully general, and correctly recovers earlier theories of piezoelectricity as a special case.

In order to address some conceptual issues (e.g. regarding the role of the surfaces, or regarding the calculation of some components of the bulk flexoelectric tensor that are presently difficult to access) we have gone a step further, and developed a fully *microscopic*

---

[y] One could equivalently choose different orthogonal axes, e.g. those corresponding to the *principal curvatures* of the surface. Since $1/\xi_y + 1/\xi_z$ is the trace of the shape operator, the result is independent of such a choice.

theory of the linear response to an inhomogeneous strain field. In this context, we have demonstrated that the use of curvilinear coordinate frames greatly facilitates the representation of the relevant physical fields and their response to mechanical deformation. The latter methodological tools are applicable well beyond the specifics of flexoelectricity, and may find application in related research areas, such as flexomagnetism.[28]

3. Application to SrTiO$_3$

In this section, we shall demonstrate the theory developed so far by applying it to SrTiO$_3$, one of the most important materials in the context of flexoelectricity, and the best known experimentally. In order to quantify the importance of surface effects, we shall consider a slab geometry, and two different lattice terminations (either of the SrO or TiO$_2$ type), as illustrated in Fig. 10.

3.1. *General Methodology*

Our goal is to calculate the total flexovoltage response of either SrTiO$_3$ slab to a bending deformation in the limit of large thickness. We shall do this by taking into account the effect of full atomic relaxation, under the initial hypothesis that our slab behaves as a

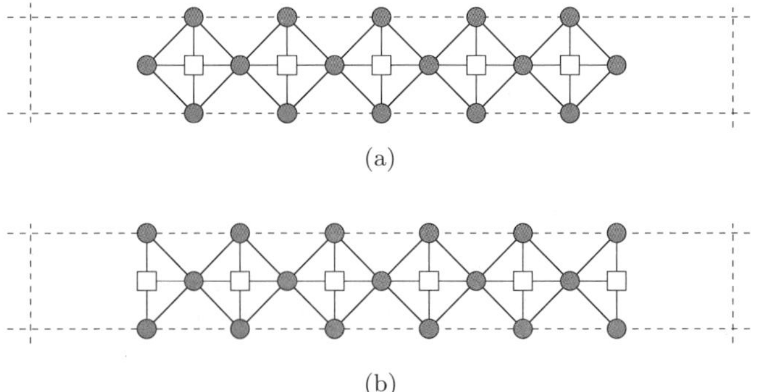

Figure 10.   Supercell models of the SrO- (a) and TiO$_2$-terminated (b) SrTiO$_3$ slabs. Ti and O atoms are represented as white squares and gray circles respectively; Sr atoms are not shown. (Adapted from Ref. [27].)

plate.[z] (We shall see in Section 3.3.3 that the beam-bending limit can easily be recovered by rescaling the plate-bending coefficient by a constant.) This requires the combination of three different computational frameworks, as detailed in the following.

3.1.1. *Bulk calculations*

Here, we perform a number of calculations on a primitive unit cell of bulk $SrTiO_3$. This is primarily aimed at calculating the bulk flexoelectric tensor via a long-wave expansion of acoustic phonons. Acoustic phonons are treated at the linear-response level by means of density-functional perturbation theory as implemented in a modified version of the ABINIT package[29] in which the contribution of the macroscopic electric fields has been removed according to the discussion of Section 2.3. In particular, We choose a small star of wave-vectors $\mathbf{q}$ surrounding the Γ point in the Brillouin zone,

$$\mathbf{q} = \frac{2\pi\tilde{q}}{a_0}(\pm 1, 0;\ \pm 1, 0;\ 0),$$

and perform a full linear-response calculation for each of these points. (In practice, we make full use of symmetries to minimize the number of actual calculations.) Next, we perform a long-wave expansion of the charge-density response and interatomic force constants and extract the second-order-in-$\mathbf{q}$ coefficients via numerical differentiation with respect to $\mathbf{q}$. We obtain: (i) the flexoelectric force-response tensor $T^{\kappa}_{\alpha\lambda,\beta\gamma}$ via equation (31), from which $C^{\kappa}_{\alpha\lambda,\beta\gamma}$ and then the internal-strain tensor $L^{\kappa}_{\alpha\lambda,\beta\gamma}$ are constructed via equations (34)–(35); and (ii) the charge-density response tensors $Q^{(1,\gamma)}_{\kappa\beta}$ and $Q^{(3,\gamma\lambda\sigma)}_{\kappa\beta}$, corresponding respectively to the Born effective charge tensor $Z^*_{\kappa,\beta\gamma}$ and the dynamical octupole tensor, via equation (45).

By combining the internal-strain tensor with the Born effective charges one can readily obtain the lattice-mediated contributions to the flexoelectric tensor as explained in Section 2.2. The octupole tensor, on the other hand, provides us with only partial information

---

[z] This means that along the direction parallel to the bending axis the system is clamped (i.e. no anticlastic bending is allowed).

on the electronic (frozen-ion) flexoelectric tensor. In particular, only the *longitudinal* component of the electronic flexoelectric tensor, $\bar{\mu}_{\hat{\mathbf{q}}}$, along an arbitrary direction $\hat{\mathbf{q}}$ can be inferred from the two linearly independent entries of $Q^{(3,\gamma\lambda\sigma)}_{\kappa\beta}$. Following Hong and Vanderbilt,[14] we define

$$\bar{\mu}_{\mathrm{L1}} = \bar{\mu}_{(100)}, \quad \bar{\mu}_{\mathrm{L2}} = 2\bar{\mu}_{(110)} - \bar{\mu}_{(100)}.$$

These are related to the components of the type-II flexoelectric tensor, μ_{L2} by[27]

$$\bar{\mu}^{\mathrm{II}}_{xx,xx} = \bar{\mu}_{\mathrm{L1}}, \tag{142}$$

$$\bar{\mu}^{\mathrm{II}}_{xx,yy} + 2\bar{\mu}^{\mathrm{II}}_{xy,xy} = \bar{\mu}_{\mathrm{L2}}. \tag{143}$$

Thus, in order to determine the transverse and shear components $\bar{\mu}^{\mathrm{II}}_{xx,yy}$ and $\bar{\mu}^{\mathrm{II}}_{xy,xy}$ independently, an additional calculation is necessary; this will be addressed shortly in Section 3.1.2.

In addition to the above calculations, which are based on the methodology described in this chapter, we also need a bulk-level calculation of some auxiliary quantities by means of more established techniques. Specifically, we extract the high-frequency dielectric constant $\bar{\epsilon}_{\mathrm{r}}$ from a separate linear-response treatment of the electric-field perturbation. At the same time we obtain a redundant set of $Z^{*}_{\kappa,\beta\gamma}$ tensor elements, which are useful for assessing the quality of the numerical differentiation at first order in $\mathbf{q}$ performed in Section 3.1.1. Similarly, we carry out an independent calculation of the elastic tensor $C_{\alpha\lambda,\beta\gamma}$ via finite differences with respect to applied strain; this allows us to check the second-order-in-$\mathbf{q}$ calculations of the force-response tensors $C^{\kappa}_{\alpha\lambda,\beta\gamma}$, since these quantities are directly related by the sum rule in equation (36).

3.1.2. *Truncated-bulk slab calculations*

Here, we carry out calculations similar to those of Section 3.1.1, but now on a slab supercell. This step is aimed at determining the transverse and shear components of the *bulk* electronic (frozen-ion) flexoelectric tensor.[aa] In fact, the two independent components of the bulk dynamical octupole tensor $Q^{(3,\gamma\lambda\sigma)}_{\kappa\beta}$ that we calculated above are not sufficient to determine the three independent entries of the bulk $\bar{\mu}^{\mathrm{II}}_{\alpha\lambda,\beta\gamma}$ tensor. We are able to circumvent this limitation by recourse to a series of calculations on a slab geometry in which we determine the charge-density response, both in the bulk and at the surface, to longitudinal, transverse, and shear strain gradients. A calculation of the flexoelectrically induced open-circuit electric field in the interior of the film, which relates [based on equation (50)] directly to the corresponding component of the bulk flexoelectric tensor in two cases out of three (longitudinal and transverse), allows us to obtain the missing component[bb] of $\bar{\boldsymbol{\mu}}^{\mathrm{II}}$. The key point here is that the missing divergence-free component of the induced polarization current, which is not currently available from bulk level calculations, manifests itself as a surface charge density, whose influence is readily apparent in the slab supercell geometry. Note that the specifics of the surface structure should not matter in these calculations. Thus, we choose the geometry that ensures the best convergence of the inner open-circuit field as a function of slab thickness, i.e. a truncated-bulk structure. (We perform such an analysis on both SrO- and TiO_2-terminated slabs, in order to verify that the results are indeed surface-independent as we expect.)

In practice, we use the same star of $\mathbf{q}$-points surrounding Γ as in the bulk calculations described above. This time, however, we neglect the information on the force constants and only focus on the charge-density response of the system. We need to analyze such a response at

---

[aa]It may seem odd to use a slab supercell to calculate a bulk-specific quantity; this is indeed a temporary work-around, which will no longer be necessary once a proper theory of the current-density response becomes available.

[bb]Strictly speaking, only the transverse component is really needed, as the longitudinal component calculated in this way is redundant with the μ_{L1} value that we already calculated at the bulk level. We shall use this as a test to assess the numerical accuracy of our calculations.

the microscopic level, by using the curvilinear-coordinate formalism of Sections 2.7 and 2.8. Of the two relevant response functions, $\rho^{\mathrm{U}}(x)$ and $\rho^{\mathrm{G}}(x)$, only the latter is really an issue, as $\rho^{\mathrm{U}}(x)$ can be straightforwardly calculated as the response to a uniform strain.[cc] The result of the second-order Taylor expansion in $\mathbf{q}$ yields $\rho^{\mathrm{G}}(x)$, and this (together with ρ^{U}) is then used to calculate the electric-field response functions $E^{\mathrm{U,G}}(x)$.

Note, however, that due to the removal of the macroscopic electric fields in the phonon calculations[13,15,27] (as required to perform the aforementioned Taylor expansions in $\mathbf{q}$, see Section 2.3), short-circuit electrical boundary conditions are enforced by construction on the calculated ρ^{G} and E^{G}. This means that there are non-vanishing macroscopic electric fields in both the vacuum and the slab interior, and these fields show an undesirable dependence on the supercell geometry (vacuum and slab thicknesses). To have a physically well-defined (and geometry-independent) value of the internal field we need to enforce open-circuit electrical boundary conditions. We do this by applying an *external* field to the system that is exactly opposite to the calculated vacuum field. To determine the charge redistribution induced in the system upon application of an external field, we perform a separate linear-response calculation of the *local* electric-field response to a macroscopic electric displacement field D. This is nothing but the local inverse dielectric permittivity of the slab supercell,

$$\frac{\partial E_x(x)}{\partial D_x} = \frac{1}{\epsilon_0} \bar{\epsilon}_{\mathrm{r}}^{-1}(x).$$

We then use

$$E_x^{\mathrm{G,OC}}(x) = E_x^{\mathrm{G,SC}}(x) - E_x^{\mathrm{G,SC}}(+\infty)\bar{\epsilon}_{\mathrm{r}}^{-1}(x),$$

[cc]We calculated $\rho^{\mathrm{U}}(x)$ separately by using standard ground-state calculations where we took finite differences in the strain. We found that this latter procedure yields slightly better accuracy than the long-wave method described above.

where $x = +\infty$ corresponds, as usual, to the vacuum region. When referring to $E^{\mathrm{G}}(x)$ in the following, we shall implicitly assume that we are speaking of the open-circuit version $E_x^{\mathrm{G,OC}}(x)$.

3.1.3. *Relaxed-ion slab calculations*

Now that we have all the necessary bulk-specific information in hand, we still need to determine the surface-specific contributions to the flexovoltage coefficient $\varphi_{xx,\mathrm{eff}}^{\mathrm{surf}}$.[dd] We shall compute φ^{surf} as the induced electrostatic potential offset upon application of a *uniform* effective strain ($\varepsilon_{yy} = \varepsilon_{\mathrm{eff}}$; $\varepsilon_{xx} = -\nu\varepsilon_{\mathrm{eff}}$) to a free-standing slab with (001) surface orientation. This quantity can be conveniently accessed by means of a standard plane-wave code; no linear-response features are needed. In particular, we take a slab supercell corresponding to a periodic lattice of alternating $SrTiO_3$ and vacuum layers, and first calculate the electronic and structural ground state by setting the in-plane lattice parameter to the equilibrium bulk value. We then apply a small positive or negative strain of the type

$$\varepsilon = \frac{\varepsilon_{\mathrm{eff}}}{2} \begin{pmatrix} -2\nu & 0 & 0 \\ 0 & 1 & 0 \\ 0 & 0 & 1 \end{pmatrix},$$

where $\varepsilon_{\mathrm{eff}}$ is a small dimensionless number, typically $\varepsilon_{\mathrm{eff}}(\pm) = \pm 0.001$. (We find it computationally advantageous to preserve the fourfold axis of the $SrTiO_3$ surface by applying an isotropic in-plane strain.)

In each perturbed configuration, we first calculate the electronic ground state with the reduced coordinates of the atoms kept fixed to their unperturbed values; the resulting electrostatic potential profile is then processed by means of macroscopic averaging[17,26] to extract the perturbed frozen-ion (FI) surface potential offsets. Next, we let the atoms relax to their new equilibrium positions in the strained

---

[dd]Recall that we need to consider, for a bent slab at mechanical equilibrium, an *effective* combination of transverse and longitudinal strain-gradient deformations, $\varepsilon_{yy,x} = \varepsilon_{\mathrm{eff},x}$; $\varepsilon_{xx,x} = -\nu\varepsilon_{\mathrm{eff},x}$, where $\nu = C_{yy,xx}/C_{xx,xx}$.

lattices, and repeat the macroscopic averaging procedure to obtain the relaxed-ion (RI) offsets. Finally, we numerically differentiate the perturbed offsets (both FI and RI) to obtain their corresponding first-order variation,

$$\varphi^{\text{surf}} = \frac{\phi(+) - \phi(-)}{2|\varepsilon_{\text{eff}}|},$$

where $\phi(\pm)$ refers to the surface potential offset at positive or negative strain.[ee] This procedure readily yields the RI and FI values of φ^{surf}. The lattice-mediated (LM) values are simply calculated as the difference of the RI and FI ones. Of course, the slab needs to be sufficiently thick in order for the inner layers to be truly bulk-like, i.e. unaffected by the atomic distortions that originate from the surface truncation of the bonding network.

3.2. *Computational Parameters*

We use the local-density approximation[30] to DFT. The interactions between valence electrons and ionic cores are described by separable norm-conserving pseudopotentials in the Troullier–Martins[31] form, generated with the fhi98PP code.[32] The $4s^2 4p^6$ and $3s^2 3p^6$ shells of Sr and Ti, respectively, are explicitly treated as valence electrons. The reference states (numbers in parentheses indicate the core radius in bohr) of the isolated neutral atom used in the pseudopotential generation are $2s(1.4)$, $2p(1.4)$, and $3d(1.4)$ for O, $4s(1.5)$, $4p(1.5)$, and $4d(2.0)$ for Sr and $3s(1.3)$, $3p(1.3)$, and $3d(1.3)$ for Ti. The local angular-momentum channel is $l = 2$ for Sr and O and $l = 0$ for Ti. The rigid-core corrections of equation (60) are not included in the presented results. The cutoff for the wavefunction plane-wave basis is set to 150 Ry in the slab calculations. (A test calculation with a 300 Ry cutoff did not show appreciable changes in the calculated

---

[ee] As a technical note, many first-principles codes use the Ewald procedure to calculate the self-consistent electrostatic potential. This involves adding to the electronic density a lattice of spherical Gaussian compensating charges, whose spurious contribution must be removed from the calculated value of φ^{surf}. See the Supplementary Notes of Ref. [27] for details.

electronic response functions; the 300 Ry cutoff was, nonetheless, necessary to ensure satisfactory accuracy in the force-response tensor at the bulk level.) The surface Brillouin zone of the slab supercell is sampled by means of a 8×8 Monkhorst–Pack grid;[33] for the bulk primitive cell we use a sampling of up to $12 \times 12 \times 12$ k-points. The finite-difference parameter in the long-wave expansion, $\tilde{q}$, is set to 0.01 (tests with $\tilde{q} = 0.02$ or $\tilde{q} = 0.03$ indicated a convergence better than 1% in the calculated electronic response functions; smaller values of $\tilde{q}$ were found to yield less accurate results because of the excessive numerical noise). The lattice parameter of the cubic cell is set to $a_0 = 7.268$ bohr, which corresponds to the calculated equilibrium value.

The supercell models are based on the schematic illustrations of Figs. 10(a) and (b). For the truncated-bulk linear-response calculations we use 5.5-unit-cell (uc) thick $SrTiO_3$ slabs alternating with vacuum layers whose thickness is set to 2.5 uc. Of course, both (slab and vacuum) thicknesses are intended as convergence parameters in our calculations, whose scope is to describe the thermodynamic limit of a macroscopic slab. Tests with thinner slabs and thicker vacuum layers (up to 3.5 uc) showed optimal convergence for the aforementioned values of these parameters (again, better than 1%). For the relaxed-ion slabs, we use 7.5 uc-thick slabs with 3.5 uc-thick vacuum layers.

3.3. *Results*

3.3.1. *Bulk calculations*

In Table 1, we report the relevant values of the force-response tensor of bulk $SrTiO_3$, calculated by using the long-wave method described in Section 2.2. In Table 2, we compare the above physical quantities to the analogous ones that were calculated in Ref. [14]. To perform the comparison we first recast the force-response components into a tensorial representation that follows the same prescriptions as equations (142) and (143),

$$C^{\prime\kappa}_{L1} = C^{\kappa}_{xx,xx}, \tag{144}$$

$$C^{\prime\kappa}_{L2} = C^{\kappa}_{xx,yy} + 2C^{\kappa}_{xy,xy}. \tag{145}$$

Table 1.   Force-response tensor $C^{\kappa}_{\alpha\lambda,\beta\gamma}$ of bulk SrTiO$_3$ in short-circuit boundary conditions. O1, O2, and O3 refer to oxygen atoms forming x-, y- or z-oriented Ti–O–Ti bonds, respectively. All values are in eV.

Atom	(xx, xx)	(xy, xy)	(xx, yy)
Sr	−24.9	7.9	−28.7
Ti	−67.9	3.8	−102.3
O1	159.3	15.3	97.4
O2	35.2	17.3	42.3
O3	35.2	−0.9	30.9

Table 2.   Summary of the linear-response data obtained from the long-wave (LW) approach at the bulk level, compared to the results of Hong and Vanderbilt[14] (here used as HV) for the same quantities. Open-circuit electrical boundary conditions are enforced on the longitudinal response functions (L1 and L2). The force response to a shear strain gradient (S) is quoted in short circuit. The oxygen modes $\xi_3 = x_{O1}$ and $\xi_4 = (x_{O2} + x_{O3})/\sqrt{2}$ are defined following Ref. 14. $\bar{\varphi}^{\text{bulk}}$ is in V; other values are reported in eV.

	L1(LW)	L1(HV)	L2(LW)	L2(HV)	S(LW)	S(HV)
$\bar{\varphi}^{\text{bulk}}$	−16.15	−16.25	−18.07	−18.17	—	—
Sr	16.3	17.0	33.2	35.7	7.9	8.4
Ti	49.1	52.3	36.3	38.9	3.8	3.0
ξ_3	67.2	68.7	24.9	13.1	15.3	15.7
ξ_4	3.0	3.6	22.7	18.2	11.6	12.0

Then, we convert the longitudinal quantities L1 and L2 from fixed-**E** or short-circuit (SC) to fixed-**D** or open-circuit (OC) boundary conditions by using [see equation (106) of Ref. [14]]

$$C^{\kappa}_{\text{L}}(\text{OC}) = C^{\kappa}_{\text{L}}(\text{SC}) - \bar{\varphi}^{\text{bulk}}_{\text{L}} Z^{*}_{\kappa}, \tag{146}$$

where Z^{*}_{κ} is the Born effective charge (calculated values are reported in Table 3), $\bar{\varphi}^{\text{bulk}}_{\text{L}}$ is the purely electronic flexovoltage coefficient, and L stands for either L1 or L2. The calculated values of $\bar{\varphi}^{\text{bulk}}_{\text{L1,L2}}$ are also reported in Table 2 for direct comparison to those reported by Hong and Vanderbilt.[14] The agreement is overall very good, especially

Table 3. Calculated Born effective charges and dielectric properties of bulk SrTiO$_3$.

Z^*_{Sr}	Z^*_{Ti}	Z^*_{O1}	Z^*_{O2}	Z^*_{O3}	$\bar{\epsilon}_r$	ϵ_r (static)
2.5548	7.2455	−5.7027	−2.0488	−2.0488	6.1785	1657

Table 4. Calculated elastic tensor of bulk SrTiO$_3$. The two rows refer to the bulk force-response calculation ("Force") and to a direct bulk calculation where we took finite differences of the calculated stress tensor while varying the strain around the equilibrium cubic configuration ("Strain"). Values are in GPa.

Method	(xx, xx)	(xy, xy)	(xx, yy)
Force	385.3	122.2	111.7
Strain	386.2	122.4	112.6

considering the different computational strategy, first-principles code and pseudopotentials that were used in Ref. [14].

As a numerical test of the calculated force-response tensor (Table 1), in Table 4 we report the elastic constants of bulk SrTiO$_3$ that we computed in two different ways: either as a first derivative of the stress with respect to the applied strain ("strain") or by using the sum rule of equation (36) ("force"). The agreement is excellent (better than 1%), confirming the high numerical quality of the calculation. Note that the choice of the electrical boundary conditions is irrelevant for this test, as the sublattice sum of the atomic forces induced by a hypothetical electric field vanishes due to the acoustic sum rule.

3.3.2. *Truncated-bulk slab calculations*

In Figs. 11(a) and (d) we plot the calculated $E_x^{U,G}(x)$, corresponding to either a SrO- or a TiO$_2$-terminated slab and to each of the three types of imposed strain gradients shown in Fig. 6 (with no internal relaxations allowed). As anticipated in Section 2.8.1, there is an important qualitative difference between the case of the longitudinal or transverse response, where the strain gradient is oriented along the

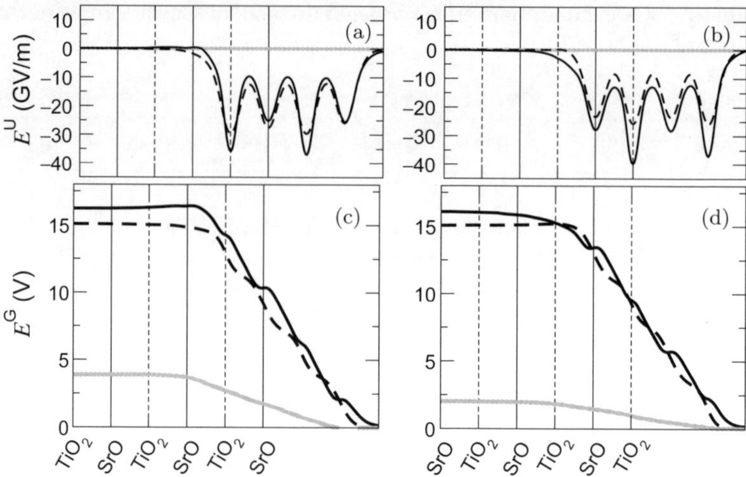

Figure 11.   Electric-field response to mechanical deformations. The E_x^{U} (a–b) and E_x^{G} (c–d) response functions are shown for a SrO- (a, c) and TiO$_2$-terminated (b, d) slab. Solid black, dashed black and solid gray curves refer to longitudinal, transverse and shear deformations, respectively. The location of the SrO (dashed) and TiO$_2$ (solid) atomic layers is indicated by vertical lines (only half of the symmetric slab is shown). (Adapted from Ref. [27].)

surface normal, and that of the shear response, where it is directed in plane.

In the former case, $E_{x,\beta\beta}^{\mathrm{U}}(x)$ (describing the **E**-field response to a *uniform* strain) yields the surface contribution to the flexovoltage coefficient, $\bar{\varphi}_{xx,\beta\beta}^{\mathrm{surf}}$, via equation (128),[ff] while the functions $E_{xx,\beta\beta}^{\mathrm{G}}(x)$ provide us with the sought-after information on the *bulk* flexovoltage coefficient of SrTiO$_3$,

$$\varphi_{xx,\beta\beta}^{\mathrm{bulk}} = -E_{xx,\beta\beta}^{\mathrm{G}}(x=0).$$

Note that the $E_{xx,\beta\beta}^{\mathrm{G}}(x)$ functions are roughly uniform inside the film, which indicates that the slab is thick enough to display bulk

---

[ff]Note, however, that here we are dealing with a truncated-bulk slab, whose surface atomic coordinates were artificially frozen to ideal bulk positions. The surface contributions that one extracts from such a geometry do not necessarily reflect, therefore, the response of a realistic system; they are quoted here mostly for illustrative purposes.

properties therein, and zero outside, consistent with the open-circuit electrical boundary conditions that were enforced. Moreover, the uniform internal field appears to be nicely *independent* of the surface termination for the longitudinal and transverse deformations, which is a further important consistency test for our computational approach.

In the shear case, however, the flexoelectric field depends on both bulk and surface-specific properties,[15] and such a termination dependence is clear from a comparison of the gray curves in Figs. 11(c) and (d). From the electric-field response functions of Figs. 11(c) and (d) we can thus only extract the *total* flexovoltage coefficient of the slab, $\varphi_{xy,xy} = -E^{\mathrm{G}}_{xy,xy}(x = 0)$. To separate $\varphi_{xy,xy}$ into bulk and surface terms it suffices, however, to complement the above data with the $\varphi^{\mathrm{bulk}}_{\mathrm{L1,L2}}$ values that we calculated at the bulk level. Indeed, by replacing the flexoelectric tensor components in equations (142) and (143) with the corresponding flexovoltage coefficients, we have

$$\varphi^{\mathrm{bulk}}_{\mathrm{L1}} = \varphi^{\mathrm{bulk}}_{xx,xx}, \tag{147}$$

$$\varphi^{\mathrm{bulk}}_{\mathrm{L2}} = \varphi^{\mathrm{bulk}}_{xx,yy} + 2\varphi^{\mathrm{bulk}}_{xy,xy}. \tag{148}$$

Equation (147) constitutes a useful consistency check of the methodology, as $\varphi^{\mathrm{bulk}}_{\mathrm{L1}}$ is redundant with the already calculated value of $\varphi^{\mathrm{bulk}}_{xx,xx}$. Equation (148), on the other hand, yields the desired value of $\varphi^{\mathrm{bulk}}_{xy,xy}$ since we already know $\varphi^{\mathrm{bulk}}_{xx,yy}$ from the slab calculations. Finally, we use $\varphi_{xy,xy} = -E^{\mathrm{slab}}_{xy,xy}$ to infer $\varphi^{\mathrm{surf}}_{xy,xy} = \varphi_{xy,xy} - \varphi^{\mathrm{bulk}}_{xy,xy}$.

Our results for the bulk, surface, and total flexovoltage coefficients of the truncated-bulk, frozen-ion deformation of a $SrTiO_3$ slab are summarized in Table 5. At the bulk level, it is interesting to note the relatively small magnitude of the shear coefficients $\varphi^{\mathrm{bulk}}_{xy,xy}$ and $\varphi^{\mathrm{surf}}_{xy,xy}$ compared to both the longitudinal and the transverse ones. Meanwhile, in the latter two cases there is a substantial cancelation between bulk and surface terms; as a result, the values of the total flexovoltage coefficients φ are all comparable in magnitude. This fact can be rationalized by observing that the linear response to atomic displacements in a ionic (or partially ionic) solid is largely dominated by the rigid displacement of an approximately spherical

Table 5. Frozen-ion flexovoltage coefficients of a truncated-bulk SrTiO$_3$ slab. To compute φ^{bulk} we used $\varphi^{\text{bulk}}_{\text{L1,L2}}$ as reported in Table 2 and $E^{\text{slab}}_{xx,yy} = 15.08$ V [extracted from Figs. 11(c) and (d)]. (L), (T), and (S) stands for longitudinal, transverse and shear, respectively. Units of Volts are used throughout.

| | | φ^{surf} | | φ (total) | |
	φ^{bulk}	SrO	TiO$_2$	SrO	TiO$_2$
xx, xx (L)	-16.15	14.36	16.95	-1.80	0.80
xx, yy (T)	-15.08	15.68	12.45	0.61	-2.63
xy, xy (S)	-1.50	-2.38	-0.51	-3.88	-2.01

charge density distribution surrounding each atom. The spherical contribution, which is typically large and negative,[12] shows up in $\varphi^{\text{bulk}}_{xx,\beta\beta}$, and with opposite sign in $\varphi^{\text{surf}}_{xx,\beta\beta}$; in the shear case neither the bulk nor the surface term are affected (see Section 2.5.1). Remarkably, the resulting values of φ depend strongly on the details of the surface, and in some cases even have opposite signs in the SrO- and TiO$_2$-terminated slabs. Such a conclusion, in fact, persists after we take into account the full relaxation of the atomic structure; we shall demonstrate this point in the following paragraphs.

3.3.3. *Relaxed-ion slab calculations*

The results of the relaxed-ion slab calculations allow us to complete the picture of the fully relaxed flexovoltage response of a SrTiO$_3$ slab in the plate-bending limit. [The beam-bending case is easily recovered by multiplying the reported values by $\tau = \mathcal{C}_{xx,xx}/(\mathcal{C}_{xx,xx} + \mathcal{C}_{xx,yy})$. By using the calculated elastic constants of bulk SrTiO$_3$, reported in Table 4, we find $\tau = 0.77$.] A summary of the results is reported in Table 6. The respective contributions of the bulk and surface are, overall, in line with the available order-of-magnitude estimates.[34] The values shown in bold font, i.e. the total flexovoltage coefficients of the two types of slab, comprise the main result of this work. Note that they depart substantially from the corresponding bulk coefficient, confirming the dramatic impact of the surface structural and electronic properties on the electromechanical response of the system. In fact, the aforementioned response coefficients are even opposite

Table 6. Flexovoltage coefficients of a relaxed $SrTiO_3$ slab. FI, lattice-mediated (LM) and total relaxed-ion (RI=FI+LM) values of the bulk, surface, and total slab response are reported. Units of Volts are used throughout.

	φ^{bulk}	φ^{surf}		φ (total)	
		SrO	TiO_2	SrO	TiO_2
FI	−10.37	13.47	6.84	3.10	−3.53
LM	−0.44	−4.93	5.34	−5.38	4.90
RI	−10.81	8.53	12.18	**−2.28**	**1.37**

in sign depending on whether a SrO- and TiO_2-terminated slab is considered. This is a remarkable result, as it means that an atomically thin surface termination layer can modify, and even reverse, the flexovoltage response of a macroscopically thick sample. This constitutes a rather drastic departure from the characteristics of other electromechanical phenomena (e.g. piezoelectricity), where the details of the surfaces typically become irrelevant in the thermodynamic limit.

It is interesting to note that the surface shows an even larger termination dependence at the frozen-ion level, but with *opposite* sign. The LM contribution to φ^{surf} is indeed large, and depends so strongly on the termination that its inclusion results in a voltage reversal, both in the TiO_2- and SrO-type slabs. (By contrast, the LM contribution to the bulk flexovoltage coefficient is relatively minor, about one order of magnitude smaller than any other value reported in the table, and has little impact on the final results.) To illustrate the reason for such a strong dependence, a microscopic analysis of the surface relaxations is provided in Fig. 12. In the SrO case, the layer-by-layer decomposition of the induced dipole shown in Fig. 12(e) has an oscillatory behavior whose amplitude decays exponentially as a function of the distance from the surface; as a consequence, the surface layer clearly dominates the overall response.[gg] Instead, for the

---

[gg]Interestingly, the structural relaxation pattern in the unperturbed state, Fig. 12(a), appears very similar to the *induced* relaxation pattern under an applied tensile strain. This suggests that the former might be, in fact, rationalized as a response of the system to a large surface stress.

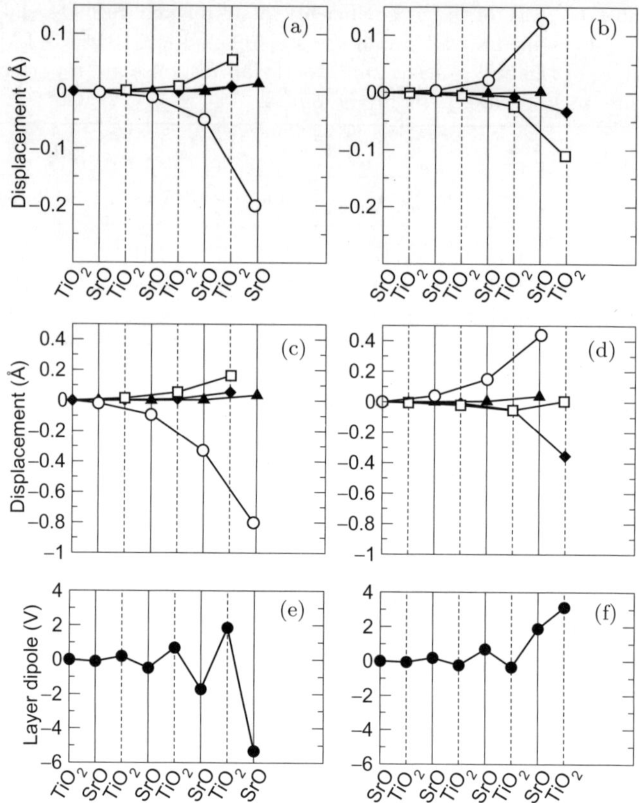

Figure 12. Static and induced ionic relaxations at the $SrTiO_3$ surface. (a–b): Ionic relaxations in the unperturbed slabs (displacements from ideal bulk-like sites). Circles, squares, diamonds and triangles correspond, respectively, to Sr, Ti, O(Ti), and O(Sr) atoms. (Cations are indicated by empty symbols, oxygen atoms by filled ones.) Negative values indicate inward displacements (i.e. towards the slab center). (c–d): Displacements induced by a uniform strain of the type $\varepsilon_{yy} - \nu\varepsilon_{xx}$; for the two oxygen atoms in the TiO_2 layers, only one value (their average displacement) is shown. (e–f) Layer-by-layer decomposition of the lattice-mediated contribution to the induced surface potential offset. Vertical lines indicate the position of the SrO (solid) and TiO_2 (dashed) atomic planes. (Adapted from Ref. [27].)

TiO_2-terminated slab shown in Fig. 12(f), the surface layer responds with a positive dipole instead of a negative one, in sharp contrast to the "underdamped" oscillatory behavior in Fig. 12(e). This behavior is probably due to the alteration of the bonding network, which we speculate to be much more profound at the TiO_2-type surface than at the SrO-type one, whereby the boundary atoms no longer behave as bulk-like but rather as a distinct chemical entity.

Apart from the obvious relevance of the above observations to the physics of $SrTiO_3$ surfaces, the analysis of Figs. 12(e) and (f) carries a general message that we have already anticipated in the above paragraphs. Any single atomic layer near the surface has a remarkably large contribution to φ^{surf}, sometimes of the same order as (or even larger than) the overall flexovoltage response of the slab. In fact, the total open-circuit voltage results from the subtle cancelation of many contributions of dissimilar physical nature. This implies that exceptional care is needed when dealing with flexoelectric phenomena, either when performing the calculations or when interpreting the experiments.

4. Conclusions and Outlook

In this chapter, we have described the main advances in the first-principles theory of flexoelectricity that have taken place during the past five years. The progress that emerges from these pages is undoubtedly impressive — we are at the stage where the full flexoelectric response of real materials, including bulk and surface effects, can be calculated *ab initio* with great accuracy. Still, much remains to be done before the field can be regarded as mature. We discuss here several research avenues that we identify as being of pivotal importance for future progress.

- **Theory of the current-density response.** The most fundamental and complete framework for the theory of flexoelectricity is the current-response formalism introduced in Section 2.2. Unlike the charge-response formalism summarized in Section 2.5, the current-response approach is capable in principle of resolving all independent components of the flexoelectric tensor. However, two issues remain to be settled in relation to this approach. First, direct methods for obtaining the current response functions $\overline{P}^{\mathbf{q}}_{\kappa\beta}$ of equation (13) by computing the linear response to a phonon of small but finite wave-vector $\mathbf{q}$ have not yet been developed and tested. Once implemented, this would allow for a finite-difference calculation of the $\overline{P}^{(2,\gamma\lambda)}_{\alpha,\kappa\beta}$ of equation (14), and thence, the electronic contribution in equation (18). Second, some aspects of the connection between the current-response theory and the theory of charge responses (including surface charges) remain to be clarified, as discussed in the context of equation (50) and following equation (101). A solution of these two issues would help put the theory of flexoelectricty on a truly sound footing.

- **Analytic derivation of the q-expansions.** The conceptual foundation of most of the material treated in this chapter is a long-wave expansion of certain physical observables as a function of the wave-vector $\mathbf{q}$ of an acoustic phonon. The calculations described in Section 3 were performed by taking such a $\mathbf{q}$-expansion numerically via finite differences, which is computationally cumbersome. Ideally, it would be best to perform the expansion analytically, i.e. to derive the DFPT equations that directly yield the wavefunction response to a strain gradient perturbation. This would also be desirable in the context of the direct current-density implementation sketched just above. When implemented in an existing DFPT code, such methods would allow for a more straightforward calculation of flexoelectric properties of materials, and thus foster a more widespread application of these techniques within the research community.

- **Application to complex materials.** Our focus in this chapter has been on materials with cubic symmetry. Clearly a proper theory that also covers crystals of lower symmetry is strongly required. The extension of the theory to such materials will require attention not just to the proliferation of independent parameters in the flexoelectric tensor, but also to subtle physical issues having to do, for example, with the anisotropic electronic screening that occurs when the symmetry is reduced. In the case of crystals that are piezoelectric (and possibly also polar), care will be needed to separate the higher-order flexoelectric from dominant piezoelectric (and possibly spontaneous) polarization response. The application to insulating ferromagnets or antiferromagnets should introduce no special difficulties in most cases, but may involve subtleties for magnetoelectric crystals or when spin-orbit coupling is strong. A first-principles theory of *flexomagnetism* has yet to be developed.

- **Compositional gradients.** An electric polarization can also arise in the presence of a *compositional* gradient, e.g. in $Ba_{1-x}Ti_xO_3$ films.[35] To our knowledge, a proper theory of such an effect is lacking. Since a compositional gradient generally also entails a strain gradient, some care will be called for in separating these effects and computing them independently before combining the contributions to make physically meaningful predictions.

- **Connection to higher-level models.** With the techniques described here, one can in principle calculate the fundamental flexoelectric properties of an arbitrary material. To use this information in real physical problems, however, one typically has to deal with many additional issues that are intractable by means of direct first-principles simulation: large samples with complex shapes, temperature effects, etc. It would be very desirable in this context to be able to extract the relevant physical parameters from the *ab initio* calculations, and incorporate them in some higher-level theory (e.g. atomistic, effective Hamiltonian, or continuum) where length- and time-scale limitations are much less stringent. A successful attempt in this sense has already been reported[36]; still, consistently incorporating the latest first-principles developments into macroscopic theories remains an open challenge. For example, it would be of crucial importance, for a realistic description of the flexoelectric effect, to extract the relevant surface-specific properties from the density-functional calculations, and incorporate them into the higher-level model. Making progress in this direction will also promote a closer interaction between different communities working on flexoelectricity (continuum numerical modeling, Landau theory, etc.), which we believe would have a strong positive impact on the field.

In summary, there has been dramatic progress in the development of a full first-principles theory of flexoelectricity. Several important challenges remain, as discussed above, but at least these have been identified, and solutions appear to be within reach. In any case, the development of the theory of flexoelectricity has already revealed many fascinating links to other, at first sight unrelated, research areas (e.g. the relationship to transformation optics, where the use of curvilinear coordinates facilitates the solution of complex electrical engineering problems). We believe that more surprises are in store, and will progressively emerge while further progress is made along the above lines. As the study of flexoelectricity touches so many subfields of condensed matter physics, we expect cross-cutting progress that will likely benefit the first-principles materials theory community at large. All in all, we look forward to the day when predictive calculations of flexoelectric responses can become a routine part of the tool-kit of first-principles computational materials theory.

Acknowledgments

We thank Jiawang Hong for useful discussions. We acknowledge support from ONR Grant N00014-12-1-1035 (D.V.), a grant from the Simons Foundation (#305025 to D.V.), MINECO-Spain Grant FIS2013-48668-C2-2-P (M.S.), and Generalitat de Catalunya Grant 2014 SGR 301 (M.S.). M.S. also acknowledges support from MINECO-Spain through the "Severo Ochoa" Programme for Centres of Excellence in R&D (SEV-2015-0496).

References

1. R. O. Jones and O. Gunnarsson. *Rev. Mod. Phys.* **61**, 689 (1989).
2. S. Baroni, S. de Gironcoli, and A. D. Corso. *Rev. Mod. Phys.* **73**, 515 (2001).
3. R. M. Martin. *Phys. Rev. B.* **5**, 1607–1613 (1972).
4. R. Resta. *Ferroelectrics.* **136**, 51–55 (1992).
5. R. D. King-Smith and D. Vanderbilt. *Phys. Rev. B.* **47**, R1651–R1654 (1993).
6. R. Resta and D. Vanderbilt. Theory of polarization: A modern approach. In: (eds.) K. M. Rabe, C. H. Ahn, and J.-M. Triscone, *Physics of Ferroelectrics: A Modern Perspective.* Springer-Verlag, Berlin Heidelberg (2007).
7. R. Resta. *J. Phys. Condensed Matter.* **22**, 123201 (2010).

8. A. K. Tagantsev. *Phys. Rev. B.* **34**, 5883 (1986).

9. A. Tagantsev. *Phase Transit.* **35**, 119–203 (1991).

10. J. Hong, G. Catalan, J. F. Scott, and E. Artacho. *J. Phys. Condens. Matter.* **22**, 478–492 (2010).

11. R. Resta. *Phys. Rev. Lett.* **105**, 127601 (2010).

12. J. Hong and D. Vanderbilt. *Phys. Rev. B.* **84**, 180101(R) (2011).

13. M. Stengel. *Phys. Rev. B.* **88**, 174106 (2013).

14. J. Hong and D. Vanderbilt. *Phys. Rev. B.* **88**, 174107 (2013).

15. M. Stengel. *Nat. Commun.* **4**, 2693 (2013).

16. P. Zubko, G. Catalan, A. Buckley, P. R. L. Welche, and J. F. Scott. *Phys. Rev. Lett.* **99**, 167601 (2007).

17. A. Baldereschi, S. Baroni, and R. Resta. *Phys. Rev. Lett.* **61**, 734–737 (1988).

18. R. Resta. *Phys. Rev. B.* **44**, 11035–11041 (1991).

19. R. Resta, L. Colombo, and S. Baroni. *Phys. Rev. B.* **41**, 12358–12361 (1990).

20. F. Gygi. *Phys. Rev. B.* **48**, 11692–11700 (1993).

21. D. R. Hamann, X. Wu, K. M. Rabe, and D. Vanderbilt. *Phys. Rev. B.* **71**, 035117 (2005).

22. W. Yan, M. Yan, Z. Ruan, and M. Qiu. *New J. Phys.* **10**, 043040 (2008).

23. U. Leonhardt and T. G. Philbin. *New J. Phys.* **8**, 247 (2006).

24. M. Stengel, N. A. Spaldin, and D. Vanderbilt. *Nat. Phys.* **5**, 304–308 (2009).

25. P. Umari, A. D. Corso, and R. Resta. *AIP Conf. Proc.* **582**, 107–117 (2001).

26. J. Junquera, M. H. Cohen, and K. M. Rabe. *J. Phys. Condens. Matter.* **19**, 213203 (2007).

27. M. Stengel. *Phys. Rev. B.* **90**, 201112(R) (2014).

28. R. Hertel. *Spin.* **3**, 1340009 (2013).

29. X. Gonze, B. Amadon, P.-M. Anglade, J.-M. Beuken, F. Bottin, P. Boulanger, F. Bruneval, D. Caliste, R. Caracas, M. Côté, T. Deutsch, L. Genovese, P. Ghosez, M. Giantomassi, S. Goedecker, D. Hamann, P. Hermet, F. Jollet, G. Jomard, S. Leroux, M. Mancini, S. Mazevet, M. Oliveira, G. Onida, Y. Pouillon, T. Rangel, G.-M. Rignanese, D. Sangalli, R. Shaltaf, M. Torrent, M. Verstraete, G. Zerah, and J. Zwanziger. *Computer Phys. Commun.* **180**, 2582–2615 (2009).

30. J. P. Perdew and Y. Wang. *Phys. Rev. B.* **45**, 13244 (1992).

31. N. Troullier and J. L. Martins. *Phys. Rev. B.* **43**, 1993–2006 (1991).

32. M. Fuchs and M. Scheffler. *Computer Phys. Commun.* **119**, 67–98 (1999).

33. H. J. Monkhorst and J. D. Pack. *Phys. Rev. B.* **13**, 5188–5192 (1976).

34. P. V. Yudin and A. K. Tagantsev. *Nanotechnology*. **24**, 432001 (2013).
35. J. Zhang, R. Xu, A. R. Damodaran, Z.-H. Chen, and L. W. Martin. *Phys. Rev. B*. **89**, 224101 (2014).
36. I. Ponomareva, A. K. Tagantsev, and L. Bellaiche. *Phys. Rev. B*. **85**, 104101 (2012).

Chapter 3

A Continuum Theory of Flexoelectricity

Q. Deng

Department of Mechanical Engineering,
University of Houston, Houston, TX 77024, USA

L. Liu

Department of Mathematics and Department
of Mechanical Aerospace Engineering,
Rutgers University, Piscataway, NJ 08854, USA

P. Sharma

Department of Mechanical Engineering,
University of Houston, Houston, TX 77024, USA

In this chapter, we present a nonlinear continuum theory of flexoelectricity. Due to the scaling of strain gradients with structural feature size, flexoelectricity is expected to show a strong size dependency in electromechanical coupling. Aside from several illustrative and pedagogical boundary value problems, we present examples that highlight the applications of flexoelectricity e.g. creation of piezoelectric materials without using piezoelectric materials, energy harvesting, soft active materials, and biological membranes.

1. Introduction

Recently, a somewhat understudied electromechanical coupling, flexoelectricity, has attracted a fair amount of attention from both fundamental and applications point of view leading to intensive experimental[1-9] and theoretical[10-22] activity in this topic. To understand flexoelectricity better, it is best first to allude to the central mathematical relation that describes piezoelectricity:

$$P_i \sim d_{ijk}\varepsilon_{jk}. \tag{1}$$

In equation (1), the polarization vector P_i is related to the second order strain tensor ε_{jk} through the third order piezoelectric material property tensor d_{ijk}. Tensor transformation properties require that under inversion-center symmetry, all odd-order tensors vanish. Thus, most common crystalline materials, e.g. Silicon, and NaCl are not piezoelectric whereas ZnO and GaAs are. Physically, however, it is possible to visualize how a non-uniform strain or the presence of strain gradients may potentially break the inversion symmetry and induce polarization even in centrosymmetric crystals.[23–27] This is tantamount to extending relation (1) to include strain gradients:

$$P_i \sim d_{ijk}\varepsilon_{jk} + f_{ijkl}\frac{d\varepsilon_{jk}}{dx_l}, \tag{2}$$

where f_{ijkl} are the components of the so-called flexoelectric tensor. While the piezoelectric property is non-zero only for selected materials, the strain gradient-polarization coupling (i.e. flexoelectricity tensor) is in principle non-zero for all (insulating) materials. This implies that under a non-uniform strain, all dielectric materials are capable of producing a polarization.[28] The flexoelectric mechanism is well-illustrated by the non-uniform straining of a graphene nanoribbon — a manifestly non-piezoelectric material (Fig. 1(a)).[22,29] As another widely studied two-dimensional soft materials, biological membranes also show flexoelectricity (Fig. 1(b)).[30–32] Flexoelectricity has been experimentally confirmed in several crystalline materials such as NaCl, ferroelectrics like Barium Titanate among others.[8,9] Recent works, some of which are summarized elsewhere in this book, have provided important insights into the atomistic origins of flexoelectricity in crystalline solids.[12,33–39] The mechanisms of flexoelectricity in polymers (while experimentally proven) still remain unclear[40–42] and atomistic modeling (being conducted by the authors) is expected to shed light on this issue in the near future. We speculate that the presence of frozen dipoles and their

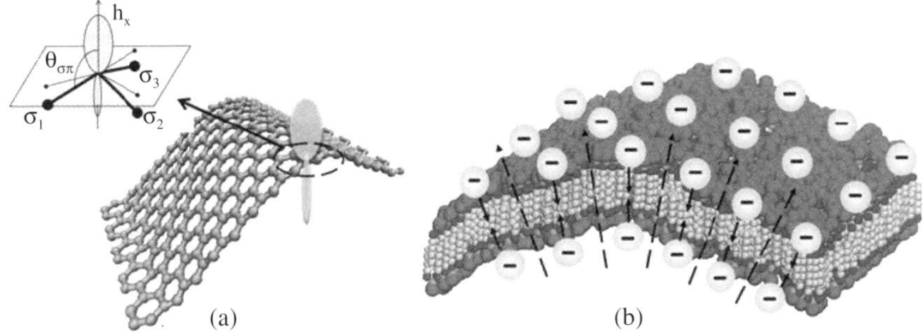

Figure 1. Flexoelectricity in membranes. (a) Bending of graphene: upon bending, the symmetry of the electron distribution at each atomic site is broken, which leads to the polarization normal to the graphene ribbon; An infinite graphene sheet is semi-metallic however finite graphene nanoribbons can be dielectric depending upon surface termination. (b) Bending of a lipid bilayer membrane: due to bending, both the charge and dipole densities in the upper and lower layers become asymmetric. This asymmetry causes the normal polarization in the bilayer membrane. Reprinted with permission from Elsevier.[43] Copyright 2014.

thermal fluctuations is the cause of flexoelectricity in soft materials, however, we cannot offer a more definitive explanation at this point and simply emphasize that this phenomenon has been experimentally confirmed[40–42] and further elucidation is a subject of future research.

Flexoelectricity results in the size-dependency of electromechanical coupling and researchers (including us) have advocated several tantalizing applications that can result through its exploitation. For example, the notion of creating piezoelectric materials without using piezoelectric materials,[9,17,18,29] giant piezoelectricity in inhomogeneously deformed nanostructures,[13,15] enhanced energy harvesting,[14,16] the origins of nanoindentation size effects,[20] renormalized ferroelectric properties,[6,10,11,44] the origins of the dead-layer effect in nano capacitors[45] among others. In fact, Chandratre and Sharma[29] have recently shown that graphene can be coaxed to behave like a piezoelectric material merely by creating holes of certain symmetry. The artificial piezoelectricity thus produced was found to be almost as strong as that of well-known piezoelectric

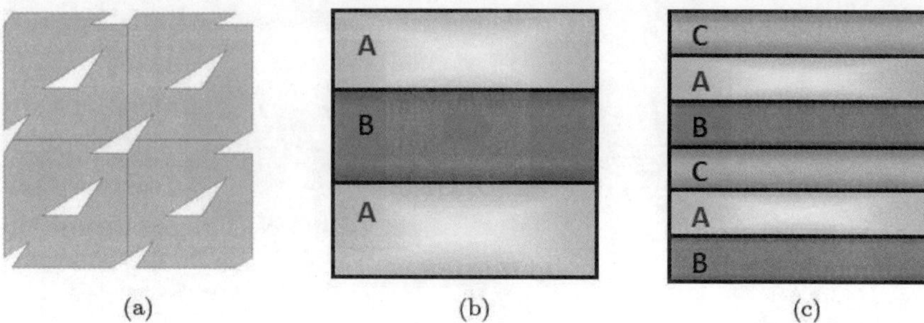

(a)   (b)   (c)

Figure 2.   Creating a piezoelectric material without using piezoelectric materials. (a) A material with a second phase, under a uniform stress, will produce local strain gradients and hence local polarization due to flexoelectricity. If the shape of the second phase is non-centrosymmetrical, the average polarization will be non-zero as well thus, from a macroscopic viewpoint, exhibiting a piezoelectric like effect. In this figure, this concept is illustrated by riddling a sheet with triangular holes. Such a sheet, with circular holes will yield a net zero average polarization. (b) The concept may also be realized by using a superlattice of differing materials. However, here care must be taken. A bilayer superlattice will result in a zero average polarization and (c) a trilayer superlattice is required to break the requisite symmetry. Nevertheless, finite bilayers (due to symmetry breaking at the free surfaces) will produce this effect. Reprinted with permission from Elsevier.[43] Copyright 2014.

substances such as quartz. Such a constructed graphene nano ribbon may be considered to be the thinnest known piezoelectric material. We briefly elaborate on this notion (Fig. 2). Consider a material consisting of two or more different non-piezoelectric dielectrics — as a concrete example that has been studied in the past we may think of a (dielectric) graphene nano ribbon impregnated with holes (Fig. 2(a)).[29] Upon the application of uniform stress, differences in material properties at the interfaces of the materials will result in the presence of strain gradients. Those gradients will induce polarization due to the flexoelectric effect. As long as certain symmetry rules are followed, the net average polarization will be non-zero. Thus, the artificially structured material will exhibit an electrical response under uniform stress behaving therefore like a piezoelectric material. The length scales must be "small" since this concept requires very large strain gradients and those for a given strain are generated easily only at the nanoscale. Here we mention that the precise scale at which this effect becomes prominent depends on the strength of the flexoelectric coefficients. For several materials studied, sub-10 nm characteristic length scales are required albeit (as this study will also show) this effect can also manifest with feature size of less than a micron. Regarding symmetry: topologies of only certain symmetries can realize the aforementioned concept. For example, circular holes distributed in a material will not yield apparently piezoelectric behavior even though the flexoelectric effect will cause local polarization fields. Due to circular symmetry, the overall average polarization is zero. A similar material but containing triangular shaped holes (or inclusions) for example, and aligned in the same direction, will exhibit the required apparent piezoelectricity. In a similar vein, a finite bilayer or multilayer configuration may also be used (Fig. 2(b)) — see discussions.[18]

In this chapter, we present a nonlinear continuum framework for flexoelectricity. It is worthwhile to mention that in finite dielectrics, because of the breaking of symmetry at the surfaces and interfaces, the so-called *surface* flexoelectric and piezoelectric effects may also play an important role.[24,46–48] However, in our current framework, we have just considered bulk flexoelectricity and ignored surface effects. In addition to the discussions by Tagantsev and others,[24,46–48] a recent work by one of us has also highlighted some aspects of this surface behavior.[49] Through the solution to various illustrative boundary value problems, we highlight the notable features and the applications of the presented framework. In Section 2, the governing equations and the associated boundary conditions are derived. In Section 3, we illustrate the notion of creating piezoelectric materials without using piezoelectric materials by solving the boundary value problem of a thin film layered superlattice structure. In Section 4, we present the solution of a truncated cone under uniaxial compression. This solution is useful in the experimental extraction of flexoelectric properties. In Section 5, we consider the bending and dynamical vibration of a beam and implications for energy harvesting are explored. In Section 6, we analyze the flexoelectric response of a biological membrane and finally, in Section 7, we consider nonlinear effects in the context of flexoelectricity in soft materials.

2. Continuum Theory of Flexoelectricity

2.1. *Continuum Kinematics*

Consider a deformable continuum body as shown in Fig. 3. Let the three-dimensional region occupied by the undeformed body be denoted Ω_R, with the boundary $\partial\Omega_R$. The position of a material point $\mathbf{A}$ in Ω_R can be described by the coordinate system X_K ($K = 1, 2, 3$). After deformation $\boldsymbol{\chi}$, the material point $\mathbf{A}(X_K)$ moves to the new position $\mathbf{a}(x_k)$, where x_k ($k = 1, 2, 3$) denotes a new coordinate system associated with the deformed body that occupies the region Ω. These two coordinate systems need not be identical. In traditional continuum mechanics, we call X_K the material or Lagrangian coordinates and x_k the spatial or Eulerian coordinates. Three base vectors for material coordinate system and the spatial coordinate system are respectively denoted by $\mathbf{I}_K$ and $\mathbf{i}_k$. The deformation $\boldsymbol{\chi}$ may be understood as a mapping from the material coordinates to the spatial coordinates, $\boldsymbol{\chi} = \mathbf{x}(\mathbf{X})$. We will also refer to the undeformed body as the reference configuration and the deformed body as the current configuration. In this chapter, all the quantities in the reference configuration are denoted by capital letters and the lower-case letters are used to describe those quantities in the current configuration.

Based on the above setup, the deformation gradient $F_{kK} \equiv x_{k,K} = \partial x_k / \partial X_K$ is introduced here. In the rest of this chapter, for

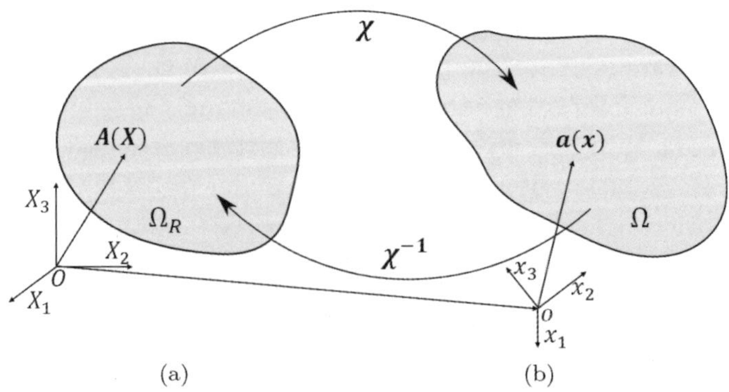

Figure 3.   (a) Reference. (b) Current configurations.

the convenience of presentation, we use ",K" and ",k" to represent the partial derivatives with respect to the coordinates X_K and x_k, respectively. Unless specified otherwise, the Einstein summation convention is applied throughout the chapter. The determinant of the deformation gradient is the so-called Jacobian which is denoted by $J \equiv det(x_{k,K})$. This scalar field describes the change of a infinitesimal volume dV after deformation through the relationship $dv = JdV$. Thus the mass density field $\rho_0(X_K)$ of the undeformed body is related to its counterpart $\rho(x_k(X_K))$ in the deformed body by $\rho_0(X_K) = J\rho(X_K)$ due to the assumption of local mass conservation. A similar relationship also holds between the charge density field $\rho_0^e(X_K)$ and $\rho^e(x_k(X_K))$, $\rho_0^e(X_K) = J\rho^e(X_K)$.

2.2. *Maxwell's Equations*

To account for energies associated with electric fields and loading devices, we shall first solve for the electric field via the Maxwell equations, i.e. the electric field $e_k = -\phi_{,k}$ in the current configuration is determined by

$$d_{k,k} = \epsilon_0 e_{k,k} + p_{k,k} + p_{k,k}^e = \rho^e, \tag{3}$$

where ϕ is the electrostatic potential, $d_k = \epsilon_0 e_k + p_k + p_k^e$ is the electric displacement field, ϵ_0 is the permittivity of free space, p_k is the intrinsic polarization, p_k^e is the external polarization, and ρ^e is the external charge density. By *external* we refer to polarization or charge density that does not change during the deformation or electrical stimulus process. In standard electrostatics textbooks, in the context of charge, this would be identical to the so-called free-charge.

Before proceeding further, it is worthwhile to examine our use of the notion of a *polarization density field*. While this is fairly standard in classical continuum mechanics (as evident in most field theory papers cited in this chapter), recent developments in the physics literature, in the context of periodic solids, have brought to light the fact that the polarization density field depends on the choice of the unit cell. The reader is referred to the paper by Resta and

Vanderbilt[50] and references therein for a detailed discussion on this matter and how the concept of the so-called Berry phase has been used to resolve this controversy. Notwithstanding this and other related works, we wish to point out a very important observation made by Marshall and Dayal[51]: If we start with a *finite* periodic solid and then examine its limit (to an infinite size) *correctly* accounting for surface and bulk bound charges, the physically relevant quantities (such as the energies and forces) are *uniquely* determined — independent of the choice of the unit cell used to compute the polarization field density. A different choice of the unit cell will lead to a different polarization density field. However, corresponding to this change in the bulk bound charges, the surface charges will change accordingly to yield exactly the same electric field and energy. Marshall and Dayal[51] draw an analogy of this situation to the definition of the strain field. Calculation of the strain field depends on the choice of the reference configuration. Nevertheless, as long as the calculations are then carried out consistently, the physically relevant quantities, such as the change in the elastic energy, are uniquely determined.

It is known that the electric field e_k, polarization p_k, and even charge density ρ^e can experience a change due to the deformation of the dielectric material. This renders the variational calculations of the field equations a bit tedious. To facilitate such calculations, it is beneficial to pull back all field quantities to the reference configuration (prior to deformation). Particularly in the case of polarization and other electrical quantities, this is a mathematical artifact but a useful one — and a standard one in continuum mechanics of electromechanical deformable media. The quantities after the pull-back are given by:

$$E_K = e_k x_{k,K}, \quad P_K = J p_k \alpha_{kK},$$

$$D_K = J X_{K,k} d_k, \quad P_K^e = J p_k^e \alpha_{kK},$$

$$\rho_0^e = J \rho^e, \tag{4}$$

where $\alpha_{kK} = \mathbf{i}_k \cdot \mathbf{I}_K$ is called the shifter which may be viewed as the rotation from the coordinate frame $\mathbf{I}_K$ to $\mathbf{i}_k$.

Upon a change of variables, in the reference configuration, we can rewrite the Maxwell equations (3) as

$$D_{K,K} = (\epsilon_0 J X_{K,k} X_{L,k} E_L + X_{K,k}\alpha_{kL}P_L + X_{K,k}\alpha_{kL}P_L^e)_{,K} = \rho_0^e, \tag{5}$$

where $D_K = J X_{K,k} d_k = \epsilon_0 J X_{K,k} X_{L,k} E_L + X_{K,k}\alpha_{kL}(P_L + P_L^e)$ is the nominal electric displacement which corresponds to the electric displacement if there is only electric field but no deformation applied to the dielectric body. The identity[52] $(J X_{K,k})_{,K} = 0$ is used to obtain the last equality of (5).

2.3. *Free Energy of an Electromechanical System*

To model flexoelectricity, we postulate that the internal/stored energy of the system is given by

$$U[x_k, P_K] = \int_{\Omega_R} \psi(F_{kK}, G_{kKL}, P_K, \Pi_{KL}), \tag{6}$$

where $G_{kKL} = x_{k,KL}$ and $\Pi_{KL} = P_{K,L}$ correspond to the gradient of deformation gradient F_{kK} and the nominal polarization P_K, respectively. Note that if we remove the dependency of the internal energy on the terms G_{kKL} and Π_{KL}, our formulation will reduce to classical elasticity of soft dielectric materials.[52,53] As will be shown later in this section, the dependence of the internal energy on $G_{kKL} = x_{k,KL}$ and $\Pi_{KL} = P_{K,L}$, leads to high order flux terms that enter both the Euler–Lagrange equations and the boundary conditions.

The boundary conditions also contribute to the free energy of the deformable dielectric system. Figure 4 depicts the schematic of the problem where the (generally possible) electrical and mechanical boundary conditions are indicated. For this system, we identify the total free energy of the system as[54]

$$F[x_k, P_K] = U[x_k, P_K] + \varepsilon^{elect}[x_k, P_K] + P^{mech}[x_k], \tag{7}$$

where

$$\varepsilon^{elect}[x_k, P_K] = \frac{\epsilon_0}{2}\int_{\Omega+V*} e_k e_k dv + \int_{\Gamma_D} \phi_b N_K D_K dA \tag{8}$$

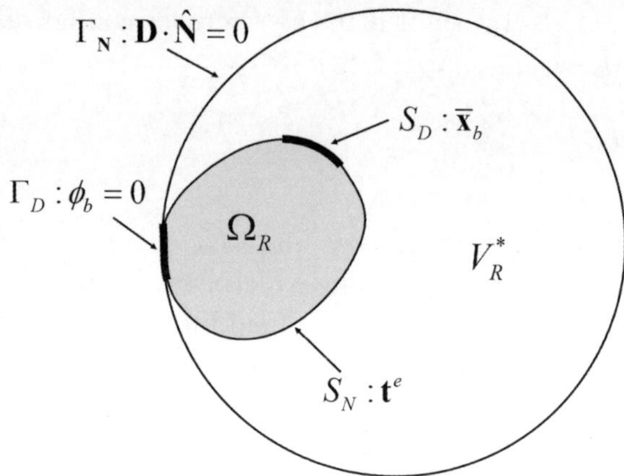

Figure 4.   A deformable dielectric system with electrical and mechanical boundary conditions ($\hat{\mathbf{N}}$ denotes the normal vector of ∂V_R^*).

is the total electric energy associated with the electric field and the boundary electric device, V^* is the space outside the body Ω, Γ_D is the Dirichlet boundary where the potential is set to ϕ_b, n_k is the normal vector of Γ_D, and

$$P^{mech}[x_k] = -\int_{S_N} t^e_K x_{k\alpha KK} dA - \int_{\Omega_R} B^e_K x_{k\alpha KK} dV \qquad (9)$$

is the potential energy of mechanical loading with t^e_K being the surface traction applied on S_N and B^e_K being the applied body force.

It is also worthwhile to mention that, for given polarization p_k, we can solve the Maxwell equation (3) with boundary conditions for a unique electric field e_k, and hence the electric energy (8) is well defined.

2.4. *Governing Equations*

By the principle of minimum free energy, the equilibrium state of the system is determined by

$$\min_{(x_k, P_K) \in \mathcal{S}} F[x_k, P_K] \qquad (10)$$

with the constraint of Maxwell equations (3) or (5), where S is a suitable space for the state variables such that sufficient smoothness and boundedness is ensured.

To find the Euler–Lagrange equations associated with a minimizer (10), we consider variations of

(1) polarization

$$x_k \to x_k, \quad P_K \to P_K + \delta Q_K, \tag{11}$$

(2) deformation

$$x_k \to x_k + \delta u_k, \quad P_K \to P_K, \tag{12}$$

where δ is a small number which controls the magnitude of the variations, Q_K and u_k are admissible variations of the field variables P_K and x_k, respectively.

We first consider the variation of polarization (11). Standard variational calculations yield:

$$\frac{d}{d\delta}F[x_k, P_K + \delta Q_K]|_{\delta=0} = \frac{d}{d\delta}U[x_k, P_K + \delta Q_K]|_{\delta=0}$$

$$+ \frac{d}{d\delta}\varepsilon^{elect}[x_k, P_K + \delta Q_K]|_{\delta=0}$$

$$= 0. \tag{13}$$

Physically, the variation of polarization P_K should result in the change of p_k, e_k, d_k, E_K, D_K, and ϕ because of the Maxwell equations and the pull back relationships (3)–(5). This implies that the variation Q_K results in:

$$p_k \to p_k + \delta \tilde{p}_k + o(\delta), \quad \phi \to \phi + \delta \tilde{\phi} + o(\delta),$$

$$e_k \to e_k + \delta \tilde{e}_k + o(\delta), \quad d_k \to d_k + \delta d_k + o(\delta),$$

$$E_K \to E_K + \delta \tilde{E}_K + o(\delta), \quad D_K \to D_K + \delta D_K + o(\delta).$$

Furthermore, the following equations are valid for their leading order terms $\tilde{p}_k$, $\tilde{e}_k$, $\tilde{d}_k$, $\tilde{E}_K$, $\tilde{D}_K$, and $\tilde{\phi}$:

$$\tilde{d}_k = \epsilon_0 \tilde{e}_k + \tilde{p}_k, \quad \tilde{d}_{k,k} = 0,$$

$$\tilde{p}_k = J^{-1}\alpha_{kK}Q_K, \quad \tilde{E}_K = \tilde{e}_k x_{k,K},$$

$$\tilde{D}_K = JX_{K,k}\tilde{d}_k, \quad \tilde{e}_k = \tilde{\phi}_{,k}.$$

Using divergence theorem, we can rewrite the first term in the RHS of (13) in the following form

$$\frac{d}{d\delta}U[x_k, P_K + \delta Q_K]|_{\delta=0} = \int_{\Omega_R}\left[\frac{\partial\psi}{\partial P_K} - \left(\frac{\partial\psi}{\partial\Pi_{KL}}\right)_{,L}\right]Q_K dV$$

$$+ \int_{\partial\Omega_R}\frac{\partial\psi}{\partial\Pi_{KL}}Q_K N_L dA, \tag{14}$$

where N_L represents the normal vector of the boundary $\partial\Omega_R$.

The Dirichlet boundary Γ_D, where the electric potential is specified, is the boundary for both Ω_R and V_R^*. By divergence theorem, the second term in the RHS of (13), the variation of electric energy, can be written as

$$\frac{d}{d\delta}\varepsilon^{elect}[x_k, P_K + \delta Q_K]|_{\delta=0} = \int_{\Omega+V*}(\epsilon_0 e_k\tilde{e}_k)dv + \int_{\Gamma_D}\phi_b N_K\tilde{D}_K dA$$

$$= \int_{\Omega+V*}(\epsilon_0 e_k\tilde{e}_k)dv + \int_{\Omega_R+V_R^*}(\phi\tilde{D}_K)_{,K}dV$$

$$= \int_{\Omega+V*}(\epsilon_0 e_k\tilde{e}_k)dv + \int_{\Omega+V*}(\phi\tilde{d}_k)_{,k}dv$$

$$= \int_{\Omega+V*}(-e_k\tilde{p}_k)dv$$

$$= \int_{\Omega}(-e_k\tilde{p}_k)dv$$

$$= \int_{\Omega_R}(-X_{L,k}E_L\alpha_{kK}Q_K)dV. \tag{15}$$

Then the first Euler–Lagrange equation and the associated boundary conditions can be obtained by substituting (14) and (15) into (13)

$$\begin{cases} \dfrac{\partial \psi}{\partial P_K} - \left(\dfrac{\partial \psi}{\partial \Pi_{KL}} \right)_{,L} - X_{L,k} E_L \alpha_{kK} = 0 & \text{in } \Omega_R, \\[4mm] \dfrac{\partial \psi}{\partial \Pi_{KL}} N_L = 0 & \text{on } \partial \Omega_R. \end{cases} \tag{16}$$

Now, we consider the variation of deformation (12). The variational calculations for this case lead to

$$\frac{d}{d\delta} F[x_k + \delta u_k, P_K]|_{\delta=0}$$

$$= \frac{d}{d\delta} U[x_k + \delta u_k, P_K]|_{\delta=0} + \frac{d}{d\delta} \varepsilon^{elect}[x_k + \delta u_k, P_K]|_{\delta=0}$$

$$+ \frac{d}{d\delta} P^{mech}[x_k + \delta u_k]|_{\delta=0} = 0. \tag{17}$$

Associated with the variation (12), there will also be changes for F_{kK}, G_{kKL}, J and F_{Kk}^{-1}

$$F_{kK} \to F_{kK} + \delta \tilde{F}_{kK}, \quad G_{kKL} \to G_{kKL} + \delta \tilde{G}_{kKL},$$

$$J \to J + \delta \tilde{J} + o(\delta), \quad F_{Kk}^{-1} \to F_{Kk}^{-1} + \delta \tilde{F}_{Kk}^{-1} + o(\delta).$$

Their leading terms are given by

$$\tilde{F}_{kK} = u_{k,K}, \quad \tilde{G}_{kKL} = u_{k,KL},$$

$$\tilde{J} = J X_{K,k} u_{k,K}, \quad \tilde{F}_{kK}^{-1} = -X_{L,k} X_{K,l} u_{l,L}. \tag{18}$$

We note that the reference quantities P_K, E_K, and D_K do not change with the variation u_k. However, the current quantities p_k, e_k, and d_k are affected by the variation of u_k because of the equation (4). For this reason, it is simpler to use P_K, E_K, and D_K in the variational calculations.

Using the divergence theorem, we first rewrite the internal energy into the following form

$$\frac{d}{d\delta}U[x_k + \delta u_k, P_K]|_{\delta=0} = \int_{\Omega_R} -\left[\frac{\partial\psi}{\partial x_{k,K}} - \left(\frac{\partial\psi}{\partial x_{k,KL}}\right)_{,L}\right]_{,K} u_k dV$$

$$+ \int_{\partial\Omega_R} \left[\frac{\partial\psi}{\partial x_{k,K}} - \left(\frac{\partial\psi}{\partial x_{k,KL}}\right)_{,L}\right] u_k N_K dA$$

$$+ \int_{\partial\Omega_R} \frac{\partial\psi}{\partial x_{k,KL}} u_{k,K} N_L dA. \tag{19}$$

It is important to note that, in equation (19), the tangential component of $u_{k,K}$ is independent of u_k due to the constraint

$$\int_{\partial\Omega_R} V_{K,L}(\delta_{KL} - N_L N_K)dA = 0,$$

for any vector V_K on the closed boundary $\partial\Omega_R$, where $\delta_{KL} = \mathbf{I}_K \cdot \mathbf{I}_L$ is the Kronecker delta which is non-zero only if $K = L$. The term $V_{K,L}(\delta_{KL} - N_L N_K)$ may also be viewed as a surface divergence of an arbitrary vector V_K on the boundary $\partial\Omega_R$.

The last integral on the rhs of (19) can be rewritten as:

$$\int_{\partial\Omega_R} \left[\frac{\partial\psi}{\partial x_{k,KL}} N_L(\delta_{KI} - N_K N_I)u_{k,I} + \frac{\partial\psi}{\partial x_{k,KL}} N_L N_K N_I u_{k,I}\right] dV$$

$$= \int_{\partial\Omega_R} \left\{\left[\frac{\partial\psi}{\partial x_{k,KL}} u_k N_L(\delta_{KI} - N_K N_I)\right]_{,I}\right.$$

$$- \left[\frac{\partial\psi}{\partial x_{k,KL}} N_L(\delta_{KI} - N_K N_I)\right]_{,I} u_k$$

$$\left.+ \frac{\partial\psi}{\partial x_{k,KL}} N_L N_K N_I u_{k,I}\right\} dV,$$

where $\frac{\partial\psi}{\partial x_{k,KL}}u_k N_L(\delta_{KI} - N_K N_I)$ is a tangent vector on $\partial\Omega_R$ since $\frac{\partial\psi}{\partial x_{k,KL}}u_k N_L(\delta_{KI} - N_K N_I)N_I = 0$. Thus, the divergence of this tangential vector is equal to its surface divergence. In other word, the integral $\int_{\partial\Omega_R}\left[\frac{\partial\psi}{\partial x_{k,KL}}u_k N_L(\delta_{KI} - N_K N_I)\right]_{,I}$ is zero.

Let $V_{kI} = \frac{\partial\psi}{\partial x_{k,KL}}N_L(\delta_{KI} - N_K N_I)$, then the equation (19) can be expressed as the following form

$$\frac{d}{d\delta}U[x_k + \delta u_k, P_K]|_{\delta=0}$$

$$= \int_{\Omega_R} -\left[\frac{\partial\psi}{\partial x_{k,K}} - \left(\frac{\partial\psi}{\partial x_{k,KL}}\right)_{,L}\right]_{,K} u_k dV$$

$$+ \int_{\partial\Omega_R} \left\{\left[\frac{\partial\psi}{\partial x_{k,K}} - \left(\frac{\partial\psi}{\partial x_{k,KL}}\right)_{,L}\right] N_K - V_{kI,I}\right\} u_k dA$$

$$+ \int_{\partial\Omega_R} \frac{\partial\psi}{\partial x_{k,KL}} N_L N_K N_I u_{k,I} dA. \tag{20}$$

Similar treatment of the higher-order boundary terms may be found in Liu[54] and Yurkov.[55]

To do the variational calculus for the electric energy ε^{elect}, by divergence theorem, we first rewrite it in the following form:

$$\varepsilon^{elect}[x_k, P_K] = \int_{\Omega+V^*} \frac{\epsilon_0}{2}e_k e_k dv + \int_{\Gamma_D} \phi_b N_K D_K dA$$

$$= \int_{\Omega+V^*} \left(\frac{\epsilon_0}{2}e_k e_k - e_k d_k\right) dv$$

$$= \int_{\Omega_R+V_R^*} \left[\frac{\epsilon_0}{2}J X_{K,k} X_{L,k} E_K E_L\right.$$

$$\left. - J X_{K,k} E_K(\epsilon_0 X_{L,k} E_L + J^{-1}\alpha_{kL} P_L)\right] dV$$

$$= \int_{\Omega_R+V_R^*} \left[-\frac{\epsilon_0}{2}J X_{K,k} X_{L,k} E_K E_L - X_{K,k}\alpha_{kL} E_K P_L\right] dV.$$

Using the relations in (18), we obtain the following form for the variation of electric field energy

$$\frac{d}{d\delta}\varepsilon^{elect}[x_k + \delta u_k, P_K]|_{\delta=0}$$

$$= \int_{\Omega_R + V_R^*} \left[-\frac{\epsilon_0}{2} J X_{K,k} u_{k,K}(X_{L,l} X_{M,l} E_L E_M) \right.$$

$$\left. + X_{K,l} u_{k,K}(\epsilon_0 J X_{L,k} X_{M,l} E_L E_M + X_{L,k} \alpha_{lM} E_L P_M) \right] dV$$

$$= \int_{\Omega_R + V_R^*} \left[-\frac{\epsilon_0}{2} J X_{K,k} u_{k,K}(X_{L,l} X_{M,l} E_L E_M) + u_{k,K} X_{L,k} E_L D_K \right] dV$$

$$= -\int_{\Omega_R} \left[-\frac{\epsilon_0}{2} J X_{K,k}(X_{L,l} X_{M,l} E_L E_M) + X_{L,k} E_L D_K \right]_{,K} u_k dV$$

$$+ \int_{\partial\Omega_R} \left[-\frac{\epsilon_0}{2} J X_{K,k}(X_{L,l} X_{M,l} E_L E_M) + X_{L,k} E_L D_K \right] u_k N_K dA$$

$$- \int_{V_R^*} \left[-\frac{\epsilon_0}{2} J X_{K,k}(X_{L,l} X_{M,l} E_L E_M) + X_{L,k} E_L E_K \right]_{,K} u_k dV.$$

$$(21)$$

Let

$$\tilde{\Sigma}_{kK}^{MW} = -\frac{\epsilon_0}{2} J X_{K,k}(X_{L,l} X_{M,l} E_L E_M) + E_L X_{L,k} D_K \qquad (22)$$

which is related to the Maxwell stress expression defined in current configuration[52,53]

$$\sigma_{kl}^{MW} = e_k d_l - \frac{\epsilon_0}{2} e_i e_i \delta_{kl} \qquad (23)$$

by

$$\Sigma_{kK}^{MW} = J X_{K,l} \sigma_{kl}^{MW}.$$

Subsequently, we refer to Σ_{kK}^{MW} as the Piola–Maxwell stress.

Note that the polarization P_K vanishes in vacuum. Thus the two Maxwell stresses reduce to

$$\tilde{\Sigma}_{kK}^{MW} = -\frac{\epsilon_0}{2} J X_{K,k}(X_{L,l}X_{M,l}E_L E_M) + E_L X_{L,k}E_K,$$

and

$$\sigma_{kl}^{MW} = e_k e_l - \frac{\epsilon_0}{2} e_i e_i \delta_{kl},$$

in the surrounding vacuum.

By substituting (20), (21), and (22) into (17), we have the second Euler–Lagrange equation and the associated boundary conditions

$$\begin{cases} \left[\left[\dfrac{\partial \psi}{\partial x_{k,K}} - \left(\dfrac{\partial \psi}{\partial x_{k,KL}}\right)_{,L} + \Sigma_{kK}^{MW}\right]_{,K} + B_K^e \alpha_{Kk} = 0 & \text{in } \Omega_R, \\[2ex] \tilde{\Sigma}_{kK,K}^{MW} = 0 & \text{in } V*, \\[2ex] \left[\dfrac{\partial \psi}{\partial x_{k,K}} - \left(\dfrac{\partial \psi}{\partial x_{k,KL}}\right)_{,L} + \Sigma_{kK}^{MW}\right] N_K = V_{kI,I} + t_K^e \alpha_{Kk} & \text{on } \partial\Omega_R, \\[2ex] \dfrac{\partial \psi}{\partial x_{k,KL}} N_K N_L = 0 & \text{on } \partial\Omega_R. \end{cases}$$

$$(24)$$

In summary, in equilibrium, the electroelastic system should satisfy the following governing equations

$$\begin{cases} \dfrac{\partial \psi}{\partial P_K} - \left(\dfrac{\partial \psi}{\partial \Pi_{KL}}\right)_{,L} - X_{L,k}E_L \alpha_{kK} = 0 & \text{in } \Omega_R, \\[2ex] \left[\dfrac{\partial \psi}{\partial x_{k,K}} - \left(\dfrac{\partial \psi}{\partial x_{k,KL}}\right)_{,L} + \Sigma_{kK}^{MW}\right]_{,K} + B_K^e \alpha_{Kk} = 0 & \text{in } \Omega_R, \\[2ex] \tilde{\Sigma}_{kK,K}^{MW} = 0 & \text{in } V*, \\[2ex] D_{K,K} = \rho_0^e & \text{in } \Omega_R. \end{cases}$$

$$(25)$$

and the boundary conditions

$$
\begin{cases}
\dfrac{\partial \psi}{\partial \Pi_{KL}} N_L = 0 & \text{on } \partial\Omega_R, \\[2ex]
\left[\dfrac{\partial \psi}{\partial x_{k,K}} - \left(\dfrac{\partial \psi}{\partial x_{k,KL}} \right)_{,L} + \Sigma_{kK}^{MW} \right] N_K = V_{kI,I} + t_K^e \alpha_{Kk} & \text{on } \partial\Omega_R, \\[2ex]
\dfrac{\partial \psi}{\partial x_{k,KL}} N_K N_L = 0 & \text{on } \partial\Omega_R.
\end{cases}
$$

$$(26)$$

2.5. *Internal Energy Density*

To model flexoelectricity, we consider the coupling between strain gradient and polarization as well as the coupling between polarization gradient and strain in the definition of the internal energy density.

The internal energy density $\psi(x_{k,K}, G_{k,KL}, P_K, \Pi_{KL})$ must be invariant under rigid translations and rotations. By Cauchy's theorem, the dependency of ψ on $x_{k,K}$ should be replaced by the dependency on $C_{KL} = x_{k,K} x_{k,L}$. It is also convenient to use the following strain tensor

$$
S_{KL} = \frac{1}{2}(C_{KL} - \delta_{KL}),
$$

instead of C_{KL}. Thus $\frac{\partial \psi}{\partial x_{k,K}}$ is always written as $\frac{\partial \psi}{\partial S_{KL}} x_{k,L}$ using the chain rule.

We will reconsider nonlinearities in Section 7 when we address soft materials. Until Section 7, we assume small strains, keep only the leading order terms in the internal energy density and ignore the difference between the reference and the current configurations. In that case, the internal energy is given by

$$
\psi(S_{kl}, G_{k,lm}, P_k, \Pi_{kl}) = \frac{1}{2} a_{kl} P_k P_l + \frac{1}{2} b_{klmn} \Pi_{kl} \Pi_{mn} + \frac{1}{2} c_{klmn} S_{kl} S_{mn}
$$

$$
+ \frac{1}{2} d_{ijklmn} G_{ijk} G_{lmn} + e_{klmn} S_{kl} \Pi_{mn}
$$

$$
+ f_{klmn} P_k G_{lmn} + g_{klm} P_k \Pi_{lm} + h_{klm} P_k S_{lm},
$$

$$(27)$$

where the second order tensor a_{kl} is the reciprocal dielectric susceptibility, the fourth order tensor b_{klmn} is the polarization gradient–polarization gradient coupling tensor and c_{klmn} is the elastic tensor, the sixth order tensor d_{ijklmn} strain gradient–strain gradient coupling tensor, the fourth order tensor e_{klmn} corresponds to polarization gradient and strain coupling introduced by Mindlin,[56] whereas f_{klmn} is the fourth order flexoelectric tensor, h_{klm} and g_{klm} are the third order piezoelectric tensor and the polarization gradient–polarization gradient coupling tensor, respectively.

3. Piezoelectric Thin Film Super Lattices Without Using Piezoelectric Materials

Consider a composite consisting of two or more different non-piezoelectric dielectric materials. Even under the application of uniform stress, differences in material properties will result in the presence of strain gradients around the interfaces. Those strain gradients will induce polarization due to the flexoelectric effect. For "properly designed" composites,[18] the net average polarization will be non-zero. Thus, the nanostructure will exhibit an overall electromechanical coupling under uniform stress behaving like a piezoelectric material. The individual constituents must be at the nanoscale since this concept requires very large strain gradients and those are generated easily only at the nanoscale.

Within the assumption of the linearized theory for centrosymmetric dielectrics, the internal energy density ψ can be assumed to be quadratic function of terms involving small strain S_{ij}, polarization P_i, polarization gradient $P_{i,j}$, and strain gradient $u_{i,jk}$

$$\psi(e_{ij}, P_i, P_{i,j}, u_{i,jk}) = \frac{1}{2}a_{kl}P_kP_l + \frac{1}{2}b_{ijkl}P_{i,j}P_{k,l} + \frac{1}{2}c_{ijkl}S_{ij}S_{kl}$$

$$+ d_{ijkl}P_{i,j}S_{kl} + f_{ijkl}P_iu_{j,kl}. \tag{28}$$

Usually we must consider another term $\frac{1}{2}g_{ijklmn}u_{i,jk}u_{l,mn}$ in the internal energy form (28) to ensure the thermodynamic stability of the system.[18,19,27] This extra term represents purely non-local elastic effects and corresponds to the so-called strain gradient elasticity theories (see also Mao and Purohit[57]). However, the contribution of this term is small for the current problem.[19] There is no qualitative change to the effective piezoelectricity with the consideration of this term and the system studied here does not lose its stability when we exclude it.[18,19]

For a thin film structure the fields vary only in the thickness direction. Let x_1 be the thickness direction. After substituting (28) in the governing equation $(25)_{1,2,4}$, we obtain the following one-dimensional governing equations for a single layer structure:

$$
\begin{cases}
c\dfrac{\partial^2 u}{\partial x_1^2} + (d-f)\dfrac{\partial^2 P}{\partial x_1^2} = 0 \\[2ex]
(d-f)\dfrac{\partial^2 u}{\partial x_1^2} + b\dfrac{\partial^2 P}{\partial x_1^2} - aP - \dfrac{\partial \phi}{\partial x_1} = 0. \\[2ex]
-\epsilon_0 \dfrac{\partial^2 \phi}{\partial x_1^2} + \dfrac{\partial P}{\partial x_1} = 0
\end{cases}
\tag{29}
$$

Under open-circuit conditions, the electric displacement is zero

$$
-\epsilon_0 \frac{\partial \phi}{\partial x_1} + P = 0.
$$

We arrive at the following equations

$$
\frac{bc - (d-f)^2}{c}\frac{\partial^2 P}{\partial x_1^2} - \left(a + \epsilon_0^{-1}\right)P = 0 \Rightarrow \frac{\partial^2 P}{\partial x_1^2} - \frac{P}{l^2} = 0,
\tag{30}
$$

where

$$
l^2 = \frac{[bc - (d-f)^2]\epsilon_0}{c\eta} \quad \text{and} \quad \eta = (1 + a\epsilon_0).
$$

Solving equation (30) for polarization, we obtain:

$$
P = A_1 e^{-x_1/l} + A_2 e^{x_1/l}.
\tag{31}
$$

The displacement field is

$$
u = A_3 + A_4 x_1 - \frac{d-f}{c} e^{-x_1/l}(A_1 + A_2 e^{2x_1/l}).
\tag{32}
$$

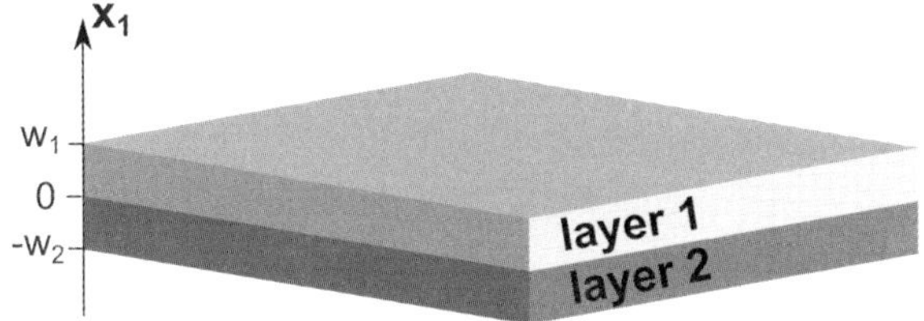

Figure 5.   A bilayer film composed of two different dielectric materials.

A_1, A_2, A_3, and A_4 are the constants of integration which may be determined by the boundary conditions. Since the coefficients d and f appear together, for conciseness, we write h instead of $(d - f)$.

For the bilayer film shown in Fig. 5, its polarization and displacement fields for each layer are given by

$$\begin{cases} P_i = A_{i1} \exp\left(-\dfrac{x_1}{l_i}\right) + A_{i2} \exp\left(\dfrac{x_1}{l_i}\right) \\[2ex] u_i = A_{i3} + A_{i4}x_1 - \dfrac{h_i}{c_i} \exp\left(-\dfrac{x_1}{l_i}\right)\left[A_{i1} + A_{i2}\exp\left(\dfrac{2x_1}{l_i}\right)\right], \end{cases}$$

$$(33)$$

where $i = 1, 2$ correspond to "layer 1" and "layer 2" shown in Fig. 5. The 8 constants of integration are determined by the following boundary conditions:

(1) Applied stress boundary conditions

$$(c_i \partial_{x_1} u_i + h_i \partial_{x_1} P_i) = \sigma.$$

(2) Continuity of stress at the interface

$$[\![\sigma]\!] = (\sigma^{(1)}|_{x_1 \to 0} - \sigma^{(2)}|_{x_1 \to 0}) = 0.$$

This condition is redundant, since in this case, the previous two (applied stress) conditions trivially ensure this continuity.

(3) Displacements at the interface are zero

$$u_i|_{x_1 \to 0} = 0.$$

(4) Electric tensor (conjugate to the polarization gradient) $\Lambda_{ij} = \frac{\partial \psi}{\partial P_{i,j}}$ is set to zero at the free boundaries

$$\Lambda_{ij}^{(1)} = (d_i \partial_{x_1} u_i + b_i \partial_{x_1} P_i)|_{x_1 \to w_i} = 0.$$

(5) The electric tensor is specified to be continuous (but not necessarily zero) at the interface

$$[\![\Lambda_{ij}]\!]|_{x_1 \to 0} = \left(\Lambda_{ij}^{(1)}|_{x_1 \to 0} - \Lambda_{ij}^{(2)}|_{x_1 \to 0}\right) = 0.$$

(6) Polarization (P) is specified to be continuous at the interface

$$[\![P]\!]|_{x_1 \to 0} = (P_1|_{x_1 \to 0} - P_2|_{x_1 \to 0}) = 0.$$

Unlike classical theory of piezoelectricity, an additional boundary condition is required at the interface on the polarization field in order to avoid the singularities of polarization gradient field.

Finally, the following results are obtained:

$$P_1 = \frac{A_1 + B_1}{C_1},$$

$$P_2 = \frac{A_2 + B_2}{C_2}, \tag{34}$$

where

$$A_1 = \sigma \epsilon_0 \cos h \left(\frac{x_1 - w_1}{l_1}\right) \left[-1 + \cos h \left(\frac{w_2}{l_2}\right)\right] c_1 h_2 l_1 \eta_1,$$

$$B_1 = \sigma \epsilon_0 c_2 h_1 \left\{\left[\cos h \left(\frac{x_1}{l_1}\right) - \cos h \left(\frac{x_1 - w_1}{l_1}\right)\right] \cos h \left(\frac{w_2}{l_2}\right) l_1 \eta_1\right.$$

$$\left. + \sin h \left(\frac{x_1}{l_1}\right) \sin h \left(\frac{w_2}{l_2}\right) l_2 \eta_2\right\},$$

$$C_1 = c_1 c_2 l_1 \eta_1 \left[\cos h \left(\frac{w_2}{l_2}\right) \sin h \left(\frac{w_1}{l_1}\right) l_1 \eta_1\right.$$

$$\left. + \cos h \left(\frac{w_1}{l_1}\right) \sin h \left(\frac{w_2}{l_2}\right) l_2 \eta_2\right],$$

$$A_2 = \sigma \epsilon_0 \cos h \left(\frac{x_1 + w_2}{l_2} \right) \left[-1 + \cos h \left(\frac{w_1}{l_1} \right) \right] c_2 h_1 l_2 \eta_2,$$

$$B_2 = \sigma \epsilon_0 c_1 h_2 \left\{ \left[\cos h \left(\frac{x_1}{l_2} \right) - \cos h \left(\frac{x_1 + w_2}{l_2} \right) \right] \cos h \left(\frac{w_1}{l_1} \right) l_2 \eta_2 \right.$$

$$\left. - \sin h \left(\frac{x_1}{l_2} \right) \sin h \left(\frac{w_1}{l_1} \right) l_1 \eta_1 \right\},$$

$$C_2 = c_1 c_2 l_2 \eta_2 \left[\cos h \left(\frac{w_2}{l_2} \right) \sin h \left(\frac{w_1}{l_1} \right) l_1 \eta_1 \right.$$

$$\left. + \cos h \left(\frac{w_1}{l_1} \right) \sin h \left(\frac{w_2}{l_2} \right) l_2 \eta_2 \right].$$

Numerical results for $BaTiO_3$–MgO bilayer are shown in Fig. 6. The calculated polarization is divided by the applied stress σ and

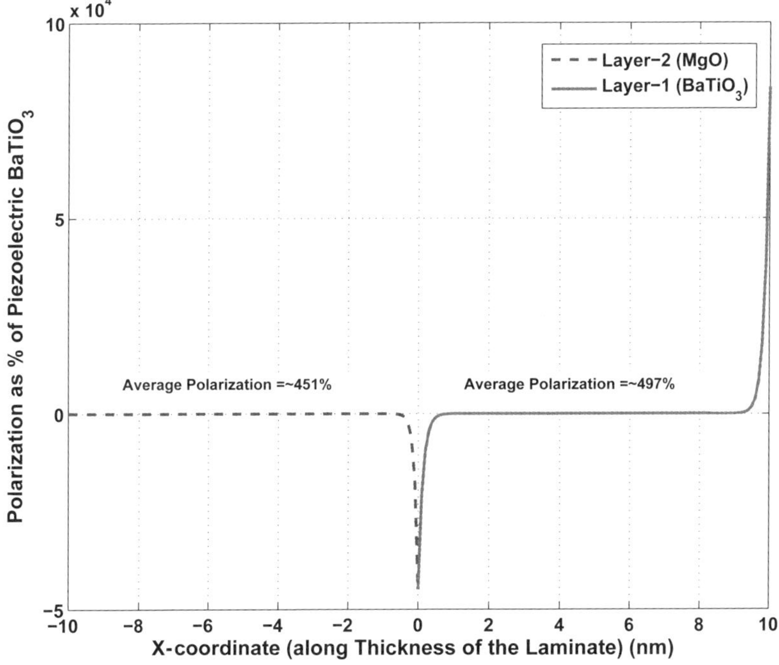

Figure 6. Polarization distribution in each layer of a MgO–$BaTiO_3$ finite bilayer. Total average polarization in the bilayer is 23% of piezoelectric $BaTiO_3$. Reprinted with permission from AIP Publishing LLC.[18] Copyright 2010.

Table 1. Material properties for $BaTiO_3$ and MgO.

	$BaTiO_3$	MgO
$a_{11}(Nm^2/C^2)$	2.824×10^7	1.298×10^{10}
$b_{11}(Nm^4/C^2)$	6.77×10^{-6}	5.67×10^{-8}
$c_{11}(N/m^2)$	1.62×10^{11}	3.00×10^{11}
$h_{11}(Nm/C)$	-1.55×10^5	1.29×10^2
$l(\text{Å})$	1.30	1.00

then normalized by the piezoelectricity p_{33} of $BaTiO_3$ which is equal to $7.8 \times 10^{-11} C/N$. For these results, we assume both layers, layer-2(MgO) and Layer-1($BaTiO_3$) to be $10\,\text{nm}$ thick subject to a unit applied stress.

The material constants used in the above calculation are listed in Table 1.

To find the effective piezoelectricity of this bilayer structure, we first calculate its average polarization $\bar{P}$ as follows:

$$\bar{P} = \frac{1}{w_1 + w_2} \left(\int_0^{w_1} P_1 dx_1 + \int_{-w_2}^0 P_2 dx_1 \right) = \frac{\bar{A}\bar{B}}{\bar{C}}, \qquad (35)$$

where

$$\bar{A} = 4\sigma \sin h \left(\frac{w_1}{2l_1} \right) \sin h \left(\frac{w_2}{2l_2} \right) \epsilon_0 (\eta_1 - \eta_2),$$

$$\bar{B} = \cos h \left(\frac{w_1}{2l_1} \right) \sin h \left(\frac{w_2}{2l_2} \right) c_1 h_2 l_1 \eta_1$$

$$+ \cos h \left(\frac{w_2}{2l_2} \right) \sin h \left(\frac{w_1}{2l_1} \right) c_2 h_1 l_2 \eta_2,$$

$$\bar{C} = c_1 c_2 (w_1 + w_2) \eta_1 \eta_2 \left[\cos h \left(\frac{w_2}{l_2} \right) \sin h \left(\frac{w_1}{l_1} \right) l_1 \eta_1 \right.$$

$$\left. + \cos h \left(\frac{w_1}{l_1} \right) \sin h \left(\frac{w_2}{l_2} \right) l_2 \eta_2 \right].$$

Then the effective piezoelectricity $d^{e\!f\!f}$ is defined as $\bar{P}/\sigma$. We note that $d^{e\!f\!f}$ directly depends on the difference between the dielectric constants of the constituent materials. Larger dielectric contrast between the two layers will lead to a larger apparent piezoelectric coefficient. As shown in Fig. 6, the effective piezoelectricity for the present MgO–BaTiO$_3$ bilayer structure is 23% of that of BaTiO$_3$

4. Compression of a Truncated Cone, Effective Piezoelectric Response, and Determination of Material Properties

The solution for the compression response of a truncated cone is an important case-study. Cross[1] profitably used this notion to show that such a geometrically graded structure behaves like an apparent piezoelectric material since even a uniform stress (e.g. compression) will result in an inhomogeneous strain and hence, due to flexoelectricity, produce a non-zero average polarization. Conversely, the solution to this problem also allows a facile way to experimentally estimate the flexoelectric properties of a material. We remark here that the expression used by Cross and co-workers (although physically intuitive) is but a crude approximation which can lead to rather large errors — recent computational work by Abdollahi *et al.*[58] attests to this. An exact solution to this problem is not possible and in this section we provide a perturbative calculation (which is an improvement) over what has been used in the past.[1]

As shown in Fig. 7, when the force F is applied to the top surface of the truncated cone, due to flexoelectricity, a current will flow through the circuit. Obviously, this is a three-dimensional problem since the cross section changes with the coordinate x_1. However, the problem can be easily converted to a one-dimensional problem if we assume that (1) after deformation, the displacement u and polarization P are both in x_1 direction and (2) u and P only vary with the coordinate x_1. These two assumptions lead to the following consequences: (1) $S_{11} = du/dx_1$ is the only non-zero strain, (2) $u_{,11} = d^2u/dx_1^2$ is the only non-zero strain gradient,

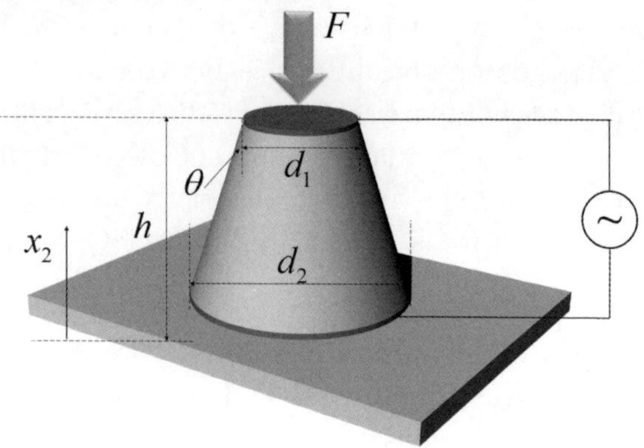

Figure 7.   A truncated cone under compression.

(3) $P_{,1} = dP/dx_1$ is the only non-zero polarization gradient. In the current example, to emphasize flexoelectricity, the contribution of polarization gradient to the internal energy is ignored. We note that omitting the polarization gradient term does not alter the governing equations (although the boundary conditions do change). In this case, the high order electric boundary conditions are replaced by the high order stress boundary conditions. We also assume that the deformation is small so that there is no difference between the current and the reference configurations — this is a reasonable assumption for hard crystalline materials. The volume integration over the truncated cone, $\int_\Omega dv$, can be written into an one-dimensional integration from 0 to L, $\int_0^L A(x)dx$, with $A(x)$ being the area of the cross section at $x_1 = x$.

Based on the setup described above, the variations (14) and (15) become

$$\frac{d}{d\delta}U[\boldsymbol{\chi}, \mathbf{P}_\delta]|_{\delta=0} = \int_0^L A(x)\psi_P P_1 dx$$

and

$$\frac{d}{d\delta}\varepsilon^{elect}[\boldsymbol{\chi}, \mathbf{P}_\delta]|_{\delta=0} = \int_0^L A(x)\left(\frac{d\phi}{dx}\right) P_1 dx.$$

Similarly, the variations (19) can be written into the following 1D form

$$\frac{d}{d\delta}U[\boldsymbol{\chi}_\delta, \mathbf{P}]|_{\delta=0} = \int_0^L \left\{ -\frac{d}{dx}\left[A(x)\frac{\partial\psi}{\partial S_{11}}\right] + \frac{d^2}{dx^2}\left[A(x)\frac{\partial\psi}{\partial u_{,11}}\right] \right\} u_1\, dx$$

$$+ \left[A(x)\frac{\partial\psi}{\partial u_{,11}}\right]_{x=0\,\&\,L}$$

$$+ \left\{ A(x)\frac{\partial\psi}{\partial S_{11}} - \frac{d}{dx}\left[A(x)\frac{\partial\psi}{\partial u_{,11}} - F\right] \right\}_{x=L}.$$

Because of the small deformation assumption, the variation (21) becomes zero so that the Maxwell stress vanishes. Then, we have the Euler–Lagrange equations for this truncated cone

$$\begin{cases} \dfrac{d}{dx}\left[A(x)\dfrac{\partial\psi}{\partial S_{11}} - \dfrac{d}{dx}\left(A(x)\dfrac{\partial\psi}{\partial u_{,11}}\right)\right] = 0 \\[3mm] \dfrac{\partial\psi}{\partial P} + \dfrac{d\phi}{dx} = 0 \end{cases} \tag{36}$$

and the associated boundary conditions

$$\begin{cases} \dfrac{\partial\psi}{\partial u_{,11}} = 0 & \text{at } x = 0 \text{ and } x = L, \\[3mm] A(x)\dfrac{\partial\psi}{\partial S_{11}} - \dfrac{d}{dx}\left[A(x)\dfrac{\partial\psi}{\partial u_{,11}}\right] = F & \text{at } x = L, \\[3mm] u = 0 & \text{at } x = 0. \end{cases} \tag{37}$$

Here, we use a reduced version of the internal energy density

$$\psi = \frac{1}{2}aP^2 + \frac{1}{2}c\left(\frac{du}{dx}\right)^2 + \frac{1}{2}g\left(\frac{d^2u}{dx^2}\right)^2 + fP\frac{d^2u}{dx^2}, \tag{38}$$

where the polarization gradient term is dropped. It is found that the governing equation remains unaltered without the consideration of this term due to the fact that[18]

$$\frac{dP}{dx}\frac{du}{dx} = \frac{d}{dx}\left(P\frac{du}{dx}\right) - P\frac{d^2u}{dx^2}$$

and $\frac{d}{dx}\left(P\frac{du}{dx}\right)$ can be converted to a boundary term by divergence theorem.

Substituting (38) into the governing equations (36) and the associated boundary conditions (37), we obtain

$$\begin{cases} -gA(x)\dfrac{d^3u}{dx^3} - g\dfrac{dA(x)}{dx}\dfrac{d^2u}{dx^2} + cA(x)\dfrac{du}{dx} \\[2ex] \qquad - fA(x)\dfrac{dP}{dx} - f\dfrac{dA(x)}{dx}P = F & \text{on } (0,h), \\[3ex] aP + f\dfrac{d^2u}{dx^2} + \dfrac{d\phi}{dx} = 0 & \text{on } (0,h), \\[3ex] \dfrac{d}{dx}\left[A(x)\left(-\dfrac{1}{\epsilon_0}\dfrac{d\phi}{dx} + P\right)\right] = 0 & \text{on } (0,h), \\[3ex] \dfrac{\partial\psi}{\partial u_{,11}} = g\dfrac{d^2u}{dx^2} + fP & \text{at } x = 0\&h, \\[2ex] u(0) = 0 \\[1ex] \phi(0) - \phi(h) = 0 \end{cases} \tag{39}$$

where $A(x) = \frac{\pi}{4}\left(d_2 - \frac{d_2-d_1}{h}x\right)^2$ denotes the cross section area at x. Note that if the half cone angle θ is very small, $a(x)$ can be approximately expressed as

$$A(x) = \pi\theta^2 x^2 + \bar{d}\theta x + a_2, \tag{40}$$

where $\bar{d} = -\pi d_2$ and $a_2 = \frac{\pi}{4}d_2^2$.

In the extreme case, when θ is zero, the truncated cone becomes a finite circular cylinder. Obviously, in that case, neglecting any edge effects, the flexoelectric response vanishes. Let $u_0(x)$ and $P_0(x)$ denote the solution for the cylinder problem. Then the solution for a truncated cone problem can be written as

$$
\begin{cases}
u(x) = u_0(x) + \theta u_1(x) + \theta^2 u_2(x) + \cdots \\[2mm]
P(x) = P_0(x) + \theta P_1(x) + \theta^2 P_2(x) + \cdots \\[2mm]
\phi(x) = \phi_0(x) + \theta \phi_1(x) + \theta^2 \phi_2(x) + \cdots
\end{cases}
\tag{41}
$$

Substitutng (40) and (41) into the governing equations and considering only zeroth order tens, we have the following equations for $u_0(x)$ and $P_0(x)$

$$
\begin{cases}
-g\dfrac{d^3 u_0}{dx^3} + c\dfrac{du_0}{dx} - f\dfrac{dP_0}{dx} = F/a_2 & \text{on } (0, h), \\[4mm]
aP_0 + f\dfrac{d^2 u_0}{dx^2} + \dfrac{d\phi_0}{dx} = 0 & \text{on } (0, h), \\[4mm]
-\epsilon_0 \dfrac{d^2 \phi_0}{dx^2} + \dfrac{dP_0}{dx} = 0 & \text{on } (0, h), \\[4mm]
g\dfrac{d^2 u_0}{dx^2} + fP_0 = 0 & \text{at } x = 0 \& h. \\[4mm]
u_0(0) = 0 \\[2mm]
\phi_0(0) - \phi_0(h) = 0
\end{cases}
\tag{42}
$$

Unsurprisingly the solution for the zeroth order equations (the case corresponding to the cylinder) is

$$
u_0(x) = \frac{F}{ca_2}x \quad \text{and} \quad P_0(x) = \phi_0(x) = 0.
$$

The first order equations are:

$$
\begin{cases}
g\dfrac{d^3 u_1}{dx^3} - c\dfrac{du_1}{dx} + f\dfrac{dP_1}{dx} = \bar{d}F/a_2^2 x & \text{on } (0, h), \\[2ex]
aP_1 + f\dfrac{d^2 u_1}{dx^2} + \dfrac{d\phi_1}{dx} = 0 & \text{on } (0, h), \\[2ex]
-\epsilon_0 \dfrac{d^2 \phi_1}{dx^2} + \dfrac{dP_1}{dx} = 0 & \text{on } (0, h), \\[2ex]
g\dfrac{d^2 u_1}{dx^2} + fP_1 = 0 & \text{at } x = 0 \& h. \\[2ex]
u_1(0) = 0 \\[1ex]
\phi_1(0) - \phi_1(h) = 0
\end{cases}
\tag{43}
$$

From $(43)_3$, we have

$$
\frac{d\phi}{dx} = \frac{1}{\epsilon_0} P_1 - \frac{C_1}{\epsilon_0},
\tag{44}
$$

where C_1 is the constant of integration. Substituting the relationship (44) into $(43)_2$, we can express the polarization P_1 in terms of the displacement u_1 as

$$
P_1 = -\frac{f\epsilon_0}{a\epsilon_0 + 1}\frac{d^2 u_1}{dx^2} + \frac{C_1}{a\epsilon_0 + 1}.
$$

A non-homogeneous ODE for $u_1(x)$ is obtained from $(43)_1$

$$
\left(g - \frac{f^2 \epsilon_0}{a\epsilon_0 + 1}\right)\frac{d^3 u_1}{dx^3} - c\frac{du_1}{dx} = \bar{d}F/a_2^2 x.
$$

The solution for this ODE is given by

$$
u_1 = A_1 e^{-x/l} + A_2 e^{x/l} - \frac{\bar{d}F}{2ca_2^2}x^2 + C_2,
\tag{45}
$$

where $l^2 = (g - \frac{f^2 \epsilon_0}{a\epsilon_0 + 1})/c$ depends on the material properties, A_1, A_2, C_1, and C_2 are constants to be determined by the following four boundary conditions

$$
\begin{cases}
(A_1 + A_2) = -\dfrac{fC_1}{c(a\epsilon_0 + 1)} = m \\[2.5em]
(A_1 e^{-h/l} + A_2 e^{h/l}) = -\dfrac{fC_1}{c(a\epsilon_0 + 1)} = m \\[2.5em]
A_1 + A_2 + C_2 = 0 \\[1.5em]
2(A_1 - A_2) + \dfrac{alh}{f}C_1 = \dfrac{\bar{d}Flh}{ca_2^2}
\end{cases},
$$

where the last equation originates from the short-circuit condition $\phi_1(0) = \phi_1(h)$. The solution for these four equations is given by

$$
\begin{cases}
A_1 = m\dfrac{e^{2h/l} - e^{h/l}}{e^{2h/l} - 1} \\[2em]
A_2 = m\dfrac{e^{h/l} - 1}{e^{2h/l} - 1} \\[2em]
C_1 = \dfrac{\bar{d}Flh}{ca_2^2} \bigg/ \left[\dfrac{alh}{f} - \dfrac{2f}{c(a\epsilon_0 + 1)}\tanh\dfrac{h}{2l}\right] \\[2em]
C_2 = -m
\end{cases}. \qquad (46)
$$

Finally, using the relationship between P_1 and u_1, the expression for P_1 can be obtained as

$$
P_1 = -\frac{f\epsilon_0}{a\epsilon_0 + 1}\left[\frac{1}{l^2}\left(A_1 e^{-x/l} + A_2\right) - \frac{\bar{d}F}{ca_2^2}\right] + \frac{C_1}{a\epsilon_0 + 1}. \qquad (47)
$$

As a simplification, if we set $A_1 = A_2 = 0$, then we have $C_2 = 0$ and $C_1 = \frac{\bar{d}Ff}{aca_2^2}$ by (46). It is also found from equation (45) that u_1 becomes a quadratic function of x. This implies that the strain gradient d^2u_1/dx^2 is a constant. Thus, the solution for P_1 reduces to

$$
P_1 = \frac{f\epsilon_0}{a\epsilon_0 + 1}\frac{\bar{d}F}{ca_2^2} + \frac{C_1}{a\epsilon_0 + 1} = \frac{\bar{d}Ff}{aca_2^2}
$$

and the overall polarization P takes the following form

$$P = P_0 + \theta P_1 = -2\left(\frac{d_2 - d_1}{d_2}\frac{d_1^2}{d_2^2}\right)\frac{f}{ach}\frac{F}{a_1},$$

where $a_1 = \frac{\pi}{4}d_1^2$ is the cross-sectional area of the top surface where the external force F is applied. Since the flexoelectric coefficient $\mu = -\frac{f}{a}$ and the applied surface traction $t^e = \frac{F}{a_1}$, the polarization can be also written as

$$P = 2\left(\frac{d_2 - d_1}{d_2}\frac{d_1^2}{d_2^2}\right)\frac{\mu}{ch}t^e.$$

Then, the effective piezoelectric coefficient is given by

$$d^{\text{eff}} = \frac{\partial P}{\partial t^e} = 2\left(\frac{d_2 - d_1}{d_2}\frac{d_1^2}{d_2^2}\right)\frac{\mu}{ch}. \tag{48}$$

This result is similar to that obtained in Cross's work[1] where an effective piezoelectric coefficient

$$d^{\text{eff}} = \mu\frac{\frac{d_2^2 - d_1^2}{d_1^2}}{ch},$$

is obtained for the truncated pyramid. Both cases show that d^{eff} is proportional to the flexoelectric coefficient μ and the reciprocals of elastic modulus c and the dimension h. A more sophisticated perturbation approach may be necessary to obtain an analytical result that matches the computational corrections shown by Abdollahi *et al.*[58]

5. Flexoelectric Beam Bending, Vibration, and Energy Harvesting

Bending a beam is perhaps the easiest way to induce strain gradients and invoke a flexoelectric response. This is also a model problem when considering energy harvesting from mechanical vibrations. As shown in Fig. 8, in this section we consider a cantilever beam subjected to a distributed load. B, h, and L correspond to the width, height, and length of the beam, respectively. $q(x_1)$ is the distributed loading applied to the beam.

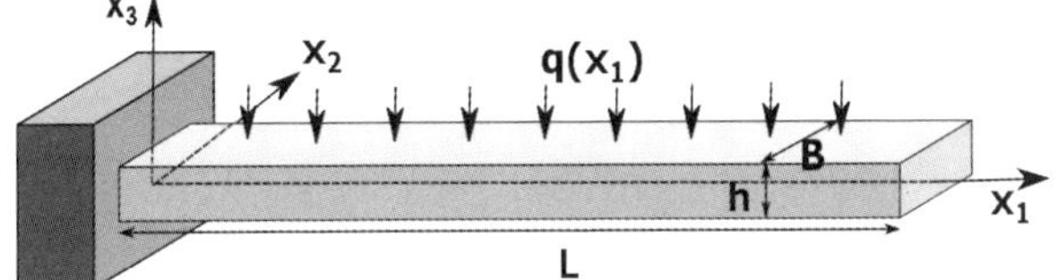

Figure 8.   A cantilever beam subjected to a distributed load.

To illustrate the central ideas of the flexoelectricity in this beam, we use the Euler–Bernoulli model. The key conclusions that we are interested in emphasizing in this work are unlikely to be affected by this assumption. The displacement field in the Euler–Bernoulli model is:

$$\mathbf{u} = \left\{ -x_3 \frac{\partial w(x_1)}{\partial x_1}, 0, w(x_1) \right\}^T, \tag{49}$$

where $w(x_1)$ is the transverse displacement of the neutral surface at point x_1. From this displacement field, the normal strain in x_1 direction is the only non-zero strain component which is given by

$$S_{11} = -x_3 \frac{\partial^2 w}{\partial x_1^2}. \tag{50}$$

The non-zero strain gradient components are

$$S_{11,1} = -x_3 \frac{\partial^3 w(x_1)}{\partial x_1^3}, \quad S_{11,3} = -\frac{\partial^2 w(x_1)}{\partial x_1^2}. \tag{51}$$

Since the thickness h of the beam is much smaller than its length L, it is found that $S_{11,1}$ is much smaller than $S_{11,3}$ for this Euler–Bernoulli beam problem. Therefore, the component $S_{11,1}$ is ignored in the present work.

Generally, the strain gradient $S_{11,3}$ will induce the separation of positive and negative charge centers. A schematic representation for the polarization induced by strain gradient is given in Fig. 9. As can be seen from Fig. 9, after deformation, the induced polarization is generated along the x_3 direction. The polarization density field

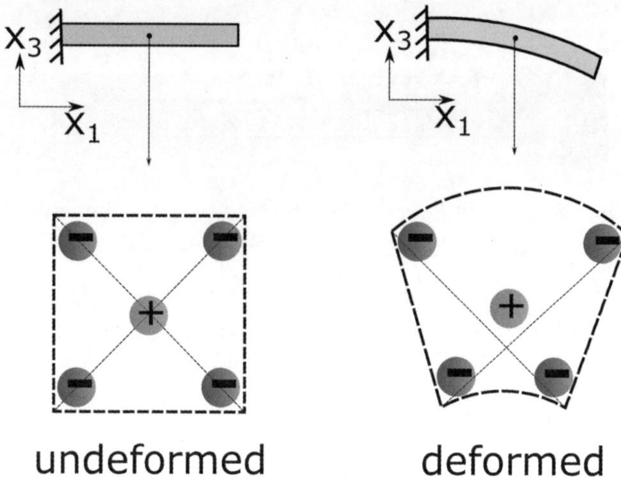

Figure 9.   Polarization due to bending of a centrosymmetric beam. Reprinted with permission from Elsevier.[59] Copyright 2014.

within the cantilever beam has the following form:

$$\mathbf{P}(x_1, x_3) = \{0, 0, P(x_1, x_3)\}^T. \tag{52}$$

For this problem, the stored energy density ψ is given by

$$\psi = \frac{1}{2}cS_{11}^2 + \frac{1}{2}gS_{11,3}^2 + \frac{1}{2}aP^2 + fS_{11,3}P$$

$$= \frac{1}{2}cx_3^2 w_{,11}^2 + \frac{1}{2}gw_{,11}^2 + \frac{1}{2}aP^2 - fw_{,11}P, \tag{53}$$

where $w_{,1}$ and $w_{,11}$ denote the first and second order derivatives of w with respect to x_1, respectively; c, g, a, and f are material properties mentioned in Section 2.

The so-called *dynamic* flexoelectric effect may strongly affect the electromechanical behavior of the sample at extremely high frequency vibrations.[24,46,47] However, in our current example, we have ignored this effect since the considered sample size is on the scale of microns and its natural frequency is still very low compare to that of the lattice vibration at which the dynamic flexoelectric effect is expected to play a role.

Once again, we stay within the linearized regime and ignore the distinction between current and reference configurations and the effect of Maxwell stress. Both these effects are important for soft materials but not for hard ones. Given these assumptions, the governing equations $(25)^{1,4}$ become

$$\begin{cases} aP - fw_{,11} + \phi_{,3} = 0 \\ -\epsilon_0 \phi_{,33} + P_{,3} = 0 \end{cases}, \tag{54}$$

where $\phi_{,3} = -E$ is the derivative of potential ϕ with respect to x_3. From (54), we further have

$$\begin{cases} P = \dfrac{f}{a} w_{,11} - \dfrac{1}{a} \phi_{,3} \\ -\left(\epsilon_0 + \dfrac{1}{a} \right) \phi_{,33} = 0 \end{cases}. \tag{55}$$

With the applied boundary conditions for electric potential ϕ: $\phi(x_1, h/2) = V$ and $\phi(x_1, -h/2) = 0$, the leading order solutions of (55) are

$$\phi = \frac{V}{h} x_3 + \frac{V}{2} \tag{56}$$

and

$$P = \frac{f}{a} w_{,11} - \frac{V}{h}. \tag{57}$$

Substituting (56) and (57) into (53) and integrating over the beam cross section, the following one-dimensional energy density ψ_b is obtained

$$\psi_b = \int_A \psi \, da = \frac{1}{2} G_E w_{,11}^2 + \frac{AV^2}{2ah^2}, \tag{58}$$

where $G_E = cI + gA - \dfrac{f^2 A}{a}$ and A denotes the cross-sectional area and I is the moment of inertia.

Free-energy minimization leads to the following equation:

$$\int_0^L \delta\psi_b dx_1 = \int_0^L G_E w_{,11}\delta w_{,11} dx_1 = \int_0^L q(x)\delta w dx_1.$$

Using the divergence theorem, we obtain the following governing equation and boundary conditions:

$$\begin{cases} G_E \dfrac{\partial^4 w(x_1)}{\partial x_1^4} = q(x_1) \\[3mm] w(0) = \dfrac{\partial w(0)}{\partial x_1} = 0 \\[3mm] \dfrac{\partial^2 w(L)}{\partial x_1^2} = \dfrac{\partial^3 w(L)}{\partial x_1^3} = 0 \end{cases} \tag{59}$$

If the distributed load is a constant q_0, then the displacement w, and polarization P are given by

$$w(x_1) = \frac{q_0}{24G_E}x_1^4 - \frac{q_0 L}{6G_E}x_1^3 - \frac{q_0}{2G_E}\left(\frac{L^2}{2} - L\right)x_1^2, \tag{60}$$

$$P(x_1) = \frac{f}{a}\left[\frac{q_0}{2G_E}x_1^2 - \frac{q_0 L}{G_E}x_1 - \frac{q_0}{G_E}\left(\frac{L^2}{2} - L\right)\right] - \frac{V}{ah}. \tag{61}$$

As shown in (61), because of flexoelectricity, both V and q_0 contribute to the polarization P. The former is due to the electric field while the latter is the result of flexoelectricity.

The example problem discussed in the preceding paragraphs shows that the bending of a dielectric beam will induce the net polarization because of flexoelectricity. One of the applications of this phenomena is energy harvesting. A proposed flexoelectric energy harvester is shown in Fig. 10. The cantilever beam is mounted to a base moving in x_3 direction. The transverse base displacement is denoted by $w_b(t)$. Due to the movement of the base, the cantilever beam undergoes bending vibrations. Dynamic strain gradient associated with vibration results in an alternating potential difference across the electrodes. The electrodes are connected to a resistive load

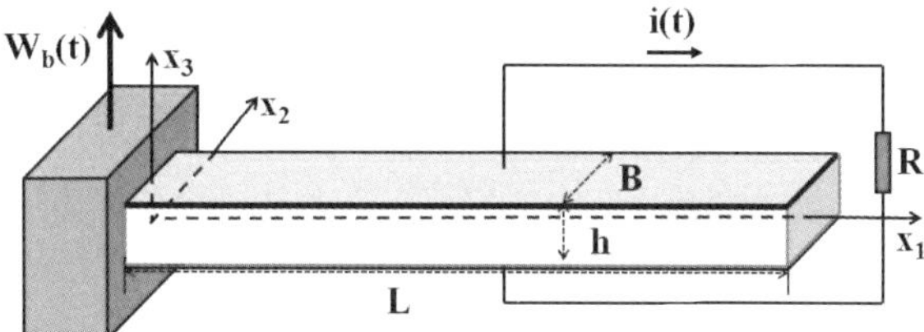

Figure 10. A centrosymmetric flexoelectric energy harvester under base excitation. Reprinted with permission from Elsevier.[59] Copyright 2014.

(R) to quantify the electrical power output. Although the internal resistance of the dielectric beam is not taken into account, it can easily be considered as a resistor connected in parallel to the load resistance.

For this vibration problem, the kinetic energy, which results in the inertial term in the governing equation, should be considered. In our recent paper,[59] from the following variational principle

$$\delta \int_{t_1}^{t_2} dt \int_V \left[\frac{1}{2}\rho \dot{u}_k \dot{u}_k - \left(\psi - \frac{1}{2}\epsilon_0 \phi_{,k}\phi_{,k} + P_k \right) \right] dV$$

$$+ \int_{t_1}^{t^2} dt \int_V (q_k \delta u_k + E_k^0 P_k)dV + \int_{t_1}^{t^2} dt \int_{\partial V} \tilde{D}\delta\phi dA = 0, \qquad (62)$$

along with the Euler–Bernoulli beam assumption (49), we obtained the following equation

$$\int_{t_1}^{t^2} dt \int_0^L \rho A(\ddot{w} + \ddot{w}_b)\delta w dx_1$$

$$+ \int_{t_1}^{t^2} dt \int_0^L \left(G_E \frac{\partial^2 w}{\partial x_1^2} - \frac{fA}{ah}V(t) \right) \delta \left(\frac{\partial^2 w}{\partial x_1^2} \right) dx_1 = 0. \qquad (63)$$

The current $i(t)$ flows through the resistor R must be equal to the time rate of change of the average electric displacement

$\tilde{D}_3 = \frac{1}{h} \int_V D_3 dV$, resulting the electric circuit equation with flexo-electric coupling

$$i(t) = \frac{V(t)}{R} = \frac{1}{h}\frac{d}{dt}\int_V \left(-\epsilon_0 \frac{V(t)}{h} + P\right) dV. \tag{64}$$

The governing equations (63) and (64) for variables $w(x)$ and $V(t)$ are numerically solved using the assumed-mode method.[60,61] The following finite series is used to represent the mechanical response of the beam:

$$w(x_1,t) = \sum_{k=1}^{N} a_k(t)\xi_k(x_1),$$

where N is the number of modes used in the series discretization, $\xi_k(x_1)$ are the kinematically admissible trial functions which satisfy the essential boundary conditions, while $a_k(t)$ are unknown generalized coordinates.

Since the focus in energy harvesting is placed on the resonance behavior (i.e. damping controlled region), it is necessary to account for structural dissipation in the system. In this work, we resort to Rayleigh damping which is proportional to the mass and the stiffness matrices. We introduce the damping matrix $\mathbf{D}$ with

$$\mathbf{D} = \mu\mathbf{M} + \gamma\mathbf{K},$$

where μ and γ are constants of proportionality which can be calculated using two modal damping ratios, ζ_1 and ζ_2 through the following equation[62]:

$$\begin{bmatrix} \gamma \\ \mu \end{bmatrix} = \frac{2\omega_1\omega_2}{\omega_1^2 - \omega_2^2}\begin{bmatrix} \dfrac{1}{\omega_2} & -\dfrac{1}{\omega_1} \\ -\omega_2 & \omega_1 \end{bmatrix}\begin{bmatrix} \zeta_1 \\ \zeta_2 \end{bmatrix},$$

where ω_1 and ω_2 are the first two nature frequencies of the beam. In the absence of other damping mechanisms, the damping ratio is related to the material quality factor ($Q = 1/2\zeta$).

Then the discrete Euler–Lagrange equations, duly considering Rayleigh damping, are given by

$$\mathbf{M}\ddot{\mathbf{a}}(t) + \mathbf{D}\dot{\mathbf{a}} + \mathbf{K}\mathbf{a} - \boldsymbol{\Theta}V(t) = \bar{\mathbf{f}}, \quad C_f\dot{V}(t) + \frac{V(t)}{R} + \boldsymbol{\Theta}^T\dot{\mathbf{a}}(t) = 0,$$

$$(65)$$

where

$$M_{kl} = \rho A \int_0^L \xi_k(x_1)\xi_l(x_1)dx_1,$$

$$K_{kl} = G_E \int_0^L \xi_k''(x_1)\xi_l''(x_1)dx_1,$$

$$D_{kl} = \mu M_{kl} + \gamma K_{kl},$$

$$\Theta_l = \frac{fA}{ah} \int_0^L \xi_l'' dx_1,$$

$$\bar{f}_l = -\ddot{w}_b(t) \int_0^L \rho A \xi_l(x_1)dx_1,$$

$$C_f = \frac{BL}{h}\left(\epsilon_0 + \frac{1}{a}\right).$$

We choose polyvinylidene difluoride (PVDF) as the model material system which has the following properties: $a = \frac{1}{\epsilon - \epsilon_0} = 1.38 \times 10^{10}\,\mathrm{Nm^2/C^2}$, $f = -a\mu_{12}' = -179\,\mathrm{Nm/C}$ ($\mu_{12}' = 1.3 \times 10^{-10}\,\mathrm{C/m}$ is the flexoelectric coefficient), $c = 3.7\,\mathrm{GPa}$, $g = 5 \times 10^{-7}\,\mathrm{N}$, and $\rho = 1.78 \times 10^3\,\mathrm{kg/m^3}$. To demonstrate the effect of scaling, here we explore the energy conversion efficiency of this flexoelectric energy harvester. The conversion efficiency is simply the ratio of the electrical power output to the mechanical power input, i.e. the power due to the shear force exerted on the beam by the base. The peak electrical power output is $|V(t)|^2/R$. The shear force exerted on the beam by the base is the shear force at $x_1 = 0$, which can be easily expressed by $cI\dfrac{d^3w(0)}{dx_1^3}$. Therefore, the power due to the shear force is the product of the shear force and the base velocity $\dfrac{de_b(t)}{dt}$. Then the mechanical-to-electric energy conversion efficiency is

$$\eta = \frac{\dfrac{|V(t)|^2}{R}}{\left| cI\dfrac{d^3w(0)}{dx_1^3} \right| \cdot \left| \dfrac{de_b(t)}{dt} \right|}.$$

We maintain the shape of the sample (in terms of the aspect ratio, 100:10:1) and vary the thickness of the beam from $3\,\mu$m through $0.3\,\mu$m. As shown in Fig. 11, for the 10 different sizes considered in this thickness range, the energy conversion efficiency monotonously increase as the decrease of the sample size. Specifically, the magnitude of the highest curve is about two orders higher than that of the lowest one. Further enhancement in the conversion efficiency can be expected as the beam thickness is reduced to nanometer scale.

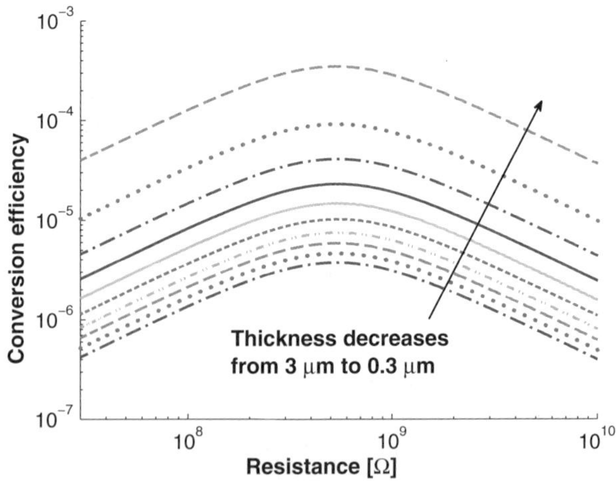

Figure 11.   Resonant energy conversion efficiency for different beam thickness levels (aspect ratio is the same: 100:10:1). Reprinted with permission from Elsevier.[59] Copyright 2014.

6. Flexoelectric Membranes

It is anticipated that the coupling between elasticity and electricity in biological membranes is important for many biological functions such as: ion transport,[63] mechanotransduction in outer hair cells,[64–66] tether formation.[67] It is also found that all the above-mentioned biological phenomena are closely related to the flexoelectricity of the biomembranes. From a mechanistic viewpoint, we can rule out the piezoelectric effect in (most) biological membranes through symmetry arguments. Therefore, the leading order coupling between strain and polarization has to be "flexoelectricity".

The flexoelectric theory of thin membranes has been developed in Mohammadi *et al.*,[68] Deng *et al.*,[43] and several other formal considerations are presented in Liu's paper.[54] In this formulation, we consider a thin membrane occupying $U \times (-h/2, h/2) \subset \mathbb{R}^3$, with $U \subset \mathbb{R}^2$ being an open bounded domain in the xy-plane and h being the thickness of the membrane. Since the thickness $h \ll 1$, the thin membrane may be idealized as a two-dimensional body; the thermodynamic state is then described by the out-of-plane displacement $w : U \to \mathbb{R}$ and the out-of-plane polarization $P : U \to \mathbb{R}$. For thin membrane, we also anticipate that bending is the predominant mode of deformation, and hence use the linearized curvature tensor (or strain gradient) of the membrane $\kappa_{ij} = -w_{,ij}$ (for this two-dimensional problem $i, j = 1, 2$) to describe the deformed state of the membrane. The polarization P is assumed to point along the normal direction throughout the deformation.

To model the flexoelectric effect, we postulate that the total internal/stored energy of an isotropic membrane is given by

$$U[w, P] = \int_U W(w_{,ij}, P), \tag{66}$$

where $W : \mathbb{R}^{2\times 2}_{sym} \times \mathbb{R} \to \mathbb{R}$ is the total internal energy density function and given by a quadratic function

$$W(w_{,ij}, P) = \frac{\kappa_b}{2}(w_{,ii})^2 + \kappa_g det(w_{,ij}) + fPw_{,ii} + \frac{1}{2}a_3 P^2, \tag{67}$$

here we notice that the elastic part, $\frac{\kappa_b}{2}(w_{,ii})^2 + \kappa_g det(w_{,ij})$, of the membrane energy coincides with the linearized Helfrich–Canham model[69] of biological membranes that are widely used by biophysicists (which is in turn identical to the Kirchhoff–Love plate theory), the term $fPw_{,ii}$ gives rise to the coupling between polarization and curvature (flexoelectric effects), and the last term $\frac{1}{2}a_3P^2$ arises from the dielectric property of the membrane. The constants κ_b, κ_g, f, and a_3 are material properties of the flexoelectric membrane and may in general depend on in-plane positions. Moreover, the stability of the membrane requires that[68]:

$$\kappa_b > 0, \quad -2\kappa_b < \kappa_g < 0, \quad a > 0, \quad \text{and} \quad \kappa_b + \frac{\kappa_g}{2} > \frac{f^2}{a}. \tag{68}$$

The boundary of the flexoelectric membrane is clamped: $w, w_{,i}|_{\partial U} = 0$. Then under the application of an external electric field $E_3 : U \to \mathbb{R}$ and a mechanical body force $b_3 : U \to \mathbb{R}$, the total free energy of the membrane is given by

$$F[w, P] = \int_U W(w_{,ij}, P) - \int_U (PE_3 + wb_3), \tag{69}$$

where the first integral is the internal energy of the flexoelectric membrane, and the second one is the potential energy arising from the interaction between the membrane and the external electric field and mechanical loading device.

In the equilibrium state, by the principle of minimum free energy the pair of (w, P) shall minimize the total free energy (69):

$$\min_{(w,P)\in\mathcal{S}} F[w, P], \tag{70}$$

where the admissible space for (w, P) is given by

$$\mathcal{S} := \left\{ (w, P) : \int_U |w_{,ij}|^2 < +\infty, \int_U |P|^2 < +\infty, w, w_{,i}|_{\partial U} = 0 \right\}. \tag{71}$$

By standard variational calculations, it can be shown that a minimizer (w, P) of the minimization problem (70) necessarily satisfies the following Euler–Lagrange equations and boundary conditions:

$$
\begin{cases}
(L_{ijkl}w_{,kl})_{,ij} + (fP)_{,ii} - b_3 = 0 & \text{on } U \\
fw_{,ii} + aP - E_3 = 0 & \text{on } U \\
w = 0, \quad w_{,i} = 0 & \text{on } \partial U
\end{cases}
\tag{72}
$$

where L_{ijkl} $(i,j,k,l = 1,2)$ represent the bending stiffness tensor which links the bending moment and the curvature of a membrane.

Using $(72)_2$ we eliminate P in $(72)_1$ and obtain

$$
\begin{cases}
\left[\left(L_{ijkl} - \dfrac{f^2}{a}\delta_{ij}\delta_{kl} \right) + \dfrac{f}{a}E_3\delta_{ij} \right]_{,ij} - b_3 = 0 & \text{on } U \\
w = 0, \quad w_{,i} = 0 & \text{on } \partial U.
\end{cases}
\tag{73}
$$

Consider a flexoelectric membrane. In the absence of applied body force $b_3 = 0$, the membrane can nevertheless be bent by an external electric field since the term $\frac{f}{a}E_3\delta_{ij}$ in $(73)_1$ serves as a "source" term for the boundary value problem (73) of w. It is of interest to investigate the effects of the external field on w. For simplicity, assume that the membrane is homogeneous and isotropic, and hence the boundary value problem can be rewritten as

$$
[\kappa_b w_{kk} + \gamma E_3]_{,ii} = 0 \quad \text{on } U,
\tag{74}
$$

where $\kappa_b = 2\mu_b + \lambda_b - (f^2/a)$ and $\gamma = f/a$. For appropriate boundary conditions as specified in $(73)_3$, it is standard to solve (74) for w. As examples, below we present a few explicit solutions, assuming infinite membrane on $\mathbb{R}^2$ and natural boundary conditions at the infinity:

$$
|w_{,ij}(x_k)| \to 0 \quad \text{as } x_k \to +\infty.
\tag{75}
$$

We remark that w is only determined within an arbitrary linear function of (x_1, x_2) by the above conditions (75). Also, appropriate decay conditions on the source term E_3 are required for (75). These simple solutions may be used for measuring the material properties in (67) and as the benchmarks of numerical schemes.

6.1. $E_3 = E_3(x_1)$

Since E_3 is independent of x_2, by symmetry we seek a solution of form $w = w(x_1)$ to (74), i.e.

$$\frac{d^2}{dx_1^2}\left[\kappa_b \frac{d^2}{dx_1^2} w(x_1) + \gamma E_3(x_1)\right] = 0 \quad \forall\, x_1 \in \mathbb{R}. \tag{76}$$

The general solution to equation (76) is given by

$$w(x_1) = -\iint \frac{\gamma}{\kappa_b} E_z(x_1) + C_0 + C_1 x_1, \tag{77}$$

where C_0 and C_1 are the integration constants and shall be determined by boundary conditions. In particular, if the external field is generated by an infinite line charge of line density q along $\mathbf{i}_2$ direction and above the membrane with distance z_0, then

$$E_3(x_1) = -\frac{qz_0}{2\pi\epsilon_0}(z_0^2 + x_1^2)^{-1}. \tag{78}$$

Substitute (78) into (77), then we have

$$w(x_1) = \frac{q\gamma}{2\pi\epsilon_0\kappa_b}\left[x \arctan\left(\frac{x_1}{z_0}\right) - \frac{z_0}{2}\log\left(x_1^2 + z_0^2\right)\right]. \tag{79}$$

Figure 12 shows that the line charge q can cause the deformation of a flexoelectric membrane. In the figure, x is normalized by the distance between the charge and the membrane z_0, $w(x_1)$ is normalized by the term $w_0 = q\gamma/\epsilon_0\kappa_b z_0$. We expect this result to have important implications for the study of lipid bilayers by ions in the surrounding electrolytes.

6.2. $E_3 = E_3(r)$

Since E_3 is only a function of $r = (x_1^2 + x_2^2)^{1/2}$, by symmetry we seek a solution $w = w(r)$ to (74)

$$\frac{d}{rdr}r\frac{d}{dr}\left[\kappa_b \frac{d}{rdr}r\frac{d}{dr}w(r) + \gamma E_3(r)\right] = 0 \quad \forall\, r > 0.$$

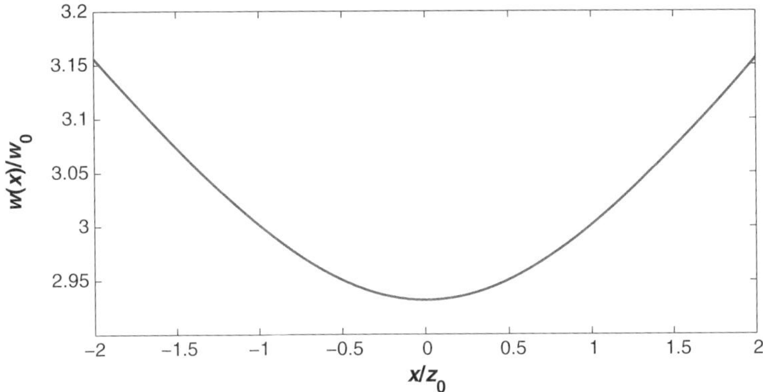

Figure 12.   Deflection of a flexoelectric membrane near a line charge. Reprinted with permission from Elsevier.[43] Copyright 2014.

Neglecting immaterial integration constants, by (75) we have

$$\frac{d}{r dr} r \frac{d}{dr} w(r) = -\frac{\gamma}{\kappa_b} E_3(r). \tag{80}$$

Upon specifying the functional form $E_3(r)$, we can integrate the RHS of (80) explicitly for $w(r)$. In particular, if the external field is generated by a point charge Q above the membrane at $x_3 = z_0$, then

$$E_3(r) = -\frac{q z_0}{4\pi\epsilon_0}(z_0^2 + r^2)^{-3/2}. \tag{81}$$

Substitute (81) into (80), then we have

$$w(r) = -\frac{q\gamma}{4\pi\epsilon_0\kappa_b}\left[\log(r) - \log\left(z_0^2 + z_0\sqrt{r^2 + z_0^2}\right)\right]. \tag{82}$$

7. Flexoelectricity in Soft Materials

7.1. *One-dimensional Formulation for Soft Materials*

In this section, our recent effort to investigate the effect of nonlinearity on flexoelectricity in soft materials[43] is presented. We now consider a one-dimensional system (a thin film) which is composed of soft materials. For this case, the strain of the system is expected to be finite and it is necessary to differentiate between the current and reference configurations. Let (X, Y, Z) be the Lagrangian coordinates of material points, $(\rho_0^e(X), P_0^e(X))$ be the external polarization (along X-direction, per unit volume) and charge density (in the reference configuration), and $b_0^e(X)$, t_0^e be the body force, surface traction (equal and opposite at top and bottom faces) in X-direction, respectively. The electrodes on the top and bottom faces are assumed to be mechanically trivial. Nevertheless, the electrodes are expected to maintain a constant electrostatic potential on both the top and bottom faces.

The electro-elastic state of the film is described by deformation and polarization — these are the independent variables in our formulation. In the presence of external charges, dipoles, applied voltage, and mechanical loads, the film is deformed; the deformation is denoted by $(x, y, z) = \chi(X, Y, Z)$ with (x, y, z) being the Eulerian coordinates in the current configuration. Let P be the intrinsic polarization in X-direction per unit volume (in the reference configuration). Since the film is thin (the thickness H is much smaller than the width L in the other two directions) and transversely isotropic, for simplicity we restrict ourselves to the following class of deformation and polarization:

$$x = X + u(X), \quad y = Y\alpha(X), \quad z = Z\beta(X), \quad P = P(X), \quad (83)$$

where the scalar functions $u, \alpha, \beta, P : (0, H) \to \mathbb{R}$ are determined by the equilibrium conditions. We remark that this kinematic assumption (83) about the possible form of deformation is the equivalent to that in the classic theory of extension. Following the standard framework of nonlinear continuum mechanics, we introduce the *stretches* in X (resp. Y, Z)-direction: $\lambda_1 = 1 + \frac{\partial u}{\partial X}$ (resp. $\lambda_2 = \frac{\partial y}{\partial Y} = \alpha, \lambda_3 = \frac{\partial z}{\partial Z} = \beta$). Thus, the Jacobian becomes $J = \lambda_1 \lambda_2 \lambda_3 = \det \mathbf{F}$. For brevity, we denote by

$$\boldsymbol{\lambda} = (\lambda_1, \lambda_2, \lambda_3) \quad \text{and} \quad \boldsymbol{\Lambda} = (\Lambda_1, \Lambda_2, \Lambda_3) = \left(\frac{d\lambda_1}{dX}, \frac{d\lambda_2}{dX}, \frac{d\lambda_2}{dX} \right).$$

Then the governing equations (25) reduce to the following one-dimensional version:

$$
\begin{cases}
\dfrac{\partial \psi}{\partial P} - \dfrac{d}{dX}\left(\dfrac{\partial \psi}{\partial \Pi}\right) + \dfrac{1}{\lambda_1}\dfrac{d\phi}{dX} = 0 & \text{on } (0, H) \\[2ex]
\dfrac{d}{dX}[\tilde{T}_i + \tilde{\Sigma}_i] + J b_i^e = 0 & \text{on } (0, H) \\[2ex]
\dfrac{d}{dX}\left[\epsilon_0 \dfrac{1}{\lambda_1}\phi_{,X} + \dfrac{P(X) + P_0^e(X)}{J}\right] = \dfrac{\lambda_1}{J}\rho_0^e(X) & \text{on } (0, H),
\end{cases}
\tag{84}
$$

where

$$
\tilde{T}_i = \frac{\partial \psi}{\partial \lambda_i} - \frac{d}{dX}\left(\frac{\partial \psi}{\partial \Lambda_i}\right) \qquad (i = 1, 2, 3)
\tag{85}
$$

and

$$
\tilde{\Sigma}_1 = -\frac{1}{\lambda_1^2}\phi_{,X}(P + P_0^e) + \frac{\epsilon_0 J}{2\lambda_1^3}\phi_{,X},
$$

$$
\tilde{\Sigma}_2 = -\frac{\epsilon_0 J}{2\lambda_1^2\lambda_2}(\phi_{,X})^2, \qquad \tilde{\Sigma}_3 = -\frac{\epsilon_0 J}{2\lambda_1^2\lambda_2}(3\phi_{,X})^2.
\tag{86}
$$

As a special case, we assume that there are no external body force and the surface traction t_0^e applied in the X direction. Then boundary conditions are given by

$$
\begin{cases}
\tilde{T}_1 + \tilde{\Sigma}_1 - t_0^e = 0 & \text{at } X = 0 \;\&\; X = H, \\[1.5ex]
\tilde{T}_2 + \tilde{\Sigma}_2 = 0 & \text{at } X = 0 \;\&\; X = H, \\[1.5ex]
\tilde{T}_3 + \tilde{\Sigma}_3 = 0 & \text{at } X = 0 \;\&\; X = H, \\[1.5ex]
\dfrac{\partial \psi}{\partial \Lambda_1} = \dfrac{\partial \psi}{\partial \Lambda_2} = \dfrac{\partial \psi}{\partial \Lambda_3} = 0 & \text{at } X = 0 \;\&\; X = H, \\[1.5ex]
\phi(0) = 0 & \phi(H) = V,
\end{cases}
\tag{87}
$$

where t_0^e is the X-component of $\boldsymbol{t}_0^e$.

To demonstrate the flexoelectric effects in soft materials, we consider isotropic flexoelectric materials with the internal energy density given by

$$\psi(\boldsymbol{\lambda}, \boldsymbol{\Lambda}, P) = W_{elast}(\boldsymbol{\lambda}) + \frac{g}{2}\Lambda_1^2 + f\Lambda_1 P + \frac{1}{2(\epsilon - \epsilon_0)J}P^2, \tag{88}$$

where $W_{elast}(\boldsymbol{\lambda})$ is the strain energy density function dictating the mechanical properties of materials, the last term implies that the dielectric constant (i.e. permittivity) ϵ of the material is independent of deformation in the absence of flexoelectric effects, i.e. the third term $f\Lambda_1 P$, the second term $\frac{g}{2}\Lambda_1^2$ guarantees that the natural state $(\boldsymbol{\lambda}, \boldsymbol{\Lambda}, P) = (0,0,0)$ is the stable equilibrium state of the film in the absence of all external electrical and mechanical loads. Here, constants $f, g > 0, \epsilon > \epsilon_0$ are material properties which can be determined by benchmark experiments.

Inserting (88) into (84), (85), and (87), then we obtain the following governing equations

$$\begin{cases} \dfrac{P}{(\epsilon - \epsilon_0)J} + f\Lambda_1 + \dfrac{1}{\lambda_1}\dfrac{d\phi}{dX} = 0 & \text{on } (0, H), \\[2ex] \left[\dfrac{\partial W_{elast}}{\partial \lambda_1} + \tilde{\Sigma}'_1 - (gu_{,XX} + fP)_{,X}\right]_{,X} + b_0^e = 0 & \text{on } (0, H), \\[2ex] \dfrac{\partial W_{elast}}{\partial \lambda_2} + \tilde{\Sigma}'_2 = 0 & \text{on } (0, H), \\[2ex] \dfrac{\partial W_{elast}}{\partial \lambda_3} + \tilde{\Sigma}'_3 = 0 & \text{on } (0, H), \\[2ex] \dfrac{d}{dX}\left[\epsilon_0\dfrac{1}{\lambda_1}\phi_{,X} + \dfrac{P(X) + P_0^e(X)}{J}\right] = \dfrac{\lambda_1}{J}\rho_0^e(X) & \text{on } (0, H), \end{cases} \tag{89}$$

and the boundary conditions

$$\begin{cases} \dfrac{\partial W_{elast}}{\partial \lambda_1} + \tilde{\Sigma}'_1 - (gu_{,XX} + fP)_{,X}] - t_0^e = 0 & \text{at } X = 0 \,\&\, X = H, \\[2ex] gu_{,XX} + fP = 0 & \text{at } X = 0 \,\&\, X = H, \\[2ex] \phi(0) = 0 & \phi(H) = V, \end{cases}$$

$$\tag{90}$$

where

$$\tilde{\Sigma}'_i = \tilde{\Sigma}_i - \frac{P^2}{2(\epsilon - \epsilon_0)J\lambda_i}.$$

We remark that if the elastic properties of the soft materials are specified, e.g. the Neo-Hookean hyperelastic model with (μ, shear modulus; κ, bulk modulus)

$$W_{elast}(\boldsymbol{\lambda}) = \frac{\mu}{2}\left[J^{-2/3}(\lambda - 1^2 + \lambda_2^2 + \lambda_3^2) - 3\right] + \frac{\kappa}{2}(J - 1)^2, \quad (91)$$

then (89) and (90) form a closed boundary value problem which can be solved to determine the state variables $(\boldsymbol{\chi}, P)$ (i.e. u, α, β, P, and ϕ).

Moreover, if the material is *incompressible* with $J = \lambda_1\lambda_2\lambda_3 = 1$, the state variables as given by (83) shall satisfy

$$J = (1 + u_{,X})\alpha\beta = 1.$$

To incorporate this incompressibility constraint, we use the Lagrange multiplier approach and add to the total free energy:

$$\int_0^H q[(1 + u_{,X})\alpha\beta - 1]dX,$$

where $q : (0, H) \to \mathbb{R}$ can be interpreted as the hydrostatic pressure as in the classic context. Taking into account the hydrostatic term and repeating the similar variational calculations as stated previously, we find the associated Euler–Lagrange equations

$$\begin{cases} \dfrac{\partial W}{\partial P} + \dfrac{1}{\lambda_1}\phi_{,X} = 0 & \text{on } (0, H), \\[2mm] \dfrac{d}{dX}\left[\tilde{T}_1 + \tilde{\Sigma}_1 + \dfrac{q}{\lambda_1}\right] + Jb_1^e = 0 & \text{on } (0, H), \\[2mm] \tilde{T}_2 + \tilde{\Sigma}_2 + \dfrac{q}{\lambda_2} = 0 & \text{on } (0, H), \\[2mm] \tilde{T}_3 + \tilde{\Sigma}_3 + \dfrac{q}{\lambda_3} = 0 & \text{on } (0, H). \end{cases} \quad (92)$$

For an incompressible flexoelectric Neo-Hookean material described by (88) and (91), based on symmetry, we note that a solution for the boundary value problem should satisfy

$$\lambda_2 = \alpha = \lambda_3 = \beta = \lambda_1^{-1/2} \quad \text{on } (0, H). \tag{93}$$

Further, we can rewrite the strain energy density as

$$W_{elast}(\lambda_1) = \frac{\mu}{2}\left(\lambda_1^2 + \frac{2}{\lambda_1} - 3\right), \tag{94}$$

which, by(85), $(92)_{3,4}$, and (93), implies

$$\tilde{T}_2 = \tilde{T}_3 = 0, \quad \tilde{\Sigma}_2 = \tilde{\Sigma}_3 = -q\lambda_1^{1/2} \quad \text{on } (0, H). \tag{95}$$

Eliminating q in $(92)_{1,2}$ by equations (95), we obtain the following explicit boundary value problem for u, ϕ, and P:

$$\begin{cases} \dfrac{P}{\epsilon - \epsilon_0} + f\lambda_{1,X} + \lambda_1^{-1}\phi_{,X} = 0 & \text{on } (0, H), \\[2mm] [\mu(\lambda_1 - \lambda_1^{-2}) + \tilde{\Sigma}_{eq} - (gu_{,XX} + fP)_{,X}]_{,X} + Jb_1^e = 0 & \text{on } (0, H), \\[2mm] [-\epsilon_0\lambda_1^{-1}\phi_{,X} + P + P_0^e]_{,X} = \lambda_1\rho_0^e & \text{on } (0, H), \\[2mm] \mu(\lambda_1 - \lambda_1^{-2}) + \tilde{\Sigma}_{eq} - (gu_{,XX} + fP)_{,X} - t_0^e = 0 & \text{at } X = 0 \& H, \\[2mm] gu_{,XX} + fP = 0 & \text{at } X = 0 \& H, \\[2mm] \phi(0) = 0, & \phi(H) = V, \end{cases}$$
$$\tag{96}$$

where $\lambda_1 = 1 + u_{,X}$, and $\tilde{\Sigma}_{eq} = \tilde{\Sigma}_1 - \lambda_1^{-3/2}\tilde{\Sigma}_2 = -\frac{1}{\lambda_1^2}\phi_{,X}(P + P_0^e) + \frac{\epsilon_0}{\lambda_1^3}(\phi_{,X})^2$.

7.2. *Flexoelectric Effects in a Soft Bilayer Structure*

Now we consider the flexoelectric effect in the context of the simple bilayer structure shown in Fig. 13. For this structure, the layers A

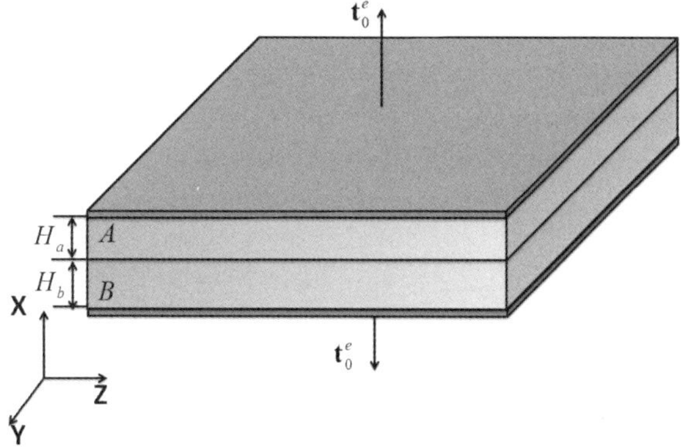

Figure 13.   A bilayer flexoelectric structure under uniform mechanical loading. Reprinted with permission from Ref. [43] Copyright 2014 by Elsevier.

and B are made of different soft materials. For this problem, the voltage difference between the two surfaces is maintained to be zero throughout the calculation however a uniform loading t_0^e is applied. Under this loading, an electric displacement will ensue in the body. The effective piezoelectricity is then calculated by $d^{\mathit{eff}} = d\tilde{D}/dt_0^e$, where $\tilde{D}$ is the average electric displacement of the bilayer structure.

In the absence of external polarization, charges and body force, i.e. $P_0^e = b_0^e = \rho_0^e = 0$, by (96) we obtain the following explicit boundary value problem for u, ϕ, and P:

$$
\begin{cases}
\dfrac{P}{\epsilon - \epsilon_0} + f\lambda_{1,X} + \lambda_1^{-1}\phi_{,X} = 0 & \text{on } (-H_b, H_a), \\[2ex]
[\mu(\lambda_1 - \lambda_1^{-2}) + \tilde{\Sigma}_{eq} \\[1ex]
\qquad - (gu_{,XX} + fP)_{,X}]_{,X} + Jb_1^e = 0 & \text{on } (-H_b, H_a), \\[2ex]
\left[-\epsilon_0\lambda_1^{-1}\phi_{,X} + P + P_0^e\right]_{,X} = 0 & \text{on } (-H_b, 0) \cup (0, H_a), \\[2ex]
\mu(\lambda_1 - \lambda_1^{-2}) + \tilde{\Sigma}_{eq} \\[1ex]
\qquad - (gu_{,XX} + fP)_{,X} - t_0^e = 0 & \text{at } X = -H_b \,\&\, H_a, \\[2ex]
gu_{,XX} + fP = 0, \quad \phi = 0 & \text{at } X = -H_b \,\&\, H_a
\end{cases}
\tag{97}
$$

with the following interfacial conditions

$$\begin{cases} [\![[\mu(\lambda_1 - \lambda_1^{-2}) + \tilde{\Sigma}_{eq} - (gu_{,XX} + fP)_{,X}]_{,X} + Jb_1^e]\!] = 0 \\ [\![[-\epsilon_0\lambda_1^{-1}\phi_{,X} + P + P_0^e]_{,X}]\!] = 0 \\ [\![gu_{,XX} + fP]\!] = 0, \quad [\![\phi]\!] = 0, \\ [\![u]\!] = 0 \quad [\![u_{,X}]\!] = 0, \end{cases} \tag{98}$$

where $[\![\,]\!] = ()|_{X=0_+} - ()|_{X=0_-}$.

For this problem, we set the displacement on the interface $u(0)$ to be zero to eliminate rigid body motion. Polypropylene cellular film and polyvinylidene fluoride (PVDF) are used for the layer A and B, respectively. The material properties of the two layers are given by

- Layer A: $\mu_a = 0.95\,\mathrm{MPa}$, $\epsilon_a = 2.35\epsilon_0$, $f_a = 46.79\,\mathrm{Nm/C}$, $g_a = 1.28 \times 10^{-8}\,\mathrm{N}$;
- Layer B: $\mu_b = 2.0\,\mathrm{GPa}$, $\epsilon_b = 9.5\epsilon_0$, $f_b = 179\,\mathrm{Nm/C}$, $g_b = 5.42 \times 10^{-10}\,\mathrm{N}$.

The shear modulus and flexoelectricity coefficient of PVDF are reported in Chu and Salem's work.[42] The shear modulus of polypropylene cellular is from (Qu and Yu, 2011).[70] Since there is no report on the flexoelectricity coefficient of polypropylene cellular film, in this work, we have assumed a reasonable value which is within the range of known values for common polymer materials. Also, the value of g has to be estimated. The reader is referred to two works[33,71] that use atomistic and microscopic considerations to determine this parameter that sets the non-local elastic length scale. We use a simple route to predict an approximate value for the polymer studied by us. As motivated by Maranganti and Sharma,[33] the characteristic non-local elastic length scale can be approximated by the radius of gyration. Accordingly, we set $\sqrt{g/3\mu} \sim R_g$, where R_g is the radius of gyration of the polymer studied.

It is not trivial to analytically solve the boundary value problem described by (97) and (98). In this section, a general purpose finite-element based partial differential equations solver (COMSOL) was used for this purpose. Because of the highly nonlinear natural of the

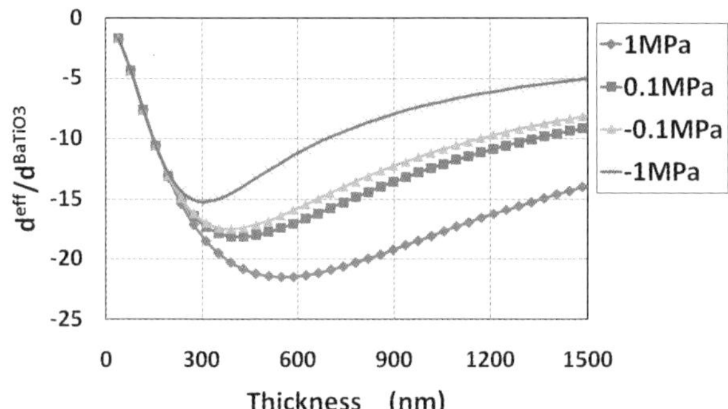

Figure 14.   Flexoelectric effect in a soft bilayer structure. Reprinted with permission from Elsevier.[43] Copyright 2014.

problem, the quartic (fourth-order) one-dimensional finite element is used for all the calculations in this section. Another reason for using the higher-order element here is that the high order continuity need to be satisfied throughout the specimen.

Figure 14 shows the size effect on the effective piezoelectricity for different stress levels. In this figure, d^{eff} is normalized by $d^{\mathrm{BaTiO_3}}$. Several observations may be made: (i) the effective piezoelectric coefficient shows size effect which can be profitably used to engineer a high electromechanical coupling by exploiting the flexoelectric effect, (ii) the resulting electromechanical coupling in soft material is quite high — reaching to 20 times that of Barium Titanate, and finally, (iii) unlike hard ceramics, the pronounced size-effect is evident even at the micron scale (as opposed to nanoscale). The latter has important ramifications in terms of experimental verification and exploitation for practical applications.

8. Concluding Remarks

In this chapter, we have attempted to provide a comprehensive account of a continuum field theory for flexoelectricity. In particular, we have emphasized a careful development of the nonlinear theory of flexoelectricity which is critical when considering soft materials. We have presented several pedagogical examples that illustrate the important applications of flexoelectricity in the fields of energy harvesting, sensing and actuation, soft active materials, and biological membranes. In particular, we strongly believe that some of the most exciting applications of flexoelectricity will emerge in the area of soft materials.

References

1. L. E. Cross. *J. Mater. Sci.* **41**, 53–63 (2006).
2. W. Ma and L. E. Cross. *Appl. Phys. Lett.* **79**(19), 4420–4422 (2001).
3. W. Ma and L. E. Cross. *Appl. Phys. Lett.* **81**(19), 3440–3442 (2002).
4. W. Ma and L. E. Cross. *Appl. Phys. Lett.* **82**(19), 3923–3925 (2003).
5. W. Ma and L. E. Cross. *Appl. Phys. Lett.* **88**, 232902 (2006).
6. G. Catalan, L. J. Sinnamon, and J. M. Gregg. *J. Phys. Condens. Mat.* **16**(13), 2253–2264 (2004).
7. G. P. Zubko, A. R. Catalan, P. Buckley, L. Welche, and J. F. Scott. *Phys. Rev. Lett.* **99**, 167601 (2007).
8. J. Y. Fu, W. Zhu, N. Li, and L. E. Cross. *J. Appl. Phys.* **100**, 024112 (2006).
9. J. Y. Fu, W. Zhu, N. Li, and L. E. Cross. *Appl. Phys. Lett.* **91**, 182910 (2007).
10. E. A. Eliseev, A. N. Morozovska, M. D. Glinchuk, and R. Blinc. *Phys. Rev. B.* **79**, 165433 (2009).
11. E. A. Eliseev, M. D. Glinchuk, V. Khist, V. V. Skorokhod, R. Blinc, and A. N. Morozovska. *Phys. Rev. B.* **84**, 174112 (2011).
12. R. Maranganti and P. Sharma. *Phys. Rev. B.* **80**, 054109 (2009).
13. M. S. Majdoub, P. Sharma, and T. Cagin. *Phys. Rev. B.* **77**, 125424 (2008).
14. M. S. Majdoub, P. Sharma, and T. Cagin. *Phys. Rev. B.* **78**, 121407(R) (2008).
15. M. S. Majdoub, P. Sharma, and T. Cagin. *Phys. Rev. B.* **79**, 119904(E) (2009).
16. M. S. Majdoub, P. Sharma, and T. Cagin. *Phys. Rev. B.* **79**, 159901(E) (2009).
17. N. D. Sharma, R. Maranganti, and P. Sharma. *J. Mech. Phys. Solids.* **55**, 2328–2350 (2007).
18. N. D. Sharma, C. M. Landis, and P. Sharma. *J. Appl. Phys.* **108**, 024304 (2010).
19. N. D. Sharma, C. M. Landis, and P. Sharma. *J. Appl. Phys.* **111**, 059901 (2012).

20. M. Gharbi, Z. H. Sun, P. Sharma, K. White, and S. El-Borgi. *Int. J. Solids Struct.* **48**, 249–256 (2011).
21. S. V. Kalinin and V. Meunier. *Phys. Rev. B.* **77**(3), 033403 (2008).
22. T. Dumitrica, C. M. Landis, and B. I. Yakobson. *Chem. Phys. Lett.* **360**(1–2), 182–188 (2002).
23. A. K. Tagantsev. *Sov. Phys. JETP.* **61**(6), 1246–1254 (1985).
24. A. K. Tagantsev. *Phys. Rev. B.* **34**, 5883–5889 (1986).
25. A. K. Tagantsev. *Phase Transit.* **35**(3–4), 119–203 (1986).
26. A. K. Tagantsev, V. Meunier, and P. Sharma. *MRS Bull.* **34**(9), 643–647 (2009).
27. R. Maranganti, N. D. Sharma, and P. Sharma. *Phys. Rev. B.* **74**, 014110 (2006).
28. T. D. Nguyen, S. Mao, Y. W. Yeh, P. K. Purohit, and M. C. McAlpine. *Adv. Mater.* **25**(7), 946–974 (2013).
29. S. Chandratre and P. Sharma. *Appl. Phys. Lett.* **100**, 023114 (2012).
30. A. G. Petrov. *Flexoelectric model for active transport.* In: *Physical and Chemical Bases of Biological Information Transfer.* Plenum Press, New York (1975).
31. A. G. Petrov. *Biochim. Biophys. Acta.* **1561**, 1–25 (2002).
32. A. G. Petrov. *Curr. Top. Membr.* **58**, 121–150 (2007).
33. R. Maranganti and P. Sharma. *J. Mech. Phys. Solids.* **55**(9), 1823–1852 (2007).
34. T. Xu, J. Wang, T. Shimada, and T. Kitamura. *J. Phys.-Condens. Matt.* **25**(41), 415901 (2007).
35. J. Hong, G. Catalan, J. F. Scott, and E. Artacho. *J. Phys.-Condens. Matt.* **22**(11), 112201 (2010).
36. J. Hong and D. Vanderbilt. *Phys. Rev. B.* **84**, 180101 (2011).
37. J. Hong and D. Vanderbilt. *Phys. Rev. B.* **88**, 174107 (2013).
38. M. Stengel. *Phys. Rev. B.* **88**, 174106 (2013).
39. M. Stengel. *Nat. Commun.* **4**, 2693 (2013).
40. S. Baskaran, X. He, Q. Chen, and J. Y. Fu. *Appl. Phys. Lett.* **98**, 242901 (2011).
41. S. Baskaran, X. He, Y. Wang, and J. J. Y. Fu. *J. Appl. Phys.* **111**, 014109 (2012).
42. B. Chu and D. R. Salem. *Appl. Phys. Lett.* **101**, 103905 (2012).
43. Q. Deng, L. Liu, and P. Sharma. *J. Mech. Phys. Solids.* **62**, 209–227 (2014).
44. G. Catalan, A. Lubk, A. Vlooswijk, E. Snoeck, C. Magen, A. Janssens, G. Rispens, G. Rijnders, D. Blank, and B. Noheda. *Nat. Mater.* **10**(12), 963–967 (2011).
45. M. S. Majdoub, R. Maranganti, and P. Sharma. *Phys. Rev. B.* **79**, 115412 (2009).

46. P. Zubko, G. Catalan, and A. K. Tagantsev. *Flexoelectric Effect in Solids*. Volume 43 of *Annual Review of Materials Research*, pp. 387–421. Annual Reviews, Palo Alto (2013).

47. P. V. Yudin and A. K. Tagantsev. *Nanotechnology*. **24**, 432001 (2013).

48. S. Shen and S. Hu. *J. Mech. Phys. Solids*. **58**, 665–677 (2010).

49. S. Dai, M. Gharbi, P. Sharma, and H. Park. *J. Appl. Phys.* **110**, 104305 (2011).

50. R. Resta and D. Vanderbilt. Theory of Polarization: A Modern Approach, in Physics of Ferroelectrics: a Modern Perspective. Edited by K. M. Rabe, C. H. Ahn, J. -M. Triscone. Springer-Verlag, Berlin (2007).

51. J. Marshall and K. Dayal. *J. Mech. Phys. Solids*. **62**, 137–162 (2014).

52. R. A. Toupin. *J. Rational Mech. Anal.* **5**(6), 849–915 (1956).

53. R. Bustamante, A. Dorfmann, and R. W. Ogden. *Z. angew. Math. Phys.* **60**, 154–177 (2009).

54. L. Liu. *J. Mech. Phys. Solids*. **63**, 451–480 (2014).

55. A. S. Yurkov. *JETP Lett.* **94**(6), 455–458 (2011).

56. R. D. Mindlin. *Int. J. Solids Struct.* **4**, 637–642 (1968).

57. S. Mao and P. Purohit. *J. Appl. Mech.* **81**(8), 081004 (2014).

58. A. Abdollahi, D. Millán, C. Peco, M. Arroyo, and I. Arias. *Phys. Rev. B.* **91**, 104103 (2015).

59. Q. Deng, M. Kammoun, A. Erturk, and P. Sharma. *Int. J. Solids Struct.* **51**(18), 3218–3225 (2014).

60. A. Erturk and D. J. Inman. *Piezoelectric Energy Harvesting*. Wiley, New Delhi (2011).

61. A. Erturk. *Comput. Struct.* **106–107**, 3218–3225 (2012).

62. R. M. Clough and J. Penzien. *Dynamics of Structures*. 2nd edn. McGraw Hill, New York (1993).

63. A. G. Petrov, B. A. Miller, K. Hristova, and P. N. R. Usherwood. *Eur. Biophys. J.* **22**, 289–300 (1993).

64. W. E. Brownell, A. A. Spector, R. M. Rapheal, and A. S. Popel. *Annu. Rev. Biomed. Eng.* **3**, 169–194 (2001).

65. W. E. Brownell, B. Farrell, and R. M. Rapheal. Membrane electromechanics at hair-cell synapses. In: Biophysics of the Cochlea: From Molecules to Models. Proceedings of the International Symposium, Titisee, Germany, 27 July–1 August, pp. 169–176 (2003).

66. R. M. Rapheal, A. S. Popel, and W. E. Brownell. *Biophys. J.* **78**(6), 2844–2862 (2000).

67. K. D. Breneman and R. D. Rabbitt. Piezo- and flexoelectric membrane underlie fast biological motors in the ear. *Materials Research Society Symposium Proceedings*, San Francisco, California, 13–17 April, 1186E, pp. 1186–JJ06-04 (2009).

68. P. Mohammadi, L. Liu, and P. Sharma. *J. Appl. Mech.* **81**(1), 011007 (2014).
69. W. Helfrich. *Z. Naturforsch C.* **28**(11), 693–703 (1973).
70. S. Qu and Y. Yu. *J. Appl. Phys.* **110**, 043525 (2011).
71. S. Nikolov, C. S. Han, and D. Raabe. *Int. J. Solids Struct.* **44**(5), 1582–1592 (2007).

Chapter 4

Mechanical Boundary Conditions for a Case When Thermodynamic Potential Depends on Strain Gradients

A. S. Yurkov

Omsk, Russia 644076

When describing elastic deformations of a body sometimes it is worth to take in account the dependence of thermodynamic potential on strain gradients. Such a dependence may be important for small body size which imply large gradients. Besides, taking into account such dependence leads to physical phenomena absent without it. An example of the latter is flexoelectricity. An remarkable fact is that while the derivation of differential equations of elastic equilibrium can be made by ordinary means in this case, the derivation of boundary conditions for them is less a trivial task. Detailed consideration of this problem is the subject of this chapter.

1. Introduction

The classical theory of elasticity is based on the thermodynamic potential dependent on the elastic strain only. However, in some cases it is important to consider dependence of the thermodynamic potential on the gradients of elastic strain.

There are two main reasons why consideration of such a thermodynamic potential may be important. First, elastic body can be of small size, small sizes imply large gradients. This becomes especially important due to rapid development of nanotechnology. Secondly, such a dependence may lead to physical phenomena absent without

169

it. An example of such phenomenon is the flexoelectric effect (see Refs. [1] and [2] and references therein).

The basic problem of the theory of elasticity is to find static equilibrium of a body. Generally speaking dynamic problems are also important, but we restrict ourselves to the static case. Generalization of the considered theory to the dynamic case is absolutely straightforward.

Even if thermodynamic potential depends on strain gradients, it remains quasi-local potential (it can be represented as integral of the corresponding density). Thus, equilibrium is described by differential equations in partial derivatives. These equations should be appended with boundary conditions on the surface of the body. Both differential equations and boundary conditions are determined by the conditions of thermodynamic potential minimum in framework of the calculus of variations.

It is important to note that in the considered case the derivation of differential equations of elastic equilibrium does not differ much from the case of the classical theory, it can be made by completely standard methods. On the contrary, the correct derivation of the boundary conditions for these equations is non-trivial task. Recent studies on the subject have been done with regard to the description of the flexoelectric effect, but they are applicable in all cases when the thermodynamic potential depends on strain gradients.

Essence of matter leading to the fact that in this case the problem of boundary conditions is non-trivial is as follows. Dependence of the thermodynamic potential of the strain gradient actually means its dependence on the second spatial derivatives of the elastic displacements u_i. When calculating the variation of the thermodynamic potential one should express it in terms of the independent variate. As usual, this is done using integration by parts. However, in the considered case surface integrals, which arises due to such an integration, contained $\delta u_{i,j}$ not just δu_i. Hereinafter index separated by comma denotes spatial derivative in respective coordinate. Sometimes such a derivative is also denoted as ∂_i. Values $\delta u_{i,j}$ are not independent variates. Thus, additional analysis and mathematical constructions are required.

Apparently a first solution to this problem has been proposed by Toupin[3,4] many years ago. This solution is based on the representation of the gradient ∂_i as a sum of "normal gradient" $n_i D$ and "tangential gradient" D_i, where n_i is unit vector normal to the surface. Operator D has been defined as $n_i \partial_i$. Hereinafter, the Einstein summation rule over repeated indices is implied. Note that formal definition of D_i has not been proposed. Certainly, if D_i is applied to a function, defined in whole three-dimensional space (at least in the body bulk), then D_i can be well-defined as difference between ∂_i and $n_i D$, no problem appears in this way. However, Toupin's boundary conditions contain not only such mathematical objects, but also $B_{ij} = D_i n_j$. Vector n_i is defined only on the surface and hence neither ∂_i nor D can be applied to it. Thus, mathematical sense of three-dimensional tensor B_{ij} has been remained unclear. Of course, this construction can be provided with some intuitive sense in Cartesian coordinate system, if a couple of axes are tangential to the surface at given point. But anyway it is not flawless mathematical definition, because tangential plane has only one point common with the curved surface. Formally, vector n_i is *not* defined in some, at least infinitesimal, region of this tangential plane. Naturally, the derivative can be defined only if differentiated expression is defined in some region.

In Refs. [3] and [4] tensor B_{ij} has been referred as "second fundamental form" of the body surface. Note that second fundamental form of the surface, at least within conventional definitions, is defined in two-dimensional space tangential to the surface, rather than in whole three-dimensional space. Thus, relation between second fundamental form and three-dimensional tensor B_{ij} remains unclear. At least additional geometric constructions and definitions are required here to make it meaningful.

In subsequent years, Toupin's construction has been used to derive boundary conditions in many papers (see Refs. [5, 6] and [7] for instance). However, mathematically flawless definition of $D_i n_j$ has not been given. Moreover, in these papers there was no mention about the relation between this tensor and the second fundamental form of the boundary surface.

In the framework of continual theory of flexoelectricity an attempt to derive the mechanical boundary conditions was also made in the paper [8]. The authors of this paper quite correctly pointed out that the mechanical boundary conditions can be obtained from the surface integral arising when the thermodynamic potential is varied in the elastic displacements. It was also correctly noticed that since the thermodynamic potential depends on strain gradients in this case, the differential equations are of the fourth order in contrary to the standard theory of elasticity, where they are of the second order. Therefore, the number of the boundary conditions should be grater than such a number in the classical theory of elasticity.

Nevertheless mechanical boundary conditions written in Ref. [8] cannot be considered as correct ones. Authors do not give detailed derivation of them, so it is not possible to specify where a mistake was done. However, erroneousness of these boundary conditions is clear already from the fact that they do not contain the vector normal to the boundary surface. This means, among others, that it is impossible to passage to the limit of the classical theory of elasticity where boundary conditions contain such a vector.

In addition it is obvious that the authors of Ref. [8] did not take into account the curvature of the boundary surface. The surface of the finite-size body cannot be flat elsewhere, further it is shown that in the considered case the surface curvature plays an important role.

Number of articles[9–11] have been published also where for the case of presence of flexoelectricity mechanical boundary conditions on the free surface of a body were declared in the form $\sigma_{ij}n_j = 0$ which is analogous to the form of the boundary conditions in the classical theory of elasticity. Here σ_{ij} is tensor of non-classical "physical stress" which enters into differential equations of equilibrium in the form $\sigma_{ij,j} = 0$. Note that this boundary conditions were simply declared without any derivation. Note also that, as mentioned above, number of boundary conditions should be greater at least.

Mathematically strict way to derive the mechanical boundary conditions in the presence of flexoelectricity was independently proposed in the paper [12]. In this paper, in contrast to earlier works, projector to the body surface, needed to decompose $\delta u_{i,j}$ to normal and tangential parts, was constructed by means of two-dimensional curvilinear coordinate system on the body surface. This two-dimensional coordinate system was introduced additionally to common three-dimensional Cartesian coordinate system in the space. Such mathematical construction does not lead to ill-defined mathematical objects. Note that although in this paper the boundary conditions has been derived in framework of the theory of flexoelectricity, in fact they can be applied to all cases where thermodynamic potential depends on strain gradients.

Later, yet another approach to the derivation of such boundary conditions was proposed in paper [13]. This approach is based on a mathematical tool which is almost identical to the one used in general theory of relativity. The only difference is that, since the space is flat, the Riemann tensor is equal to zero and covariant derivatives commute.

Note that for solution of certain problems the curvilinear coordinates[a] x^{α} are more convenient than Cartesian ones, especially if one choose a coordinate system, in which the equation of the body surface is $x^3 = x_S^3$, where x_S^3 is some constant. In such coordinate system transformation of surface integrals needed for derivation of discussed boundary conditions is performed directly without any mathematical tricks. That is exactly what was used in Ref. [13].

Detailed derivation of the boundary conditions in curvilinear coordinates has not been given in Ref. [13], there was only a brief note about the method of derivation. The proof that this boundary conditions are completely equivalent to the boundary conditions derived in Ref. [12] was also omitted. This equivalence was only mentioned without any evidence. Consideration was too brief in Ref. [13].

Thus, detailed description of the problem of mechanical boundary conditions for a case where the thermodynamic potential depends on strain gradients is desirable. It is presented here and in somewhat more general form in Refs. [12] and [13].

---

[a] Note that index distinguishing curvilinear coordinates conventionally written as superscript instead of subscript.

2. General Description of the Mechanical Equilibrium of the Elastic Body

In the most general case, the mechanical equilibrium of an elastic body can be found by minimization of thermodynamic potential H. In the framework of the theory of continuum this implies the use of calculus of variations. The minimization should be performed with respect to variations of the elastic displacements u_i.

Thermodynamic potential also depends on variables other than u_i. Particularly, in the theory of flexoelectricity it also depends on electric polarization at least. However, here we consider only the mechanical aspects of the theory, so that electric polarization and all other variables, which differ from u_i, will be considered as given.

In quasi-local case the thermodynamic potential of the body is the sum of bulk part H_B and surface part H_S. The bulk part is volume integral of the bulk density of the thermodynamic potential:

$$H_B = \int \mathcal{H}_B dV. \tag{1}$$

Analogously the surface part is an integral over the body surface of the surface density:

$$H_S = \oint \mathcal{H}_S dS. \tag{2}$$

Note that the decomposition of the thermodynamic potential to volume and surface parts can be non-unique. If $\mathcal{H}_B$ contains a term which is divergence then the volume integral of this term can be converted to a surface integral. The reverse transformation is also possible if $\mathcal{H}_S$ contains a suitable term.

As indicated above to find the body equilibrium it is necessary to calculate the variation of the thermodynamic potential and equate it to zero. The result of this calculation essentially depends on the type of the functional dependences of $\mathcal{H}_B$ and $\mathcal{H}_S$ on u_i.

In the classical theory of elasticity $\mathcal{H}_B$ depends only on the symmetrized strain tensor u_{ij} which is the symmetric part of the displacement gradients tensor (non-symmetrized strain tensor):

$$u_{ij} = \frac{1}{2}(u_{i,j} + u_{j,i}). \tag{3}$$

In the case considered here $\mathcal{H}_B$ depends on u_{ij} and also depends on $u_{ij,k}$. Note that in the general reasoning it is quite possible to assume that $\mathcal{H}_B$ depends on $u_{i,j}$ and $u_{i,j,k}$ directly. Real dependence only on u_{ij} and $u_{ij,k}$ is automatically ensured due to the specific form of this dependence in this way, in fact it is ensured by symmetry properties of the material tensors. Keeping this in mind further we use non-symmetrized strain tensor $u_{i,j}$.

As for $\mathcal{H}_S$ in the classical case it depends on u_i only, particularly this dependence describes interaction with the external forces acting on the body surface. In the case considered here one should take into account also the dependence of $\mathcal{H}_S$ on $u_{i,j}$. Dependence $\mathcal{H}_S$ on $u_{i,j}$ is of the same order of smallness as the dependence $\mathcal{H}_B$ on $u_{i,j,k}$. This is clear already from the fact that volume integral of $\mathcal{H}_B$ being a divergence can be transformed to surface integral, order of derivatives is reduced at this transformation.

For such $\mathcal{H}_B$ and $\mathcal{H}_S$ the variation δH can be represented in terms of following quantities:

$$T_{ij} = \frac{\partial \mathcal{H}_B}{\partial u_{i,j}}, \tag{4}$$

$$\Theta_{ijk} = \frac{\partial \mathcal{H}_B}{\partial u_{i,j,k}}, \tag{5}$$

$$F_i = \frac{\partial \mathcal{H}_S}{\partial u_i}, \tag{6}$$

$$\Sigma_{ij} = \frac{\partial \mathcal{H}_S}{\partial u_{i,j}}. \tag{7}$$

In this notation following equations can be written immediately:

$$\delta H_B = \int T_{ij}\delta u_{i,j}dV + \int \Theta_{ijk}\delta u_{i,j,k}dV, \tag{8}$$

$$\delta H_S = \oint F_i\delta u_i dS + \oint \Sigma_{ij}\delta u_{i,j}dS. \tag{9}$$

Variations δu_i are independent but $\delta u_{i,j}$ and $\delta u_{i,j,k}$ are not. Thus, integrals in (8) should be transformed using integration by parts. This yields

$$\delta H_B = -\int (T_{ij,j} - \Theta_{ijk,j,k})\delta u_i dV$$

$$+ \oint (T_{ij} - \Theta_{ijk,k})\delta u_i n_j dS + \oint \Theta_{ijk}\delta u_{i,j} n_k dS, \qquad (10)$$

where n_i is unit vector normal to the surface of the body. Thus, surface terms appear which are analogous to the terms in right-hand side of (9).

Variations δu_i are absolutely arbitrary. Particularly δu_i and $\delta u_{i,j}$ can be vanishing on the surface. In such a case all surface integrals disappear, remaining volume integral yields differential equations of equilibrium:

$$T_{ij,j} - \Theta_{ijk,j,k} = 0. \qquad (11)$$

Note that if one defines a "physical stress" σ_{ij} as follows

$$\sigma_{ij} = T_{ij} - \Theta_{ijk,k}, \qquad (12)$$

then (11) can be rewritten in the manner analogous to the case of the classical theory of elasticity:

$$\sigma_{ij,j} = 0. \qquad (13)$$

Boundary conditions for equations (13) should be derived from the nullification of sum of surface integrals for δu_i and $\delta u_{i,j}$ not vanishing on the surface. Thus, the boundary conditions can be derived from the integral equation

$$\oint (\sigma_{ij} n_j + F_i)\delta u_i dS + \oint (\Sigma_{ij} + \Theta_{ijk} n_k)\delta u_{i,j} dS = 0, \qquad (14)$$

where we use (12) to rewrite the first integral.

In the classical theory of elasticity $\Theta_{ijk} = 0$, $\Sigma_{ij} = 0$ and hence there is no second integral in (14) which contains $\delta u_{i,j}$. Therefore, in this theory there is no problem to derive the boundary conditions from (14) at ones. One needs to only equate to zero the expression in parentheses in the first integral.

In the case when the thermodynamic potential depends on $u_{i,j,k}$ the situation becomes more complicated. The fact is that δu_i and $\delta u_{i,j}$ cannot be considered as completely independent. Therefore, to obtain the boundary conditions one should transform the second integral in left-hand side of (14). This transformation is non-trivial, in order to make it one needs a special mathematical tool which is considered in the following sections. Only after this transformation mechanical boundary conditions can be obtained.

3. Additional Surface Coordinate System and Boundary Conditions in Cartesian Tensor Components

As stated above, to derive the boundary conditions one must transform the second integral in the left-hand side of (14). One method of this transformation based on the use of additional surface coordinate system was proposed in Ref. [12]. In this section, we consider this method with some generalization relatively to the paper [12]. This generalization is quite direct but it can be useful sometimes.

Besides the usual Cartesian coordinate system in the space, let there be an additional two-dimensional coordinate system x^α on the surface of the body, $\alpha = 1, 2$ here. Generally the closed surface of a body is curved. Certainly body surface can be flat everywhere except the sharp edges where flat parts of surface intersect. Such a situation can be considered as the limit of the body with smoothed edges for which the curvature of the surface is present only on this smoothed edges. It is what further implied.

In this way a two-dimensional non-Euclidean geometry appears on the surface of a body. This is why the surface coordinate system should be described as curvilinear and one should distinguish corresponding tensor indices, denoted by Greek letters, as superscripts and subscripts. Indices corresponding three-dimensional Cartesian coordinates are denoted by Latin letters, they are always subscripts.

If only two-dimensional tangential to the surface tensors were necessary then the well-known mathematics of differential Riemannian geometry would be sufficient. However, equation (14) contains three-dimensional tensors on the surface rather than two-dimensional tensors. This is why the standard technique of Riemannian geometry should be modified here. Such a modification is quite natural if the frame vector formalism is used.

Let us introduce the frame vectors

$$e_{i\alpha} = \frac{\partial r_i}{\partial x^\alpha} = r_{i,\alpha},\tag{15}$$

where r_i are Cartesian coordinates of a point on the surface. Consequently, if a point moves along the surface then

$$(d\mathbf{r})^2 = e_{i\alpha}e_{i\beta}dx^\alpha dx^\beta.\tag{16}$$

Thus, metric tensor of the surface coordinate system $g_{\alpha\beta}$ can be represented as follows

$$g_{\alpha\beta} = e_{i\alpha}e_{i\beta}.\tag{17}$$

Since generally the coordinates x^α are curvilinear, the basis set $e_{i\alpha}$ is neither orthogonal nor normalized. Therefore, it is necessary to introduce a mutual basis set e_i^α, which is defined by relations

$$e_i^\alpha e_{i\beta} = \delta_\beta^\alpha,\tag{18}$$

where δ_β^α is Kronecker delta in Greek indices. By means of the reverse metric tensor $g^{\alpha\beta}$ mutual reference vectors e_i^α can be represented in terms of $e_{i\alpha}$:

$$e_i^\alpha = g^{\alpha\beta}e_{i\beta}.\tag{19}$$

Thus, by means of $g_{\alpha\beta}$ and $g^{\alpha\beta}$ curvilinear indices can be raised and lowered in the standard manner. Surface integrals can be represented in the form of integration over x^α using obvious relation: $dS = \sqrt{g}d^2x$, where $d^2x = dx^1 dx^2$, $g = \det g_{\alpha\beta}$.

The key observation needed further is that $e_i^\alpha e_{j\alpha}$ is the three-dimensional projector onto the plane tangential to the surface of the body. Indeed, $e_i^\alpha e_{j\alpha}e_j^\beta e_{k\beta} = e_i^\alpha \delta_\alpha^\beta e_{k\beta} = e_i^\alpha e_{k\alpha}$. Moreover, any vector

v_j convolved with $e_i^\alpha e_{j\alpha}$ yields a linear combination of vectors e_i^α which are tangential to the body surface. These two facts prove our assertion. It is also obvious that $n_i n_j$ is the projector onto normal of the surface. Thus, the unit operator (Kronecker delta in Latin indices) can be represented as follows:

$$\delta_{ij} = n_i n_j + e_i^\alpha e_{j\alpha}. \tag{20}$$

Equation (20) allows us to convert $\delta u_{i,j}$ in such manner: $\delta u_{i,j} = \delta u_{i,l} n_l n_j + \delta u_{i,l} e_{l\alpha} e_j^\alpha$. From equation (15) it follows that $\delta u_{i,l} e_{l\alpha} = (\delta u_i)_{,\alpha}$, where $(\ldots)_{,\alpha}$ means the partial derivative in x^α. Therefore,

$$\oint (\ldots) \delta u_{i,j} dS$$

$$= \oint (\ldots) \delta u_{i,l} n_l n_j \sqrt{g} d^2 x + \oint (\ldots) e_j^\alpha (\delta u_i)_{,\alpha} \sqrt{g} d^2 x. \tag{21}$$

The second term in the right-hand side of this equation can be integrated by parts in straightforward manner. It yields

$$\oint (\ldots) \delta u_{i,j} dS$$

$$= \oint (\ldots) \delta u_{i,l} n_l n_j dS - \oint [(\ldots) e_j^\alpha \sqrt{g}]_{,\alpha} \delta u_i \, g^{-1/2} dS. \tag{22}$$

Identity (22) allows transforming the surface integral containing an independent variations $\delta u_{i,j}$ to the surface integrals containing δu_i and $\delta u_{i,l} n_l$ which are independent. Indeed, δu_i are independent initially. As for $\delta u_{i,l} n_l$ it is nothing but the gradient in the direction normal to the surface. Setting δu_i on the surface one can continue them arbitrary along the normal to the surface. Thus, near surface $\delta u_{i,l} n_l$ are arbitrary independent values.

Thus, by applying (22) to (14) and taking into account the independence and arbitrariness of δu_i and $\delta u_{i,l} n_l$, we obtain the boundary conditions in the form

$$\Sigma_{ij} n_j + \Theta_{ijk} n_j n_k = 0, \tag{23}$$

$$\sigma_{ij} n_j + F_i - (\Sigma_{ij} e_j^\alpha g^{1/2})_{,\alpha} g^{-1/2} - (\Theta_{ijk} n_k e_j^\alpha g^{1/2})_{,\alpha} g^{-1/2} = 0. \tag{24}$$

Naturally, these equations are valid only on the body surface. They are boundary conditions for differential equations of elastic equilibrium (13).

Using the Leibniz rule, the identity $\Theta_{ijk,\alpha} = \Theta_{ijk,l} r_{l,\alpha}$, equations (15) and (20), the last term in left-hand side of (24) can be represented in the form used in the paper [12]:

$$(\Theta_{ijk} n_k e_j^\alpha g^{1/2})_{,\alpha} g^{-1/2} = \Theta_{ijk,j} n_k - \Theta_{ijk,l} n_l n_j n_k + \Theta_{ijk} \gamma_{jk}, \qquad (25)$$

where

$$\gamma_{jk} = (n_k e_j^\alpha g^{1/2})_{,\alpha} g^{-1/2}. \qquad (26)$$

Generally an analogous transformation of $(\Sigma_{ij} e_j^\alpha g^{1/2})_{,\alpha} g^{-1/2}$ may be meaningless because Σ_{ij} may be defined only on the surface rather than space. However, there are special case when Σ_{ij} can be represented as convolution:

$$\Sigma_{ij} = S_{ijl} n_l, \qquad (27)$$

where tensor S_{ijl} is well-defined in the bulk of the body. In particular, such a possibility appears when the surface contribution to the thermodynamic potential arises due to the transformation of volume contribution by integration by parts. For such cases $(\Sigma_{ij} e_j^\alpha g^{1/2})_{,\alpha} g^{-1/2}$ can be transformed absolutely similarly as $(\Theta_{ijk} n_k e_j^\alpha g^{1/2})_{,\alpha} g^{-1/2}$.

Equations (23) and (24) with the equalities (25), (26) are nothing but some generalization of the boundary conditions obtained in Ref. [12]. Generalization is as follows. First, we have considered the general case of the thermodynamic potential depended on the strain gradient (see previous section) rather than a special thermodynamic potential used in Ref. [12]. Second, we have taken into account the surface contribution to this potential, it has led to additional term $\Sigma_{ij} n_j$ in (23) and additional terms F_i and $-(\Sigma_{ij} e_j^\alpha g^{1/2})_{,\alpha} g^{-1/2}$ in (24). Except for these generalizations boundary conditions derived above coincide with ones derived in Ref. [12].

Note that boundary conditions (23) and (24) do satisfy the condition that in the limit case they should correspond to the classical theory of elasticity. Indeed in the classical limit $\Sigma_{ij} = 0$,

$\Theta_{ijk} = 0$. Thus, (23) becomes an identity while (24) takes the classical form $\sigma_{ij}n_j + F_i = 0$.

An remarkable fact is that the curvature of the body surface plays an important role in the mechanical boundary conditions derived above. To see it, let us rewrite (24) by means (25) omitting surface contribution contained Σ_{ij}. It turns out:

$$\sigma_{ij}n_j + F_i - \Theta_{ijk,j}n_k + \Theta_{ijk,l}n_l n_j n_k - \Theta_{ijk}\gamma_{jk} = 0. \qquad (28)$$

Tensor γ_{jk} is a purely geometric construction (see also Appendix A). It does not depend on the elastic displacements or somewhat else of non-geometrical nature, actually it characterizes only the geometry of the body surface. Note also that this tensor is invariant with respect to the change of the surface coordinate system. It is clear from the fact that it is a scalar convolution in Greek indices. If some part of a body surface is flat then suitable Cartesian coordinates can be chosen as surface coordinates on this part. But in such surface coordinates all values under derivative in (26) are constants. Thus, $\gamma_{jk} = 0$ on the flat parts of the body surface. Otherwise $\gamma_{jk} \neq 0$ on the curved parts of the body surface, it is absolutely obvious. Thus, the last term in left-hand side of (28) appears or disappears depending on whether the body surface is curved or flat.

4. Boundary Conditions within Covariant Formalism

In the previous section, the mechanical boundary conditions were derived in terms of tensor components related to the three-dimensional Cartesian coordinate system. To do this, one has to introduce an additional curvilinear two-dimensional coordinate system on the surface of the body.

One can go the other way and just introduce curvilinear coordinate system x^α in the whole space. If it is required that the equation of the body surface in this curvilinear coordinate system has a simple form $x^3 = x_S^3$, where x_S^3 is some constant, then the derivation of the boundary conditions can be done in a more direct way. In addition, such a coordinate system is most convenient for the practical solution of differential equations of elastic equilibrium.

Although the space is flat, to work with tensor objects at these case is most convenient in framework of the mathematical tool of Riemann geometry well known from the general theory of relativity. The fact that space is flat only leads to that the Riemann tensor is identically zero. Main points of such mathematical formalism (so-called covariant formalism) are described in Appendix B for convenience. It is also needed for references.

To transform equations (12)–(14) from Cartesian to covariant form one should change partial derivatives to covariant derivatives and provide an index balance (only superscript can be convolved with subscript). Indices can be lowered using metric tensor $g_{\alpha\beta}$, they can be raised using reverse metric tensor $g^{\alpha\beta}$. Covariant derivative is denoted by lower index separated by semicolon, it is defined as follows:

$$A^{\alpha_1\ldots\alpha_n}{}_{;\beta} = A^{\alpha_1\ldots\alpha_n}{}_{,\beta} + A^{\gamma\alpha_2\ldots\alpha_n}\Gamma^{\alpha_1}_{\gamma\beta} + \cdots + A^{\alpha_1\ldots\alpha_{n-1}\gamma}\Gamma^{\alpha_n}_{\gamma\beta},$$
$$(29)$$

$$A_{\alpha_1\ldots\alpha_n;\beta} = A_{\alpha_1\ldots\alpha_n,\beta} - A_{\gamma\alpha_2\ldots\alpha_n}\Gamma^{\gamma}_{\alpha_1\beta} - \cdots - A_{\alpha_1\ldots\alpha_{n-1}\gamma}\Gamma^{\gamma}_{\alpha_n\beta}.$$
$$(30)$$

Here, Christoffel symbols $\Gamma^{\gamma}_{\alpha\beta}$ are determined as follows:

$$\Gamma^{\gamma}_{\alpha\beta} = \frac{1}{2}g^{\gamma\delta}\left(g_{\delta\alpha,\beta} + g_{\delta\beta,\alpha} - g_{\alpha\beta,\delta}\right). \tag{31}$$

Elementary volume is $\sqrt{g}d^3x$, where g is determinant of $g_{\alpha\beta}$, $d^3x = dx^1dx^2dx^3$. If a surface is determined by equation $x^3 = \text{const.}$, then elementary part of this surface is $\sqrt{\mathbf{g}}d^2x$, where $\mathbf{g}$ is determinant of 2×2 matrix $g_{(\alpha)(\beta)}$, $d^2x = dx^1dx^2$. Hereafter, index in parentheses means index whose values are 1 and 2 only. Covariant vector normal to such surface has the only component n_3.

Thus, in framework of the covariant formalism equations (12)–(14) take the form

$$\sigma^{\alpha\beta} = T^{\alpha\beta} - \Theta^{\alpha\beta\gamma}{}_{;\gamma}, \tag{32}$$

$$\sigma^{\alpha\beta}{}_{;\beta} = 0, \tag{33}$$

$$\oint \left(\sigma^{\alpha\beta} n_\beta + F^\alpha \right) \delta u_\alpha \sqrt{\mathbf{g}}\, d^2 x$$

$$+ \oint \left(\Sigma^{\alpha\beta} + \Theta^{\alpha\beta\gamma} n_\gamma \right) \delta u_{\alpha;\beta} \sqrt{\mathbf{g}} d^2 x = 0. \tag{34}$$

One should stress that equation of the body surface of the form $x^3 = x_S^3 = \text{const}$ is implied here, otherwise integration over x^1 and x^2 is *not* the integration over *body* surface. It should also be noted that the differentiation of scalar in covariant tensor components yields contravariant tensor components. Thus,

$$T^{\alpha\beta} = \frac{\partial \mathcal{H}_B}{\partial u_{\alpha;\beta}}, \tag{35}$$

$$\Theta^{\alpha\beta\gamma} = \frac{\partial \mathcal{H}_B}{\partial u_{\alpha;\beta;\gamma}}, \tag{36}$$

$$F^\alpha = \frac{\partial \mathcal{H}_S}{\partial u_\alpha}, \tag{37}$$

$$\Sigma^{\alpha\beta} = \frac{\partial \mathcal{H}_S}{\partial u_{\alpha;\beta}}. \tag{38}$$

Transformation of the second integral in left-hand side of (34) can be provided quite directly. One should represent $\delta u_{\alpha;\beta}$ in terms of partial derivative and Γ-term

$$\delta u_{\alpha;\beta} = \delta u_{\alpha,\beta} - \delta u_\gamma \Gamma^\gamma_{\alpha\beta}, \tag{39}$$

and separate the case of $\beta = 3$ in the terms contained $\delta u_{\alpha,\beta}$ as follows:

$$(\dots^{\alpha\beta}) \delta u_{\alpha,\beta} = (\dots^{\alpha 3}) \delta u_{\alpha,3} + (\dots^{\alpha(\beta)}) \delta u_{\alpha,(\beta)}. \tag{40}$$

Integral contained $\delta u_{\alpha,(\beta)}$ can be integrated by parts without any tricks. This way it turns out:

$$\oint L^\alpha \delta u_\alpha \sqrt{\mathbf{g}} d^2 x + \oint (\Sigma^{\alpha 3} + \Theta^{\alpha 3\gamma} n_\gamma) \delta u_{\alpha,3} \sqrt{\mathbf{g}} d^2 x = 0, \tag{41}$$

where

$$L^\alpha = \sigma^{\alpha\beta}n_\beta + F^\alpha - \Sigma^{\delta\varepsilon}\Gamma^\alpha_{\delta\varepsilon} - (\Sigma^{\alpha(\beta)}\mathbf{g}^{1/2})_{,(\beta)}\mathbf{g}^{-1/2}$$

$$- \Theta^{\delta\varepsilon\gamma}n_\gamma\Gamma^\alpha_{\delta\varepsilon} - (\Theta^{\alpha(\beta)\gamma}n_\gamma\mathbf{g}^{1/2})_{,(\beta)}\mathbf{g}^{-1/2}. \tag{42}$$

Note that since (B.21) and (B.23) L^α can be rewritten using g rather then $\mathbf{g}$ in last term:

$$L^\alpha = \sigma^{\alpha\beta}n_\beta + F^\alpha - \Sigma^{\delta\varepsilon}\Gamma^\alpha_{\delta\varepsilon} - (\Sigma^{\alpha(\beta)}\mathbf{g}^{1/2})_{,(\beta)}\mathbf{g}^{-1/2}$$

$$- \Theta^{\delta\varepsilon\gamma}n_\gamma\Gamma^\alpha_{\delta\varepsilon} - (\Theta^{\alpha(\beta)\gamma}g^{1/2})_{,(\beta)}g^{-1/2}n_\gamma. \tag{43}$$

From (41) the boundary conditions are obvious and have the following form

$$\Sigma^{\alpha3} + \Theta^{\alpha3\gamma}n_\gamma = 0, \tag{44}$$

$$L^\alpha = 0. \tag{45}$$

Note that since n_α has the only component n_3, (44) can be rewritten in the form

$$\Sigma^{\alpha\beta}n_\beta + \Theta^{\alpha\beta\gamma}n_\beta n_\gamma = 0, \tag{46}$$

which corresponds to (23). As for (45) with (42) or (43) it looks different from (24). Nevertheless it can be proving that (45) is equivalent to (24) (see Appendix C).

Also it is worth to transform the obtained boundary conditions (44) and (45) to the forms used in papers [13] and [14]. In these papers only the case of $F^\alpha = 0$ and $\Sigma^{\alpha\beta} = 0$ are considered, the same condition is assumed while the transformation. Thus, by means of (B.23) equation (44) can be rewritten as

$$\Theta^{\alpha33} = 0. \tag{47}$$

Using the Leibniz rule, equation (B.15) and the representation of partial derivative in terms of covariant derivative, the last term in right-hand side of (43) can be transformed as follows

$$\left(\Theta^{\alpha(\beta)\gamma}g^{1/2}\right)_{,(\beta)}g^{-1/2}n_\gamma = \Theta^{\alpha\beta\gamma}{}_{;\beta}n_\gamma - \Theta^{\delta\beta\gamma}\Gamma^\alpha_{\delta\beta}n_\gamma$$
$$- \Theta^{\alpha\beta\delta}\Gamma^\gamma_{\delta\beta}n_\gamma - \left(\Theta^{\alpha3\gamma}g^{1/2}\right)_{,3}g^{-1/2}n_\gamma. \tag{48}$$

With this identity (45) takes the form

$$\sigma^{\alpha\beta}n_\beta - \Theta^{\alpha\beta\gamma}{}_{;\beta}n_\gamma + \Theta^{\alpha\beta\delta}\Gamma^\gamma_{\delta\beta}n_\gamma + \left(\Theta^{\alpha3\gamma}g^{1/2}\right)_{,3}g^{-1/2}n_\gamma = 0. \tag{49}$$

This coincides with (12) from Ref. [13]. By means of (B.23) and (47) it can be also rewritten in the form

$$\sigma^{\alpha3} - \Theta^{\alpha\beta3}{}_{;\beta} + \Theta^{\alpha(\beta\delta)}\Gamma^3_{(\delta\beta)} + \Theta^{\alpha33}{}_{,3} = 0, \tag{50}$$

where parentheses in $(\delta\beta)$ mean that δ and β cannot be equal to 3 simultaneously. Boundary conditions (47) and (50) are exactly the ones used in Ref. [14].

5. Boundary Conditions in Local Coordinate System

Boundary conditions described above within general formalism are quite complicated. However, they can be essentially simplified using special coordinate system. It should be emphasized that, in general, this coordinate system is different for different points on the boundary surface, this is why we refer it as local coordinate system. Nevertheless, sometimes it can be convenient to find boundary conditions in the local coordinate system and then, if necessary, convert them to any other coordinate system by conventional means. This approach may be useful, particularly, for the computer calculations.

Local Cartesian coordinate system is determined by the condition that z-axis is orthogonal to the boundary surface at the considered point of the surface. Let

$$z = \zeta(x, y), \tag{51}$$

be the equation of the boundary surface in this coordinate system. It can be shown (see Appendix D for details) that the boundary conditions can be written in the following form

$$\Sigma_{iz} + \Theta_{izz} = 0, \tag{52}$$

$$\sigma_{iz} + F_i - \Sigma_{i(j),(j)} - \Theta_{i(j)z,(j)} + \Theta_{i(j)(k)}\zeta_{,(j),(k)} = 0, \tag{53}$$

where Cartesian indices in parentheses run only the values of x and y, it is implied that body bulk corresponds to the condition $z < \zeta(x,y)$. Note that (53) contains the values $\zeta_{,x,x}$, $\zeta_{,y,y}$, and $\zeta_{,x,y}$. By rotating the local coordinate system around z-axis it is possible to achieve that $\zeta_{,x,y}$ become zero. In this case remaining $\zeta_{,x,x}$ and $\zeta_{,y,y}$ are nothing but the inverse radii of the body surface curvature.

6. Conclusion

It is described above how one can generalize classical theory of elastic continuum to a case when thermodynamic potential depends on the strain and the strain gradients. In this case, in fact, one gets such a type of classical (non-quantum) field theory where function which determinates field equations ("Lagrangian") depends on the first and second derivatives of the initial field, particularly the elastic displacement field u_i. Certainly in the static case "Lagrangian" is nothing but the thermodynamic potential.

At the same time it should be noted that in essence we did not use the fact that the field u_i is a field of elastic displacements. All this is also true in cases where the vector field u_i has another physical nature if only "Lagrangian" has the form that we have expected. It is not a significant problem to add also a "mass term" (in terms of the quantum field theory) $a_{ij}u_iu_j$ where a_{ij} is some constant tensor. It will not change much, only differential equations will be slightly (and obviously) modified while the boundary conditions will remain the same. Such opportunities should be kept in mind, but practically considered theory is needed to describe namely the elastic fields, and mostly in relation to the flexoelectric effect.

It is shown in this chapter that in theories of this type, particularly in generalized theory of elasticity, differential equations can be derived in quite conventional manner. The only difference with respect to ordinal case is that while varying one should integrate by parts twice. As for the derivation of the boundary conditions for these differential equations, the situation is more complicated. We have considered two ways by which such derivation can be done. The first way uses a mathematical trick based on the use of additional surface coordinate system. The second way is more direct, but it involves the use of a covariant tensor analysis and a curvilinear coordinate system of a special class. In the coordinate systems of this class equation of boundary surface should have the form $x^3 = x_S^3$ where x_S^3 is some constant. For a complex surface shape it may be difficult to find a single coordinate system of this type, but in this case it is possible to divide the surface to such parts, for which the finding of the coordinate system of the desired type is not very difficult. It is, however, necessary to ensure that these parts are sewed sufficiently smoothly.

It should be noted also that in the considered case the differential equations and especially boundary conditions are very complicated. With regard to the calculation of the flexoelectric deformations corresponding boundary problem was solved for a homogeneously polarized ball of an isotropic material.[13] Even in this simple geometry one has to use a representation of the radial functions in the form of power series. To solve more complex boundary problems it is even more difficult. But for the calculation of specifically flexoelectric deformations it was proposed as an approximate method[14] which radically simplifies the problem. Corresponding theory is considered in detail in another chapter.[15]

Acknowledgments

A.K. Tagantsev is acknowledged for reading the manuscript.

Appendix A. Relation between Tensor γ_{ij} and Gauss Second Fundamental Form

By means of Leibniz rule equation (26), which defines the tensor γ_{ij}, can be rewritten as follows:

$$\gamma_{ij} = n_{j,\alpha} e_i^\alpha + n_j e_{i,\alpha}^\alpha + n_j e_i^\alpha \frac{1}{2g} g_{,\alpha}. \tag{A.1}$$

It can been shown (see Appendix B) that

$$g_{,\alpha} = g g^{\beta\gamma} g_{\beta\gamma,\alpha}. \tag{A.2}$$

Representing metric tensor and reverse metric tensor in terms of surface frame vectors, we find:

$$g_{,\alpha} = 2g e_l^\beta e_l^\gamma e_{k\gamma} e_{k\beta,\alpha}. \tag{A.3}$$

This equation can be essentially simplified. By means of the relation $e_l^\gamma e_{k\gamma} = \delta_{kl} - n_l n_k$, it turns out

$$g_{,\alpha} = 2g \left(e_k^\beta e_{k\beta,\alpha} - e_l^\beta e_{k\beta,\alpha} n_l n_k \right). \tag{A.4}$$

Any surface frame vector is orthogonal to vector n_l normal to the surface. Thus, second term in parentheses is identically zero and it turns out to be:

$$g_{,\alpha} = 2g e_k^\beta e_{k\beta,\alpha}. \tag{A.5}$$

Thus, (A.1) can be rewritten as follows:

$$\gamma_{ij} = n_{j,\alpha} e_i^\alpha + n_j e_{i,\alpha}^\alpha + n_j e_i^\alpha e_k^\beta e_{k\beta,\alpha}. \tag{A.6}$$

Second term in right-hand side of (A.6) can be transformed in such manner that it appears to cancel out the third term. First, we transform it as follows:

$$n_j e_{i,\alpha}^\alpha = n_j \delta_{ik} e_{k,\alpha}^\alpha = n_j e_i^\beta e_{k\beta} e_{k,\alpha}^\alpha + n_i n_j n_k e_{k,\alpha}^\alpha. \tag{A.7}$$

From the fact that $e_{k\beta} e_{k\beta}^\alpha = \delta_\beta^\alpha$ does not depend on x^α, and obvious relation $e_{k\beta,\alpha} = e_{k\alpha,\beta}$, it follows that $e_{k\beta} e_{k,\alpha}^\alpha = -e_{k\alpha,\beta} e_k^\alpha$. Thus, it turns out:

$$n_j e_{i,\alpha}^\alpha = n_i n_j n_k e_{k,\alpha}^\alpha - n_j e_i^\alpha e_{k\beta,\alpha} e_k^\beta. \tag{A.8}$$

Last term in right-hand side of this equation obviously cancels out the third term in right-hand side of (A.6). Thus,

$$\gamma_{ij} = n_{j,\alpha} e_i^\alpha + n_i n_j n_k e_{k,\alpha}^\alpha. \tag{A.9}$$

The first term in right-hand side of (A.9) can be related to Gauss second fundamental form $b_{\alpha\beta}$, which is defined as follows:

$$b_{\alpha\beta} = n_k e_{k\beta,\alpha}. \tag{A.10}$$

Since $n_k e_{k\beta} = 0$, $n_{k,\alpha} e_{k\beta} = -n_k e_{k\beta,\alpha}$. Convolving this equation with e_j^β and taking into account that $e_{k\beta} e_j^\beta = \delta_{jk} - n_j n_k$, we find:

$$n_{j,\alpha} - n_{k,\alpha} n_j n_k = -n_k e_{k\beta,\alpha} e_j^\beta. \tag{A.11}$$

From $n_k n_k = 1$, it follows that $n_{k,\alpha} n_k = 0$. Thus,

$$n_{j,\alpha} = -n_k e_{k\beta,\alpha} e_j^\beta = -b_{\alpha\beta} e_j^\beta. \tag{A.12}$$

In the theory of surfaces this equation is known as Weingarten equation. By means of it we find:

$$n_{j,\alpha} e_i^\alpha = -B_{ij}, \tag{A.13}$$

where

$$B_{ij} = b_{\alpha\beta} e_i^\alpha e_j^\beta. \tag{A.14}$$

It should be emphasized that three-dimensional tensor B_{ij} is not a tensor of second fundamental form $b_{\alpha\beta}$ itself, although it is related to $b_{\alpha\beta}$.

The second term in right-hand side of (A.9) can be also represented in terms of B_{ij}. Obviously $B_{ii} = b_{\alpha\beta} g^{\alpha\beta}$. Next,

$$n_i n_j n_k e_{k,\alpha}^\alpha = n_i n_j n_k g^{\alpha\beta}{}_{,\alpha} e_k^\beta + n_i n_j n_k g^{\alpha\beta} e_{k\beta,\alpha}. \tag{A.15}$$

The first term in right-hand side of this equation is zero identically, because $n_k e_k^\beta = 0$. It remains to be the only term, which contains $n_k e_{k\beta,\alpha} = b_{\alpha\beta}$. This eventually yields the trace of the tensor B_{ij}. Finally it turns out:

$$\gamma_{ij} = n_i n_j B_{kk} - B_{ij}. \tag{A.16}$$

Appendix B. Curvilinear Coordinates and Covariant Formalism

It is convenient to assume that besides curvilinear coordinate system there is Cartesian coordinate system also. We introduce frame vectors $e_{i\alpha}$ as follows

$$e_{i\alpha} = \frac{\partial r_i}{\partial x^\alpha} = r_{i,\alpha}, \tag{B.1}$$

where r_i are Cartesian coordinates of a point of volume, Greek indices distinguished curvilinear coordinates run three values: 1, 2, and 3. Consequently,

$$(d\mathbf{r})^2 = e_{i\alpha}e_{i\beta}dx^\alpha dx^\beta. \tag{B.2}$$

Thus, metric tensor $g_{\alpha\beta}$ can be represented in the following form

$$g_{\alpha\beta} = e_{i\alpha}e_{i\beta}. \tag{B.3}$$

Coordinates x^α are curvilinear, so that the basis set $e_{i\alpha}$ is neither orthogonal nor normalized. Therefore, one should introduce a mutual basis set e_i^α, which is defined by relations

$$e_i^\alpha e_{i\beta} = \delta_\beta^\alpha, \tag{B.4}$$

where δ_β^α is Kronecker delta in Greek indices. By means of the reverse metric tensor $g^{\alpha\beta}$ mutual reference vectors e_i^α can be represented in terms of $e_{i\alpha}$:

$$e_i^\alpha = g^{\alpha\beta}e_{i\beta}. \tag{B.5}$$

Using $g_{\alpha\beta}$ and $g^{\alpha\beta}$ curvilinear indices can be raised and lowered in the standard manner.

Convolving (B.4) with $e_{j\alpha}$ we get that $e_i^\alpha e_{j\alpha}e_{i\beta} = e_{j\beta}$. This identity should be valid for any $e_{j\beta}$. This is why

$$e_i^\alpha e_{j\alpha} = \delta_{ij}. \tag{B.6}$$

The curvilinear contravariant tensor components are defined by Cartesian tensor components as follows:

$$A^{\alpha_1 \ldots \alpha_n} = A_{i_1 \ldots i_n} e_{i_1}^{\alpha_1} \ldots e_{i_n}^{\alpha_n}. \tag{B.7}$$

The curvilinear covariant tensor components are defined analogously:

$$A_{\alpha_1\ldots\alpha_n} = A_{i_1\ldots i_n} e_{i_1\alpha_1} \cdots e_{i_n\alpha_n}. \tag{B.8}$$

By analogy it is clear how mixed tensor components are defined. By means of (B.6) it is obvious how curvilinear tensor components can be transformed back to Cartesian tensor components. Say for a tensor in covariant components this transformations looks as follows:

$$A_{i_1\ldots i_n} = A_{\alpha_1\ldots\alpha_n} e_{i_1}^{\alpha_1} \cdots e_{i_n}^{\alpha_n}. \tag{B.9}$$

Tensor convolutions are transformed by means of (B.6). For example

$$A_{\ldots i\ldots} B_{\ldots i\ldots} = A_{\ldots i\ldots} \delta_{ij} B_{\ldots j\ldots} = A_{\ldots i\ldots} e_i^\alpha e_{j\alpha} B_{\ldots j\ldots}. \tag{B.10}$$

Thus it is obtained automatically that only upper and lower Greek indices can be convolved.

Gradients of tensors are transformed in a special way. Since the frame vectors are not constant they cannot be moved through the derivative in straightforward manner, additional terms appear during this motion. For example gradient of vector is transformed as follows:

$$A_{i,j} e_{i\alpha} e_{j\beta} = A_{\alpha,\beta} - A_\gamma \Gamma^\gamma_{\alpha\beta}, \tag{B.11}$$

where

$$\Gamma^\gamma_{\alpha\beta} = e_i^\gamma e_{i\alpha,j} e_{j\beta} = e_i^\gamma e_{i\alpha,\beta}. \tag{B.12}$$

The expression in the right-hand side of (B.11) is called the covariant derivative and is denoted by index separated by a semicolon:

$$A_{\alpha;\beta} = A_{\alpha,\beta} - A_\gamma \Gamma^\gamma_{\alpha\beta}. \tag{B.13}$$

Set of values $\Gamma^\gamma_{\alpha\beta}$ is called the Christoffel symbols. They can be expressed not only in terms of frame vectors (B.12) but also in terms of metric tensor

$$\Gamma^\gamma_{\alpha\beta} = \frac{1}{2} g^{\gamma\delta} (g_{\delta\alpha,\beta} + g_{\delta\beta,\alpha} - g_{\alpha\beta,\delta}). \tag{B.14}$$

It is easy to see that this Christoffel symbols are symmetric in the lower indices. There is also useful equation

$$\Gamma^{\beta}_{\alpha\beta} = \frac{g_{,\alpha}}{2g} = (\ln \sqrt{g})_{,\alpha}, \tag{B.15}$$

which can be obtained if one differentiate $\ln g = \mathrm{Sp} \ln g_{\alpha\beta}$, $g = \det g_{\alpha\beta}$ hereafter.

The covariant derivative of contravariant vector should be defined in another manner. The reason is that frame vector e^{α}_i should be moved through the derivative rather than $e_{i\alpha}$. Note that $e^{\alpha}_i e_{i\beta}$ is a constant (see equation (B.4)), so that its partial derivative is zero. Therefore, $e^{\alpha}_{i,\gamma} e_{i\beta} = -e^{\alpha}_i e_{i\beta,\gamma}$ and from this identity and simple algebra it turns out:

$$A_{i,j} e^{\alpha}_i e_{j\beta} = A^{\alpha}_{;\beta} = A^{\alpha}_{,\beta} + A^{\gamma}\Gamma^{\alpha}_{\gamma\beta}. \tag{B.16}$$

From the above it is quite clear how to determine the covariant derivative of higher rank tensors. The terms containing the Christoffel symbols (so-called Γ-terms) appear due to frame vectors motion through the derivative. If the tensor is of the highest rank then several frame vectors should be moved through derivative. Obviously several Γ-terms should appear in this way. For example covariant derivative of two-rank tensor with lower indices is

$$A_{\alpha\beta;\gamma} = A_{\alpha\beta,\gamma} - A_{\delta\beta}\Gamma^{\delta}_{\alpha\gamma} - A_{\alpha\delta}\Gamma^{\delta}_{\beta\gamma}. \tag{B.17}$$

What happens in the other cases is clear by analogy. Covariant derivative of a scalar is determined as a usual partial derivative.

It is easy to show that the covariant derivative of the product obeys the Leibniz rule

$$(A_{...}B_{...})_{;\alpha} = A_{...;\alpha}B_{...} + A_{...}B_{...;\alpha}. \tag{B.18}$$

It can be also proved in straightforward manner that the covariant derivative of frame vectors is identically zero. Since metric tensor can be represented as the convolution (in Cartesian indices) of frame vectors, its covariant derivative is identically zero also. It is also worth that if some tensor is constant in Cartesian components (say it is a material tensor of homogeneous media) then in curvilinear components its covariant (but not usual partial) derivatives are zero.

In the general case of curved space covariant derivatives do not commute, their commutator is proportional to the Riemann curvature tensor. However, here we are interested only in the case of curvilinear coordinates in flat space. In the flat space Riemann tensor is identically zero. Therefore, the covariant derivatives commute.

To complete this brief overview of covariant tensor analysis it remains to say few words about integration. As for the volume integration everything is quite simple. It is easy to see that the Jacobian of the transformation from Cartesian coordinates to curvilinear ones is the determinant of $e_{i\alpha}$. However, it is more convenient and generally accepted to express this Jacobian in terms of metric tensor. It is clear from (B.3) that the square of the Jacobian is equal to g. Thus,

$$dV = \sqrt{g}\, d^3x, \tag{B.19}$$

where $d^3x = dx^1 dx^2 dx^3$. This formula is enough for volume integration, at least in the part that we need.

As for the integration over surface, the situation is somewhat more complicated. Here, we restrict ourselves to the case when the equation of the surface has a simple form $x^3 = \text{const}$. The more general case, we do not need.

If the equation of the surface has the simple form mentioned above then the integration over the surface reduces to integration over x^1 and x^2. Let us introduce additional notation: Greek indices in parentheses have the values 1 and 2 only. In this notation it turns out

$$dS = \sqrt{\mathbf{g}}\, d^2x, \tag{B.20}$$

where $d^2x = dx^1 dx^2$, $\mathbf{g}$ is determinant of 2×2 matrix $g_{(\alpha)(\beta)}$. Note that according to the known representation of the inverse matrix by algebraic adjuncts

$$\mathbf{g} = g^{33} g. \tag{B.21}$$

Surface integrals often contain an unit vector normal to the surface. Thus, we also need the expression for such a vector. It is clear geometrically that in the considered special coordinate system the vectors e_{i1} and e_{i2} are tangential to the surface. Therefore, in accordance with (B.4) the vector e_i^3 is normal to the surface. By making the normalization of this vector, we obtain the unit vector normal to the surface in Cartesian components:

$$n_i = \frac{e_i^3}{\sqrt{e_i^3 e_i^3}} = \frac{e_i^3}{\sqrt{g^{33}}}. \tag{B.22}$$

Conversion of this vector to the curvilinear components by means of general equations yields the following:

$$n_\alpha = n_i e_{i\alpha} = \frac{e_i^3 e_{i\alpha}}{\sqrt{g^{33}}} = \frac{\delta_\alpha^3}{\sqrt{g^{33}}}, \tag{B.23}$$

$$n^\alpha = g^{\alpha\beta} n_\beta = \frac{g^{\alpha 3}}{\sqrt{g^{33}}}. \tag{B.24}$$

The above brief consideration of the covariant formalism is sufficient for the purposes of this chapter. For more details one should refer to any textbook on general relativity.

Appendix C. Equivalence of Boundary Conditions Derived by Two Different Approaches

It was mentioned in main text that equivalence of (44) to (23) is obvious. Otherwise, equivalence of (45) to (24) is not obvious and require a proof. It is described here.

Before the description of the proof itself we should discuss some geometrical matter. The curvilinear coordinate system used in this section is such that equation of body surface is $x^3 = $ const. Thus, first two coordinates $x^{(\alpha)}$ of this coordinate system may be well used as surface coordinates system used in Section 3. Nevertheless, in this case one should distinguish some objects depending on whether $x^{(\alpha)}$ are used as first two coordinates of spatial coordinate system or as the surface coordinate system in manner of Section 3. For such a distinction here we add additional tilde to objects related to the surface coordinate system.

Note that $\tilde{e}_{i(\alpha)} = e_{i(\alpha)}$ but $\tilde{e}_i^{(\alpha)} \neq e_i^{(\alpha)}$. This is a consequence of the fact that $\tilde{g}^{(\alpha)(\beta)} \neq g^{(\alpha)(\beta)}$. By means of simple geometrical reasoning one can see that

$$\tilde{e}_i^{(\alpha)} = e_i^{(\alpha)} - \frac{g^{(\alpha)3}}{g^{33}} e_i^3. \tag{C.1}$$

Note also that

$$\tilde{g} = \mathbf{g}. \tag{C.2}$$

Now one can turn to the proof itself. Let us denote last two terms in (24) as ϕ_i. By means of (C.1) and (C.2) we can rewrite it in terms of three-dimensional coordinates:

$$\phi_i = - \left(\Sigma_{ij} e_j^{(\alpha)} \mathbf{g}^{1/2} \right)_{,(\alpha)} \mathbf{g}^{-1/2} + \left(\Sigma_{ij} \frac{g^{(\alpha)3}}{g^{33}} e_j^3 \mathbf{g}^{1/2} \right)_{,(\alpha)} \mathbf{g}^{-1/2}$$

$$- \left(\Theta_{ijk} n_k e_j^{(\alpha)} \mathbf{g}^{1/2} \right)_{,(\alpha)} \mathbf{g}^{-1/2} + \left(\Theta_{ijk} n_k \frac{g^{(\alpha)3}}{g^{33}} e_j^3 \mathbf{g}^{1/2} \right)_{,(\alpha)} \mathbf{g}^{-1/2}. \tag{C.3}$$

Second and fourth terms in right-hand side of this equation can be combined into one term

$$\left[\left(\Sigma_{ij} e_j^3 + \Theta_{ijk} n_k e_j^3 \right) \frac{g^{(\alpha)3}}{g^{33}} \mathbf{g}^{1/2} \right]_{,(\alpha)} \mathbf{g}^{-1/2}. \tag{C.4}$$

Since $e_i^3 \sim n_i$ the expression in parentheses is zero due to (23). Moreover, since it is zero everywhere on the surface (i.e. for every $x^{(\alpha)}$), its derivative in $x^{(\alpha)}$ is zero also. Thus, (C.4) is zero i.e. second and fourth terms in (C.3) cancel each other out completely. Thus,

$$\phi_i = - \left(\Sigma_{ij} e_j^{(\beta)} \mathbf{g}^{1/2} \right)_{,(\beta)} \mathbf{g}^{-1/2} - \left(\Theta_{ijk} n_k e_j^{(\beta)} \mathbf{g}^{1/2} \right)_{,(\beta)} \mathbf{g}^{-1/2}. \tag{C.5}$$

Further one can easily calculate contravariant representation of vector ϕ_i. It turns out:

$$\phi^\alpha = e^\alpha_{i,(\beta)} e^{(\beta)}_j \left(\Sigma_{ij} + \Theta_{ijk} n_k\right)$$
$$- \left(\Sigma^{\alpha(\beta)} \mathbf{g}^{1/2}\right)_{,(\beta)} \mathbf{g}^{-1/2} - \left(\Theta^{\alpha(\beta)\gamma} n_\gamma \mathbf{g}^{1/2}\right)_{,(\beta)} \mathbf{g}^{-1/2}.$$

$$(C.6)$$

In the first term of right-hand side of this equation range of the index β can be expanded to complete set 1, 2, 3. This is a consequence of (23) and the fact that $e^3_i \sim n_i$. Further one should note that

$$e^\alpha_{i,\beta} e^\beta_j = e^\alpha_{l,\beta} \delta_{li} e^\beta_j = e^\alpha_{l,\beta} e_{l\delta} e^\delta_i e^\beta_j = -\Gamma^\alpha_{\beta\delta} e^\delta_i e^\beta_j. \qquad (C.7)$$

With this identity equation (C.6) takes the following form

$$\phi^\alpha = -\Gamma^\alpha_{\delta(\beta)} \left(\Sigma^{\delta(\beta)} + \Theta^{\delta(\beta)\gamma} n_\gamma\right)$$
$$- (\Sigma^{\alpha(\beta)} \mathbf{g}^{1/2})_{,(\beta)} \mathbf{g}^{-1/2} - (\Theta^{\alpha(\beta)\gamma} n_\gamma \mathbf{g}^{1/2})_{,(\beta)} \mathbf{g}^{-1/2}. \qquad (C.8)$$

Note that range of index β in first term of right-hand side of this equation can be extended to complete set 1, 2, 3. This is a consequence of the equation (44). With this extension further elementary transformations yield exactly (45). Proof is completed.

Appendix D. Some Specific Coordinate System

Here, we apply general theory described in Section 4 to the case when curvilinear coordinates have specific form. In this case, equations, which present boundary condition, can be essentially simplified. Moreover, in this case surface curvature enter into equations explicitly.

Let

$$z = \zeta(x, y), \qquad (D.1)$$

be the equation of body surface in Cartesian coordinates x, y, z. We introduce curvilinear coordinate system x^1, x^2, x^3 as follows:

$$x = x^1, \quad y = x^2, \quad z = x^3 + \zeta(x^1, x^2). \qquad (D.2)$$

In this curvilinear coordinates equation of body surface is $x^3 = 0$. Thus, these coordinates are appropriate for the theory described in Section 4.

By direct calculations in the framework of the theory described in Appendix B one can easily get the following equations:

$$
\begin{aligned}
&e_{x1} = 1, &&e_{x2} = 0, &&e_{x3} = 0, \\
&e_{y1} = 0, &&e_{y2} = 1, &&e_{y3} = 0, \\
&e_{z1} = \zeta_{,x}, &&e_{z2} = \zeta_{,y}, &&e_{z3} = 1, \\
&g_{11} = 1 + \zeta_{,x}^2, &&g_{22} = 1 + \zeta_{,y}^2, &&g_{33} = 1, \\
&g_{23} = \zeta_{,y}, &&g_{13} = \zeta_{,x}, &&g_{12} = \zeta_{,x}\zeta_{,y}, \\
&g = 1, &&\mathbf{g} = 1 + \zeta_{,x}^2 + \zeta_{,y}^2, \\
&g^{11} = 1, &&g^{22} = 1, &&g^{33} = 1 + \zeta_{,x}^2 + \zeta_{,y}^2, \\
&g^{23} = -\zeta_{,y}, &&g^{13} = -\zeta_{,x}, &&g^{12} = 0, \\
&e_x^1 = 1, &&e_x^2 = 0, &&e_x^3 = -\zeta_{,x}, \\
&e_y^1 = 0, &&e_y^2 = 1, &&e_y^3 = -\zeta_{,y}, \\
&e_z^1 = 0, &&e_z^2 = 0, &&e_z^3 = 1.
\end{aligned}
\tag{D.3}
$$

By means of these equations it is easy to see that there are only three non-zero Christoffel symbols:

$$
\Gamma_{11}^3 = \zeta_{,x,x}, \quad \Gamma_{22}^3 = \zeta_{,y,y}, \quad \Gamma_{12}^3 = \Gamma_{21}^3 = \zeta_{,x,y}.
\tag{D.4}
$$

Now, note that if we consider the point of the surface, where $\zeta_{,x} = \zeta_{,y} = 0$ then equations (D.3) become much simpler. Moreover, in this point we cannot distinguish curvilinear covariant, curvilinear contravariant and Cartesian tensor components. All of them are the same. However, it should be emphasized that latter *does not mean* that in curvilinear coordinates covariant derivatives can be replaced by usual partial derivatives, components of partial derivatives in Cartesian coordinates are equal to components of *covariant* derivatives in curvilinear coordinates. Note also that in any representation (covariant, contravariant or Cartesian) unit vector normal to the surface has the only component $n_3 = n^3 = n_z = 1$. We imply here that body bulk corresponds to the condition $z < \zeta(x, y)$.

Without loss of generality, we can consider only the point where $\zeta_{,x} = \zeta_{,y} = 0$. Indeed, for any point of the surface it is always possible to find a Cartesian coordinate system in which z-axis is normal to the surface at this point. In this case it is obvious that $\zeta_{,x} = \zeta_{,y} = 0$. Such coordinate system we further refer as local coordinate system. After finding the boundary conditions in the local coordinate system, they can be converted to any other coordinate system by standard way. Sometimes this approach may be convenient in practice. Moreover, in this case it is possible to associate Christoffel symbols, which enter into the boundary conditions, with the curvature of the body surface. Indeed, by rotating the local coordinate system around z-axis, it is always possible to achieve vanishing of $\zeta_{,x,y}$. In this case only Γ^3_{11} and Γ^3_{22} are non-zero and they are obviously equal to pair of inverse radii of body surface curvature.

It is easy to see that in local coordinate system $g_{,(\beta)} = \mathbf{g}_{,(\beta)} = 0$. Thus, general equation (43) can be rewritten as follows:

$$L^\alpha = \sigma^{\alpha\beta} n_\beta + F^\alpha - \Sigma^{\delta\varepsilon} \Gamma^\alpha_{\delta\varepsilon} - \Sigma^{\alpha(\beta)}{}_{,(\beta)} - \Theta^{\delta\varepsilon\gamma} \Gamma^\alpha_{\delta\varepsilon} n_\gamma - \Theta^{\alpha(\beta)\gamma}{}_{,(\beta)} n_\gamma.$$

$$(D.5)$$

However, in order to be able to use the above-mentioned direct correspondence between Cartesian and curvilinear components of the tensors, it is necessary to express the usual partial derivatives in terms of covariant derivatives and Christoffel symbols. Keeping in mind that only $\Gamma^3_{(\alpha)(\beta)}$ are non-zero, by direct calculations we easily get following equations:

$$\Sigma^{\delta\varepsilon} \Gamma^\alpha_{\delta\varepsilon} + \Sigma^{\alpha(\beta)}{}_{,(\beta)} = \Sigma^{\alpha(\beta)}{}_{;(\beta)}, \tag{D.6}$$

$$\Theta^{\delta\varepsilon\gamma} \Gamma^\alpha_{\delta\varepsilon} n_\gamma + \Theta^{\alpha(\beta)\gamma}{}_{,(\beta)} n_\gamma = \Theta^{\alpha(\beta)\gamma}{}_{;(\beta)} n_\gamma - \Theta^{\alpha(\beta)(\delta)} \Gamma^\gamma_{(\beta)(\delta)} n_\gamma.$$

$$(D.7)$$

Thus, in local coordinate system boundary condition $L^\alpha = 0$ can be rewritten as follows:

$$\sigma^{\alpha\beta} n_\beta + F^\alpha - \Sigma^{\alpha(\beta)}{}_{;(\beta)} - \Theta^{\alpha(\beta)\gamma}{}_{;(\beta)} n_\gamma + \Theta^{\alpha(\beta)(\delta)} \Gamma^\gamma_{(\beta)(\delta)} n_\gamma = 0.$$

$$(D.8)$$

The remaining boundary conditions

$$\Sigma^{\alpha\beta} n_\beta + \Theta^{\alpha\beta\gamma} n_\beta n_\gamma = 0, \tag{D.9}$$

are quite simple and does not require conversion.

Using the above-mentioned correspondence between the tensor components in Cartesian and curvilinear coordinates as well as the expressions for $\Gamma^3_{(\alpha)(\beta)}$ and n_α, equations (D.8) and (D.9) can be easily rewritten in Cartesian tensor component as follows:

$$\sigma_{iz} + F_i - \Sigma_{i(j),(j)} - \Theta_{i(j)z,(j)} + \Theta_{i(j)(k)}\zeta_{,(j),(k)} = 0, \tag{D.10}$$

$$\Sigma_{iz} + \Theta_{izz} = 0. \tag{D.11}$$

Here, Cartesian indices in parentheses run only the values of x and y. In this equation, values $\zeta_{,(j),(k)}$, characterizing surface curvature, are included explicitly.

References

1. P. V. Yudin and A. K. Tagantsev. *Nanotechnology.* **24**, 432001–432036 (2013).
2. P. Zubko, G. Catalan, and A. K. Tagantsev. *Ann. Rev. Mater. Res.* **43**, 387–421 (2013).
3. R. Toupin. *Arch. Ration. Mech. Anal.* **11**(1), 385–414 (1962).
4. R. Toupin. *Arch. Ration. Mech. Anal.* **17**(2), 85–112 (1964).
5. R. D. Mindlin and N. N. Eshel. *Int. J. Solid Struct.* **4**, 109–124 (1968).
6. P. Germain. *SIAM J. Appl. Math.* **25**(3), 556–575 (1973).
7. H. S. Ling and S. S. Ping. *Sci. China Phys. Mech. Astron.* **53**(8), 1497–1504 (2010).
8. V. L. Indenbom, E. B. Loginov, and M. A. Osipov. *Kristallografiya.* **26**(6), 1157–1162 (1981).
9. R. Maranganti, N. D. Sharma, and P. Sharma. *Phys. Rev. B.* **74**, 014110 (2006).
10. N. D. Sharma, R. Maranganti, and P. Sharma. *J. Mech. Phys. Solids.* **55**, 2328–2350 (2008).
11. M. S. Majdoub, P. Sharma, and T. Cagin. *Phys. Rev. B.* **77**, 125424 (2008).
12. A. S. Yurkov. *JETP Lett.* **94**(6), 455–458 (2011).
13. A. S. Yurkov. Flexoelectric deformation of a homogeneously polarized ball. *arXiv:1304.1868* [*cond-mat.mtrl-sci*]. (2013).

14. A. S. Yurkov. *JETP Lett.* **99**, 214–218 (2014).
15. A. S. Yurkov. Flexoelectric deformations of finite-size bodies in framework of the theory of continuum. In *Flexoelectricity in Solids: From Theory to Applications*. Ed. by A. K. Tagantsev and P. V. Yudin. World Scientific, Singapore (2016).

Chapter 5

Flexoelectric Deformations of Finite-Size Bodies in Framework of the Theory of Continuum

A. S. Yurkov

Omsk, Russia 644076

In many cases the correct theoretical description of flexoelectricity requires the consideration of the finite size of a body and is reduced to the solution of boundary problems for partial differential equations. Generally speaking, in this case one should solve jointly the equations of polarization equilibrium and equations of elastic equilibrium. However, due to the fact that typically flexoelectric moduli are very small, usually one can consider the solution of polarization equilibrium equations at a given elastic strain (direct flexoelectric effect) or the solution of elastic equilibrium equations at given polarization (converse flexoelectric effect). Derivation of the polarization equilibrium equations and boundary conditions for them can be made in the quite usual way. Solution of these equations usually is not too difficult problem. On the contrary, description of converse flexoelectric effect is a more complicated problem. *Inter alia* effective solution of corresponded boundary problems requires the development of special mathematical methods. The subject of this chapter is a detailed discussion of the relevant theory focusing on the converse flexoelectric effect. Also considered are some particular examples illustrating the application of the general theory.

1. Introduction

A body in which flexoelectrical phenomena are observed always has finite size. In some cases the finite sample size is not significant, and the sample can be described as infinite medium. Examples are the influence of the flexoelectric effect on the dispersion of acoustic phonons[1] or its influence on structure of ferroelectric domain wall.[2] In other cases, such as flexoelectric bending of a plate,[3] finite size of a sample plays fundamental role.

Obviously, the effects associated with the finite size of the sample are caused by the presence of the sample boundary. In particular, this raises the surface piezoelectricity which, as it turns out, leads to the effects of the same order as the flexoelectric effect.[4]

Less trivial fact is that even without surface piezoelectricity, in the presence of flexoelectricity classical theory of elasticity should be modified. In this case it turns out that, generally speaking, one should use elastic boundary conditions of non-classical form. These boundary conditions are consistent only if one takes into account the contribution to the thermodynamic potential which is a bilinear in elastic strain gradients.[5] The latter leads to not only the boundary conditions, but also that the differential equations of elastic equilibrium should be non-classical.

Equations of elastic equilibrium are needed to describe the converse flexoelectric effect (mechanical response to the polarization). As for direct flexoelectric effect (polarization response to an inhomogeneous elastic strain), to describe it the equations of polarization equilibrium are needed. It is essential that these equations and boundary conditions for them are modified by flexoelectricity not so radically as the equations and boundary conditions describing the elastic response (converse flexoelectric effect). This is related to the fact that the flexoelectric contributions to the thermodynamic potential contains only the first spatial derivatives of the polarization, while it contains the second spatial derivatives of the elastic displacements. Equations of polarization equilibrium and boundary conditions for them in the presence of flexoelectricity were derived in paper [6].

Thus, non-trivial features of flexoelectricity in finite samples reveal themselves mainly in the converse flexoelectric effect. This is why we do not discuss equations of polarization equilibrium, focusing on the converse flexoelectric effect at a given polarization. Moreover, in the particular examples (but not in the general theory) we assume homogeneous polarization at which non-trivial features of the flexoelectric effect in finite samples reveal itself most dramatically.

The case of homogeneous polarization is interesting in the sense that a naive analysis based on constitutive equations leads to the (erroneous) conclusion that the homogeneous polarization does not cause flexoelectric deformation of a body. The consequence of this conclusion is the ability to create a sensor which does not behave as an actuator.[7,8] Such an ability contradicts the general principles of thermodynamics. But in reality violations of the principles of thermodynamics is not happening, homogeneous polarization does lead to flexoelectric deformation of the body.[4,5]

In the paper[4] the conclusion that homogeneous polarization leads to flexoelectric deformation of a thin plate was made on the basis of the particular ansatz for elastic displacement distribution over the sample and direct minimization of the thermodynamic potential. A more rigorous analysis of this problem requires the solution of differential equations of elastic equilibrium with appropriate boundary conditions. Such a solution for the case of homogeneously polarized ball was presented in paper.[9] As was expected the solution confirmed the presence of flexoelectric strains caused by homogeneous polarization.

It is worth mentioning that in the presence of flexoelectricity the exact (within the framework of the continuum media theory) equations of elastic equilibrium are too complex to be used for a case different from a simple case of a ball. Even for a ball the solution is very cumbersome and requires an introduction of non-standard functions presented by the series in powers of radial coordinate. Thus, the development of approximate method is worth. Such a method was developed in the paper.[10] In this paper, it has been shown that elastic displacements, produced by flexoelectricity in a finite-size body, is approximately equal to the sum of two parts. The first part, which was called non-classical, can be found by solution of fairly simple system of one-dimensional equations. This part is concentrated near the surface of the body and decays exponentially inside the body. The second, classical part can be found by means of the equations of the classical theory of elasticity. Boundary conditions for these equations have a standard classical form of the boundary conditions for the body under the external force on the surface. Physically there is no force on surface, these forces are formal in nature, they describe the interaction between classical and non-classical part of the elastic displacement and are proportional to the latter.

Papers on the subject[9,10] are too brief. More detailed description is presented here. General aspects of derivation of the equations of elastic equilibrium and the boundary conditions for them in the case where the thermodynamic potential depends on higher derivatives discussed in another chapter.[12] Here we apply these results to the case of flexoelectricity and develop the theory further. Besides, it is presented as a comparison of the exact solution[9] and the solution of the same problem within the approximate theory.[10] Some additional examples of approximate theory application are also presented.

2. Thermodynamic Potential, Differential Equations of Elastic Equilibrium, and Boundary Condition

It is well known[11] that within the theory of continuum, flexoelectricity is described by a contribution to thermodynamic potential density:

$$\mathcal{H}_{flx} = -f^{(1)}_{klij} P_k u_{i,j,l} - f^{(2)}_{klij} P_{k,l} u_{i,j}. \tag{1}$$

Here P_k is polarization, u_i is elastic displacement, $f^{(1)}_{klij}$ and $f^{(2)}_{klij}$ are material tensors, $(\dots)_{,i} = \partial(\dots)/\partial x_i$, and the Einstein summation rule is adopted. Note that usually symmetrized strain tensor $u_{ij} = (u_{i,j} + u_{j,i})/2$ is used instead of displacement gradients $u_{i,j}$ (non-symmetrized strain tensor). But since $u_{i,j}$ is convolved with material tensors which are symmetrical in corresponding indices, it does not matter to write down $u_{i,j}$ or u_{ij}. Note also that here we use definition of $f^{(a)}_{klij}$ slightly different from that adopted in Ref. [11], pairs of indices ij and kl are swapped. This definition corresponds to Refs. [5, 9] and [10], it is also applicable. Note that in the case of cubic material it is no difference what definition to use, $f^{(a)}_{klij}$ are symmetrical with respect to such a swap.

Thermodynamic potential of a body H is integral over body volume V from the thermodynamic potential density. Using integration by parts its flexoelectric contribution can be represented in the form:

$$H_{flx} = \frac{f_{klij}}{2} \int \left(P_{k,l} u_{i,j} - P_k u_{i,j,l} \right) dV + \oint d_{ijk} u_{i,j} P_k dS, \qquad (2)$$

where $f_{klij} = f^{(1)}_{klij} - f^{(2)}_{klij}$ is so-called flexocoupling tensor, d_{ijk} is surface piezoelectric tensor of special form:

$$d_{ijk} = -\frac{1}{2} \left(f^{(1)}_{klij} + f^{(2)}_{klij} \right) n_l. \qquad (3)$$

Hereinafter n_l is unit vector normal to body surface S. Note that since d_{ijk} depends on n_l, generally it is different in different points of the surface.

Thus, flexoelectric part of thermodynamic potential density (1) can be reduced to the Lifshitz-type form

$$\mathcal{H}_{flx} = \frac{1}{2} f_{klij} \left(P_{k,l} u_{i,j} - P_k u_{i,j,l} \right), \qquad (4)$$

together with additional surface piezoelectricity described by piezoelectric tensor (3). Further effects of surface piezoelectricity are not described so that Lifshitz-type form (4) is used.

Certainly, besides flexoelectric potential (4) one should take into account the elastic energy. It will be clear from what follows that besides conventional elastic energy $c_{ijkl} u_{i,j} u_{k,l}/2$ higher elasticity term $v_{ijklnm} u_{i,j,n} u_{k,l,m}/2$ should be added. Thus, total volume density of the thermodynamic potential $\mathcal{H}_B$ is

$$\mathcal{H}_B = \frac{1}{2} \left[c_{ijkl} u_{i,j} u_{k,l} + v_{ijklnm} u_{i,j,n} u_{k,l,m} + f_{klij} \left(P_{k,l} u_{i,j} - P_k u_{i,j,l} \right) \right]. \qquad (5)$$

Possible surface contribution to the thermodynamic potential is not considered here, so that volume integral from (5) is the total thermodynamic potential of a body.

Starting from the equation (5) it is possible to write down the equations of elastic equilibrium and the boundary conditions for them. In fact, one just needs to calculate the tensors

$$T_{ij} = \frac{\partial \mathcal{H}_B}{\partial u_{i,j}} = c_{ijkl}u_{k,l} + \frac{1}{2}f_{klij}P_{k,l}, \tag{6}$$

$$\Theta_{ijk} = \frac{\partial \mathcal{H}_B}{\partial u_{i,j,k}} = v_{ijnlkm}u_{n,l,m} - \frac{1}{2}f_{nkij}P_n, \tag{7}$$

and use the results of Ref. [12]. In this way it turns out that "physical stresses" σ_{ij} which obey the equations of equilibrium

$$\sigma_{ij,j} = 0 \tag{8}$$

are determined by equation:

$$\sigma_{ij} = T_{ij} - \Theta_{ijk,k} = c_{ijkl}u_{k,l} - v_{ijnlkm}u_{n,l,m,k} + f_{klij}P_{k,l}. \tag{9}$$

Boundary conditions for (8) can be written in different equivalent forms, one should only substitute (6) and (7) to equations from Ref. [12]. Particularly they can be written in the form which coincides with that presented in Ref. [5]:

$$\Theta_{ijk}n_j n_k\big|_S = 0, \tag{10}$$

$$\sigma_{ij}n_j - \Theta_{ijk,j}n_k + \Theta_{ijk,l}n_l n_j n_k - \Theta_{ijk}\gamma_{jk}\big|_S = 0, \tag{11}$$

where tensor γ_{jk} describes curvature of the body surface (see Ref. [12]). Naturally the terms contained Σ_{ij} and F_i, which are present in the equations from Ref. [12], do not appear because surface contributions to thermodynamic potential is not described.

It is obvious from (10) and (7) that if $v_{ijklnm} \to 0$ and $f_{ijkl}P_i\big|_S \neq 0$ then $u_{i,j,k} \to \infty$. Thus, the limit $v_{ijklnm} \to 0$ is singular. This fact explains the above statement that generally in the presence of flexoectricity one should take into account the higher elasticity described by v_{ijklnm}. Higher elasticity term can be omitted only in very special (and physically doubtful) case when $P_i = 0$ on the surface of a body. In this case, (10) becomes an identity and term $\Theta_{ijk}\gamma_{jk}$ in (11) disappears. Note that generally it remains second and third term in (11) which, in framework of classical theory of elasticity, can be considered as an external force on the surface.

However, if polarization changes only in the direction normal to the surface then this terms cancels each other out.

3. Exact Solution for Homogeneously Polarized Ball of Isotropic Dielectric

When solving a particular problem it is convenient to use curvilinear coordinates x^α within covariant formalism (see Ref. [12]). Note that indices numbering curvilinear coordinates usually are written as superscripts rather then subscripts. Such indices we denote by Greek letters.

To transform equations from Cartesian form to covariant form one should change partial derivatives to covariant derivatives and provide an index balance (only superscript can be convolved with subscript). Thus, equations of mechanical equilibrium and boundary conditions for them take the form:

$$\sigma^{\alpha\beta}{}_{;\beta} = 0, \tag{12}$$

$$\Theta^{\alpha\beta\gamma} n_\beta n_\gamma \Big|_S = 0, \tag{13}$$

$$\sigma^{\alpha\gamma} n_\gamma - \Theta^{\alpha\beta\gamma}{}_{;\beta} n_\gamma + \Theta^{\alpha\beta\gamma}{}_{;\delta} n^\delta n_\beta n_\gamma - \Theta^{\alpha\beta\gamma} \gamma_{\beta\gamma} \Big|_S = 0. \tag{14}$$

Here index separated by semicolon means covariant derivative, n_α is covariant representation of unit vector normal to body surface S, tensors $\sigma^{\alpha\beta}$ and $\Theta^{\alpha\beta\gamma}$ are determined as follows:

$$\sigma^{\alpha\beta} = c^{\alpha\beta\gamma\delta} u_{\gamma;\delta} - v^{\alpha\beta\gamma\delta\varepsilon\zeta} u_{\gamma;\delta;\zeta;\varepsilon} + f^{\gamma\delta\alpha\beta} P_{\gamma;\delta}, \tag{15}$$

$$\Theta^{\alpha\beta\gamma} = v^{\alpha\beta\varepsilon\delta\gamma\zeta} u_{\varepsilon;\delta;\zeta} - \frac{1}{2} f^{\delta\gamma\alpha\beta} P_\delta. \tag{16}$$

It is convenient to use curvilinear coordinate system of special class in which equation of body surface has the form $x^3 = $ const. In this coordinate system tensor $\gamma^{\alpha\beta}$ can be presented in terms of Christoffel symbols $\Gamma^\alpha_{\beta\gamma}$ (see Ref. [12]):

$$\gamma_{\beta\gamma} = \Gamma^\varepsilon_{\gamma\delta} n^\delta n_\beta n_\varepsilon + \Gamma^\varepsilon_{\beta\delta} n^\delta n_\varepsilon n_\gamma - \Gamma^\delta_{\gamma\beta} n_\delta. \tag{17}$$

Christoffel symbols can be obtained by differentiating the metric tensor $g_{\alpha\beta}$ as follows:

$$\Gamma^{\gamma}_{\alpha\beta} = \frac{1}{2}g^{\gamma\delta}\left(g_{\delta\alpha,\beta} + g_{\delta\beta,\alpha} - g_{\alpha\beta,\delta}\right). \tag{18}$$

Vector n_{α}, normal to the surface in such coordinate system has the only component:

$$n_3 = \frac{1}{\sqrt{g^{33}}}. \tag{19}$$

For the case of a ball it is natural to use spherical coordinate system:

$$\begin{cases} x = r\sin\theta\cos\psi \\ y = r\sin\theta\sin\psi \ . \\ z = r\cos\theta \end{cases} \tag{20}$$

Here x, y, and z are Cartesian coordinates, $\psi = x^1$, $\theta = x^2$, and $r = x^3$ are curvilinear ones. Equation of the ball surface has the form $r = R$, so that these curvilinear coordinates belong to the special class mentioned above.

Instead of curvilinear indices 1, 2, and 3 further the indices ψ, θ, and r respectively are also used. So the letters r, ψ, and θ are excluded from notation of "running" indexes. Particular Cartesian indices are further denoted as x, y, and z.

Metric tensor components $g_{\alpha\beta}$, its determinant g, and Cristoffel symbols $\Gamma^{\alpha}_{\beta\gamma}$ can be derived directly from (20):

$$\begin{cases} g_{\psi\psi} = r^2\sin^2\theta, \quad g_{\theta\theta} = r^2, \quad g_{rr} = 1, \quad g = r^4\sin^2\theta \\ \Gamma^{\psi}_{\psi\theta} = \Gamma^{\psi}_{\theta\psi} = \cot\theta, \quad \Gamma^{\psi}_{\psi r} = \Gamma^{\psi}_{r\psi} = r^{-1}, \quad \Gamma^{\theta}_{\psi\psi} = -\sin\theta\cos\theta \\ \Gamma^{\theta}_{\theta r} = \Gamma^{\theta}_{r\theta} = r^{-1}, \quad \Gamma^{r}_{\psi\psi} = -r\sin^2\theta, \quad \Gamma^{r}_{\theta\theta} = -r. \end{cases}$$

$$\tag{21}$$

Other components are zero. Since tensor $g_{\alpha\beta}$ is diagonal, inverse tensor $g^{\alpha\beta}$ is also diagonal and corresponding components are simply equal to the inverse diagonal components of $g_{\alpha\beta}$. Unit vector normal to the surface has the only component $n_r = 1$.

Further one should specify the form of the material tensors for isotropic media. It is well-known that elastic tensor of such a media is defined by two parameters, say its components c_{12} and c_{44} (Voigt notation is used), and can be expressed as follows:

$$c_{ijkl} = c_{12}\delta_{ij}\delta_{kl} + c_{44}(\delta_{ik}\delta_{jl} + \delta_{il}\delta_{jk}). \tag{22}$$

It can be shown that flexocoupling tensor of isotropic media can be expressed analogously:

$$f_{ijkl} = f_{12}\delta_{ij}\delta_{kl} + f_{44}(\delta_{ik}\delta_{jl} + \delta_{il}\delta_{jk}). \tag{23}$$

For six-rank tensor of higher elastic moduli of isotropic media further is used an expression

$$\begin{aligned}
v_{ijlknm} = {} &v_1(\delta_{ij}\delta_{lm}\delta_{nk} + \delta_{ij}\delta_{nl}\delta_{mk} + \delta_{ij}\delta_{nm}\delta_{lk} + \delta_{ik}\delta_{jn}\delta_{lm} \\
&+ \delta_{il}\delta_{jn}\delta_{km} + \delta_{im}\delta_{jn}\delta_{kl} + \delta_{in}\delta_{jk}\delta_{lm} + \delta_{in}\delta_{jl}\delta_{km} \\
&+ \delta_{in}\delta_{jm}\delta_{kl}) + v_2(\delta_{ik}\delta_{jl}\delta_{nm} + \delta_{ik}\delta_{jm}\delta_{nl} + \delta_{il}\delta_{jk}\delta_{nm} \\
&+ \delta_{il}\delta_{jm}\delta_{nk} + \delta_{im}\delta_{jk}\delta_{nl} + \delta_{im}\delta_{jl}\delta_{nk}),
\end{aligned} \tag{24}$$

where v_1 and v_2 are constant parameters.

Above, the expressions of material tensors are presented in Cartesian components. To convert them to curvilinear components one should only change Kronecker delta to metric tensor with corresponding location (up or bottom) of indices.

If the ball is polarized along z-axis with polarization P then polarization vector has two covariant components:

$$P_r = P\cos\theta, \tag{25}$$

$$P_\theta = -rP\sin\theta. \tag{26}$$

The equations above mathematically completely determine the problem under study. To solve this problem one must first separate the variables. In the case of spherical geometry this is done by expansion of the vector fields u_α and P_α in the series of spherical vectors $Y_{lm|\alpha}^{(n)}$. This expansion looks as

$$u_\alpha(\psi,\theta,r) = \sum_{nlm} Y_{lm|\alpha}^{(n)}(\psi,\theta)\,\mathcal{U}_{nlm}(r), \qquad (27)$$

for the elastic field and similarly for polarization field. $\mathcal{U}_{nlm}(r)$ are radial functions which obey one-dimensional (radial) differential equations.

Note that, since the covariant formalism is used here, spherical vectors do not coincide with their usual form.[13] However, the required form of these vectors can be easily obtained from conventional form. Covariant spherical vectors can be defined as follows

$$\begin{cases} Y_{lm|\theta}^{(1)} = \dfrac{r}{\sqrt{l(l+1)}}(Y_{lm})_{,\theta}, \quad Y_{lm|\psi}^{(1)} = \dfrac{r}{\sqrt{l(l+1)}}im Y_{lm} \\[2ex] Y_{lm|\theta}^{(2)} = \dfrac{r}{\sqrt{l(l+1)}}\cdot\dfrac{-im}{\sin\theta}Y_{lm}, \quad Y_{lm|\psi}^{(2)} = \dfrac{r}{\sqrt{l(l+1)}}\sin\theta(Y_{lm})_{,\theta}, \\[2ex] Y_{lm|r}^{(3)} = Y_{lm} \\[2ex] Y_{lm|r}^{(1)} = 0, \quad Y_{lm|r}^{(2)} = 0, \quad Y_{lm|\theta}^{(3)} = 0, \quad Y_{lm|\psi}^{(3)} = 0 \end{cases}$$

$$(28)$$

where Y_{lm} is scalar spherical functions expressed in the usual manner in terms of associated Legendre polynomials.

In the case considered here, i.e. when the polarization is homogeneous, expansion in spherical vectors is simplified radically. Indeed, it is easy to see that in such expansion of the polarization field P_α there are only the terms with $l = 1$, $m = 0$ and top indices 1, 3. Functions $Y_{10|\theta}^{(1)}$ and $Y_{10|r}^{(3)}$ are proportional to $\sin\theta$ and $\cos\theta$ respectively (compare with (25) and (26)). Since the theory is linear, it follows that the expansion of the field u_α contains only the similar terms. As a result, this leads to the fact that $u_r \sim \cos\theta$, $u_\theta \sim \sin\theta$, $u_\psi = 0$. This is why one can use a substitution:

$$u_\psi = 0, \qquad (29)$$

$$u_\theta = -rf_2(r)\sin\theta, \qquad (30)$$

$$u_r = f_1(r)\cos\theta. \qquad (31)$$

Factor r is added into (30) to make function $f_2(r)$ analytical at $r = 0$. The consequence of using the covariant formalism is that u_θ is not a physical displacement in the meridional direction, the physical displacement is $u_\theta \sqrt{g^{\theta\theta}}$, while $g^{\theta\theta} = r^{-2}$. Certainly the physical component of the displacement should not have a singularity at the ball center. Note that covariant form of $Y^{(1)}_{lm|\theta}$ contains the same factor r.

Thus, the problem is reduced to finding the functions $f_1(r)$ and $f_2(r)$. To find the corresponding differential equations and boundary conditions one should substitute (29)–(31) in the above equations and make the very cumbersome but absolutely straightforward algebraic transformations. As a result, it yields the system of two differential equations

$$v_3(\xi^4 f_1'''' + 4\xi^3 f_1''' - 8\xi^2 f_1'' + 16 f_1) + v_3(8\xi^2 f_2'' - 16 f_2)$$

$$+ v_4(\xi^4 f_1'''' + 4\xi^3 f_1''' - 6\xi^2 f_1'' + 12 f_1)$$

$$- v_4(2\xi^3 f_2''' - 2\xi^2 f_2'' - 4\xi f_2' + 12 f_2)$$

$$= c_{44}\xi^2(2\xi^2 f_1'' + 4\xi f_1' - 6 f_1) - c_{44}\xi^2(2\xi f_2' - 6 f_2)$$

$$+ c_{12}\xi^2(\xi^2 f_1'' + 2\xi f_1' - 2 f_1) - c_{12}\xi^2(2\xi f_2' - 2 f_2), \qquad (32)$$

$$v_3(\xi^4 f_2'''' + 4\xi^3 f_2''' - 4\xi^2 f_2'' + 8 f_2) + v_3(4\xi^2 f_1'' - 8 f_1)$$

$$- v_4(2\xi^2 f_2'' - 4 f_2) + v_4(\xi^3 f_1''' + 4\xi^2 f_1'' - 2\xi f_1' - 4 f_1)$$

$$= c_{44}\xi^2(\xi^2 f_2'' + 2\xi f_2' - 4 f_2) + c_{44}\xi^2(\xi f_1' + 4 f_1)$$

$$- c_{12}\xi^2 2 f_2 + c_{12}\xi^2(\xi f_1' + 2 f_1), \qquad (33)$$

and four boundary condition for it

$$2v_1(f_2''' + f_2'' - 18 f_2' + 34 f_2 + 8 f_1'' + 14 f_1' - 34 f_1)$$

$$+ 2v_2(2 f_2''' + 2 f_2'' - 20 f_2' + 36 f_2 + 4 f_1'' + 16 f_1' - 36 f_1)$$

$$- 2c_{44}R^2(f_2' - f_2 + f_1)\big|_{\xi=1} = f_{12}R^2 P, \qquad (34)$$

$$R^2[c_{12}(2f_1 - 2f_2) + (c_{12} + 2c_{44})f_1']$$

$$- v_1(9f_1''' + 18f_1'' - 46f_1' + 36f_1 - 14f_2'' + 32f_2' - 36f_2)$$

$$- 2v_2(3f_1''' + 6f_1'' - 22f_1' + 28f_1 - 2f_2'' + 16f_2' - 28f_2)\big|_{\xi=1}$$

$$= R^2 f_{12} P, \tag{35}$$

$$(18v_1 + 12v_2)f_1'' + 12v_1(3f_1' - 4f_1 - 2f_2' + 4f_2)\big|_{\xi=1}$$

$$= (f_{12} + 2f_{44})R^2 P, \tag{36}$$

$$2v_1(f_2'' + 2f_2' - 6f_2 + 2f_1' + 6f_1)$$

$$+ 4v_2(f_2'' - 2f_2' + 2f_2 + 2f_1' - 2f_1)\big|_{\xi=1} = f_{44}R^2 P. \tag{37}$$

Here $\xi = r/R$ is a dimensionless radial coordinate, prime denotes the derivative in ξ, $v_3 = (v_1 + 2v_2)R^{-2}$, $v_4 = (8v_1 + 4v_2)R^{-2}$.

In general, the system of two linear fourth-order differential equations (32) and (33) has eight linear independent particular solutions (basis functions) $f_i(\xi) = \mathcal{B}_{ki}(\xi)$. Here k is the number of particular solution. Since general solution is an linear combination of basis functions with some coefficients C_k

$$f_i(\xi) = \sum C_k \mathcal{B}_{ki}(\xi), \tag{38}$$

it may seem that to determine C_k the number of boundary conditions should be also eight, while we have only four boundary conditions (34)–(37). However, one should additionally require that the solution should be analytic at zero. This condition reduces the number of linear independent solutions and, respectively, reduces the number of the required boundary conditions (see below for details).

To find only that basis functions $\mathcal{B}_{ki}(\xi)$ which are analytic at zero, we express them in the form of the series in non-negative powers of ξ:

$$\mathcal{B}_{ki}(\xi) = \sum_{n=0}^{\infty} a_{kin}\xi^n. \tag{39}$$

The constant coefficients a_{kin} are found from the condition that $f_i(\xi) = \mathcal{B}_{ki}(\xi)$ obey the system of differential equations (32) and (33). This condition leads to the following linear algebraic equations for the coefficients (index k which is numbering the solutions is omitted):

$$\begin{cases} [v_3(n^4 - 2n^3 - 9n^2 + 10n + 16) \\ \quad + v_4(n^4 - 2n^3 - 7n^2 + 8n + 12)]a_{1\,n} \\ \quad + [v_3(8n^2 - 8n - 16) - v_4(2n^3 - 8n^2 + 2n + 12)]a_{2\,n} \\ = [c_{44}(2n^2 - 6n - 2) + c_{12}(n^2 - 3n)]a_{1\,n-2} \\ \quad - [c_{44}(2n - 10) + c_{12}(2n - 6)]a_{2\,n-2} \\ [v_3(4n^2 - 4n - 8) + v_4(n^3 + n^2 - 4n - 4)]a_{1\,n} \\ \quad + [v_3(n^4 - 2n^3 - 5n^2 + 6n + 8) - v_4(2n^2 - 2n - 4)]a_{2\,n} \\ = [c_{44}(n + 2) + c_{12}n]a_{1\,n-2} + [c_{44}(n^2 - 3n - 2) - c_{12}2]a_{2\,n-2} \end{cases} \tag{40}$$

From (40) one can see that the system of linear algebraic equations, even though it is infinite, has the characteristic structure: coefficients with greater n are determined by coefficients with smaller n. Therefore, one can find a_{in} for any finite n step by step, it is only needed to analyze the case of small n and define some of a_{in} for these small n.

Naturally a small n system of equations (40) is degenerate, some a_{in} with small n can be defined arbitrarily. This arbitrariness leads to that it may be several functions $\mathcal{B}_{ki}(\xi)$ and their normalization may be arbitrary. Detailed analysis shows that the system of equations (40) is degenerate for $n = 1, 2, 4$ only. It turns out five arbitrary constants, and so it is the same number of basis functions. However, one of these functions is physically meaningless and should be omitted (for more on this see below).

From an abstract point of view, arbitrary constants can be defined as anything, it is just needed that obtained basis functions be linearly independent. However, there is an important question: is it possible to choose these constants in such manner that the series representing at least some of the basic functions cut short, would constitute the finite sums? It turns out that this choice of constants is really possible.

From the system of equations (40) it is clear that in order to series cut short, it should be $a_{3i} = a_{4i} = 0$. For this to be possible, in any case, right-hand sides of (40) should be zero for $n = 3, 4$. Together with (40) for $n = 0, 1, 2$ these conditions lead to a system of linear algebraic equations which is quite amenable to analysis. The result is that there may be two basic functions which are finite sums.

One of them is determined by the conditions $a_{10} = a_{20} = \text{const.} \neq 0$, $a_{n>0\,i} = 0$. This function corresponds to the displacement of the ball as a whole. Such a displacement is physically meaningless and this basis function should be omitted. Certainly, such a displacement does not change the thermodynamic potential, so that this function indeed is a solution of the equations derived from a stationary condition of this potential. But it is meaningless solution.

Another basis function, which is finite sum, is determined by condition $(c_{12} - c_{44})a_{22} = (3c_{44} + 2c_{12})a_{12}$, other coefficients are zero. Thus, first basis function is determined as follows:

$$
\begin{cases}
\mathcal{B}_{1i} : \\[4pt]
a_{12} = 1, \quad a_{22} = \dfrac{3c_{44} + 2c_{12}}{c_{12} - c_{44}} a_{12},
\end{cases}
\tag{41}
$$

other a_{in} are zero.

It should be stressed that the function $\mathcal{B}_{1i}$ obeys not only the equations (32) and (33) but also the equations of the classical theory of elasticity, which can be obtained from (32) and (33) by setting $v_3 = v_4 = 0$. It can be verified by direct substitution.

Further analysis shows that there are three additional basic functions $\mathcal{B}_2$–$\mathcal{B}_4$ which are expressed by infinite series and can be chosen, for example, as follows.

$$
\begin{cases}
\mathcal{B}_{2i} : \\[6pt]
a_{12} = 1, \quad a_{22} = \dfrac{3c_{44} + 2c_{12}}{c_{44} - c_{12}} a_{12} \\[10pt]
a_{24} = 0 \\[6pt]
(40v_3 + 60v_4)a_{14} \\[4pt]
= (6c_{44} + 4c_{12})a_{12} + (2c_{44} - 2c_{12})a_{22} \\[4pt]
a_{i0} = a_{i1} = a_{i3} = 0
\end{cases}
\tag{42}
$$

coefficients with $n \geq 5$ are calculated by means of (40).

$$\begin{cases} \mathcal{B}_{3i} : \\ a_{11} = 1, \quad a_{21} = \dfrac{4v_3 + 3v_4}{4v_3 + 2v_4} a_{11}. \\ a_{i0} = a_{i2} = a_{i4} = 0 \end{cases} \qquad (43)$$

Coefficients with $n = 3$ and $n \geq 5$ are calculated by means of (40).

$$\begin{cases} \mathcal{B}_{4i} : \\ a_{14} = 1, \quad a_{24} = \dfrac{2v_3 + 3v_4}{v_4 - 4v_3} a_{14}, \\ a_{i0} = a_{i1} = a_{i2} = a_{i3} = 0 \end{cases} \qquad (44)$$

coefficients with $n \geq 5$ are calculated by means of (40).

Graphs of the basis functions $\mathcal{B}_{ik}$ defined above for some set of parameters are shown in Fig. 1. Turn our attention that in contrast

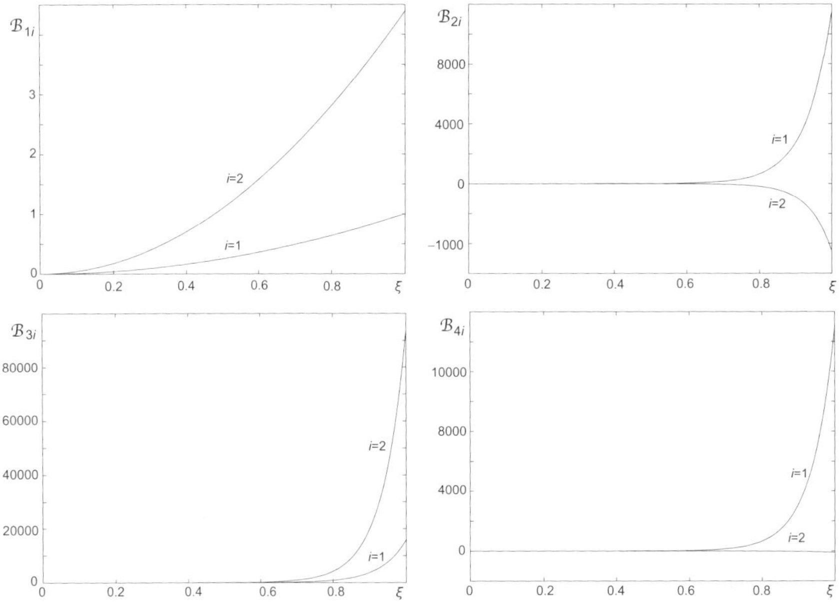

Figure 1. Basis functions for $R = 1 \cdot 10^{-5}$, $P = 1$, $c_{44} = 1.1 \cdot 10^{12}$, $c_{12} = 3.4 \cdot 10^{12}$, $f_{44} = f_{12} = 1 \cdot 10^{-3}$, $v_1 = 0.2$, $v_2 = 0.1$.

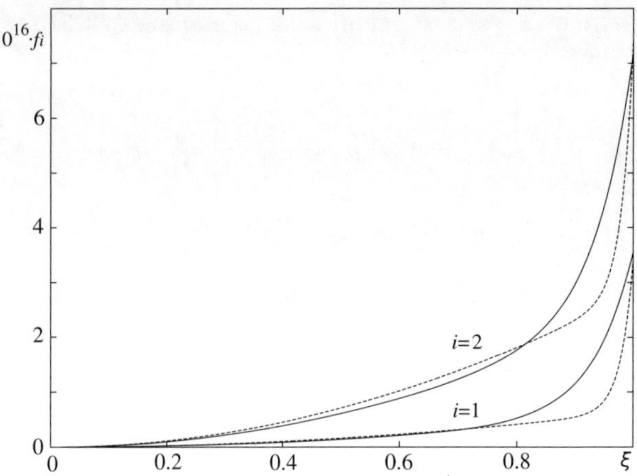

Figure 2. Result of the calculation for $R = 1 \cdot 10^{-5}$, $P = 1$, $c_{44} = 1.1 \cdot 10^{12}$, $c_{12} = 3.4 \cdot 10^{12}$, $f_{44} = f_{12} = 1 \cdot 10^{-3}$. Solid line is for $v_1 = 2 \cdot 10^{-1}$, $v_2 = 1 \cdot 10^{-1}$, dashed line is for $v_1 = 2 \cdot 10^{-2}$, $v_2 = 1 \cdot 10^{-2}$.

with $\mathcal{B}_{1i}$ the functions $\mathcal{B}_{2i}$–$\mathcal{B}_{4i}$ are concentrated near the surface and decay rapidly in the ball volume.

Above there is explicit representation of four basic functions $\mathcal{B}_1$–$\mathcal{B}_4$. It only remains to substitute (38) in the boundary conditions (34)–(37), and from the resulting system of four linear algebraic equations to find numerically C_k. Thus, the problem is solved. Result of numerical calculations for some parameters is presented in the Fig. 2 in terms of functions f_1 and f_2. In order to determine the elastic displacements corresponding to these functions one should use equations (29)–(31).

Note that the graphs shows that, except for a thin layer near $\xi = 1$, the curves are parabolic, the more so the less v_1 and v_2 are. In the bulk of the ball the only function $\mathcal{B}_{1i}(\xi)$ remains with a good approximation. Thus, for small v_i, except for a thin surface layer, solution approximately obeys the equations of classical elasticity theory (see discussion of function $\mathcal{B}_1$ properties above). This observation can be used to construct an approximate method of flexoelectric deformations calculations (see next section). This method is needed to describe more complicated geometry than spherical geometry of a ball.

4. Approximate Method to Calculate Flexoelectric Deformations of Finite Size Bodies

Exact (within the framework of a continuum media theory) equations of elastic equilibrium used in previous section are partial differential equations of the fourth order. They are too complex to be used in the cases interesting for applications. Even for a homogeneously polarized ball of isotropic media solution is cumbersome and requires the functions expressed by the series in powers of radial coordinate. Thus, to find flexoelectric deformation in more complicated cases one needs an approximate method. Such a method was proposed in Ref. 10 and it is briefly described in this section. More detailed description is given in Appendix A.

When solving the problem for a ball (see previous section) we have seen that the solution is a sum of two parts. The first part corresponding base function $\mathcal{B}_{1i}$ obeys the equations of the classical theory of elasticity, the second part (contribution of remaining basis functions) in the case of small $v^{\alpha\beta\gamma\delta\varepsilon\zeta}$ is concentrated near the surface and rapidly decays inside the body. Hereafter the first part is called a volume or classical part, and the second one — a non-classical or surface part.

It is natural to assume that this property should be preserved in more general cases. It is possible to use it to construct an approximation which is useful for solution of more general problems than the problem for a ball. Small parameter provides the possibility of this approximation is the ratio of the characteristic length, over which the non-classical part decays, to the radius of curvature of the body surface.

According to the observation mentioned above the solution of the elastic equilibrium equations is presented in the form $u_\gamma = \tilde{u}_\gamma + \hat{u}_\gamma$, where $\tilde{u}_\gamma$ obey the classical equations:

$$c^{\alpha\beta\gamma\delta}\tilde{u}_{\gamma;\delta;\beta} + f^{\gamma\delta\alpha\beta}P_{\gamma;\delta;\beta} = 0. \tag{45}$$

Thus, $\tilde{u}_\gamma$ is the classical part of elastic displacements. Respectively, $\hat{u}_\gamma$ is non-classical part. It can be shown (see Appendix A for details) that for sufficiently smooth distribution of polarization and small

$v^{\alpha\beta\gamma\delta\varepsilon\zeta}$ non-classical part of elastic displacements obeys following differential equations:

$$c^{\alpha 3\gamma 3}\hat{u}_\gamma - v^{\alpha 3\gamma 333}\hat{u}_{\gamma,3,3} = 0. \tag{46}$$

Naturally, it is implied here that equation of body surface is $x^3 = x_S^3$, where x_S^3 is a constant. Equations (46) should be appended by boundary conditions, which have the form:

$$v^{\alpha 3\varepsilon 333}\hat{u}_{\varepsilon,3,3}\big|_S = \frac{1}{2}f^{\delta 3\alpha 3}P_\delta\big|_S. \tag{47}$$

Additionally it should be required that $\hat{u}_\gamma$ exponentially decay inside the body.

Equations (46) are the system of three ordinal differential equations, the dependence on x^1 and x^2 is parametric here. Moreover, in the thin layer near the body surface one can assume that the coefficients do not depend on x^3, they are approximately equal to the surface values. Solution of such a system is easy, it is obvious that $\hat{u}_\gamma$ can be represented by an exponent decaying over the characteristic length of the order of $\sqrt{v/c}$. Here v and c are characteristic values of the corresponding tensors components. Thus, the non-classical part of elastic displacements can be found completely in explicit form.

Classical part of elastic displacements can be found by means of equations (45) appended by boundary conditions:

$$c^{\alpha 3\gamma\delta}\tilde{u}_{\gamma;\delta}\Big|_S$$
$$= s^\alpha - f^{\gamma\delta\alpha 3}P_{\gamma;\delta} - \frac{1}{2}f^{\delta 3\alpha(\beta)}P_{\delta;(\beta)} + \frac{1}{2}f^{\delta(\gamma)\alpha(\beta)}P_\delta\Gamma^3_{(\beta)(\gamma)}\Big|_S, \tag{48}$$

where

$$s^\alpha = c^{\alpha(\beta)\gamma 3}\hat{u}_{\gamma,(\beta)} - c^{\alpha\beta\gamma\delta}\Gamma^3_{\delta\beta}\hat{u}_\gamma - c^{\alpha\beta\gamma 3}\Gamma^\varepsilon_{\gamma\beta}\hat{u}_\varepsilon - h^{\alpha\beta}\hat{u}_{\beta,3,3}, \tag{49}$$

$$h^{\alpha\beta} = v^{\alpha(\gamma)\beta 3(\delta)3}\Gamma^3_{(\gamma)(\delta)} - v^{\alpha 3\gamma 333}\Gamma^\beta_{\gamma 3} - 2v^{\alpha 3\beta\delta 33}\Gamma^3_{\delta 3} - v^{\alpha\varepsilon\beta 3\delta 3}\Gamma^3_{\delta\varepsilon}, \tag{50}$$

the indices enclosed in parentheses runs only the values 1 and 2.

Remember that in the used coordinate system the unit vector normal to the body surface n_α has the only component $n_3 = 1/\sqrt{g^{33}}$. Thus, with respect to the classical part of strain, $c^{\alpha3\gamma\delta}\tilde{u}_{\gamma;\delta}\big|_S$ actually defines the external force density $\sigma^{\alpha\beta}n_\beta$ on the body surface, one should only multiply it by $1/\sqrt{g^{33}}$. This is why differential equations (45) appended by (48) correspond to a standard problem of classical theory of elasticity for a body under external forces on its surface. Methods of solution of such problems are well-known and do not require special consideration.

It should be emphasized that physically there are no external forces on the surface. Forces mentioned above are purely formal in nature and describe the interaction of non-classical and classical parts of deformation.

5. Comparison of Exact and Approximate Solutions for Homogeneously Polarized Ball

It is worth to find flexoelectric deformations of a homogeneous polarized ball by an approximate method described in the previous section and compare this deformations with ones known from exact solution. This is a test of approximate method.

For isotropic material the matrices $c^{\alpha3\gamma3}$ and $v^{\alpha3\gamma333}$ are diagonal. Thus, the solution of (46) can be found at once:

$$\begin{cases} \hat{u}_r = \dfrac{f_{12} + 2f_{44}}{2(c_{12} + 2c_{44})} P \cos\theta\, e^{\lambda_r R(\xi-1)} \\[2em] \hat{u}_\theta = -R\dfrac{f_{44}}{2c_{44}} P \sin\theta\, e^{\lambda_\theta R(\xi-1)}, \\[2em] \hat{u}_\psi = 0 \end{cases} \tag{51}$$

where

$$\lambda_\theta = \sqrt{\frac{c_{44}}{v_1 + 2v_2}}, \tag{52}$$

$$\lambda_r = \sqrt{\frac{c_{12} + 2c_{44}}{9v_1 + 6v_2}}. \tag{53}$$

For comparison with the exact solution it is useful to present $\hat{u}_\gamma$ in terms of the functions $\hat{f}_1$ and $\hat{f}_2$:

$$\hat{f}_1 = \frac{\hat{u}_r}{\cos\theta} = \frac{f_{12} + 2f_{44}}{2(c_{12} + 2c_{44})} Pe^{\lambda_r R(\xi-1)}, \tag{54}$$

$$\hat{f}_2 = -\frac{\hat{u}_\theta}{r\sin\theta} = \frac{1}{\xi} \cdot \frac{f_{44}}{2c_{44}} Pe^{\lambda_\theta R(\xi-1)}. \tag{55}$$

Next step is to find $\tilde{\sigma}^{\alpha 3} n_3 = c^{\alpha 3\gamma\delta}\tilde{u}_{\gamma;\delta} n_3$ on the ball surface. To do this, one should just use the equations of the previous section. It turns out

$$\tilde{\sigma}^{33} n_3\big|_S = \frac{2P\cos\theta}{R} \cdot \frac{c_{44}f_{12} - c_{12}f_{44}}{c_{12} + 2c_{44}}, \tag{56}$$

$$\tilde{\sigma}^{23} n_3\big|_S = \frac{P\sin\theta}{R^2} \cdot \frac{c_{44}f_{12} - c_{12}f_{44}}{c_{12} + 2c_{44}}. \tag{57}$$

Certainly $\tilde{\sigma}^{13} = 0$. This is obviously from symmetry but can be also obtained by direct calculations. Configuration of forces determined by equations (56) and (57) is shown on Fig. 3.

To solve the differential equations (45) actually is not necessary here. It is clear that in the terms of functions $\tilde{f}_1 = \tilde{u}_r/\cos\theta$ and $\tilde{f}_2 = -\tilde{u}_\theta/(r\sin\theta)$ the solution is proportional to the previously

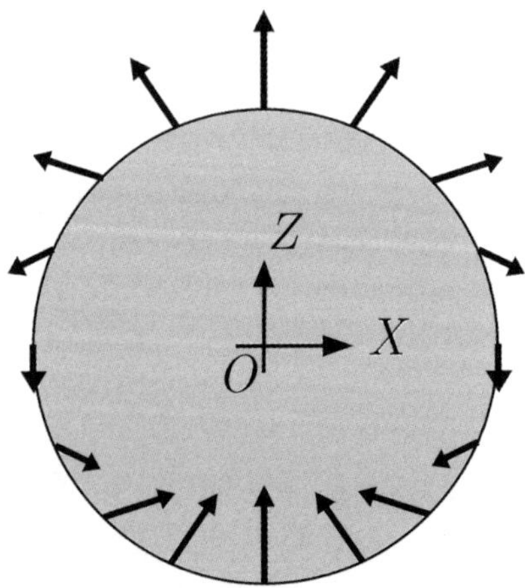

Figure 3.   Formal forces acting on the ball surface. *OXZ* cross-section.

defined function $\mathcal{B}_{1i}$. Eventually the validity of this statement can be verified by direct substitution. Thus, one just needs to find the coefficient of proportionality, it is easily done by using (56) or (57). One can use any of these conditions, they both give the same result:

$$\tilde{f}_1 = \frac{P(c_{12}f_{44} - c_{44}f_{12})(c_{12} - c_{44})}{c_{44}(c_{12} + 2c_{44})(3c_{12} + 2c_{44})}\xi^2, \tag{58}$$

$$\tilde{f}_2 = \frac{P(c_{12}f_{44} - c_{44}f_{12})(2c_{12} + 3c_{44})}{c_{44}(c_{12} + 2c_{44})(3c_{12} + 2c_{44})}\xi^2. \tag{59}$$

It is also useful to derive the expression for the classical part of the elastic displacement in Cartesian components as the functions of the Cartesian coordinates. Direct conversions yield

$$\begin{cases} \tilde{u}_z = a_1 z^2 + a_2(x^2 + y^2) \\ \tilde{u}_y = (a_1 - a_2)yz \\ \tilde{u}_x = (a_1 - a_2)xz \end{cases}, \tag{60}$$

where

$$a_1 = \frac{P(c_{12}f_{44} - c_{44}f_{12})(c_{12} - c_{44})}{R^2 c_{44}(c_{12} + 2c_{44})(3c_{12} + 2c_{44})}, \tag{61}$$

$$a_2 = \frac{P(c_{12}f_{44} - c_{44}f_{12})(2c_{12} + 3c_{44})}{R^2 c_{44}(c_{12} + 2c_{44})(3c_{12} + 2c_{44})}. \tag{62}$$

Now we can find $f_i = \hat{f}_i + \tilde{f}_i$ and compare it with the results of exact calculations. This comparison is shown in Fig. 4. This figure shows a good agreement between the approximate and the exact solution, the smaller higher elastic moduli match those better. It should also be emphasized that the latest version, when the difference is badly distinguishable, corresponds to most physically reasonable values of higher elastic moduli. Thus, the approximate method works very well at least for a homogeneously polarized ball with the given parameters.

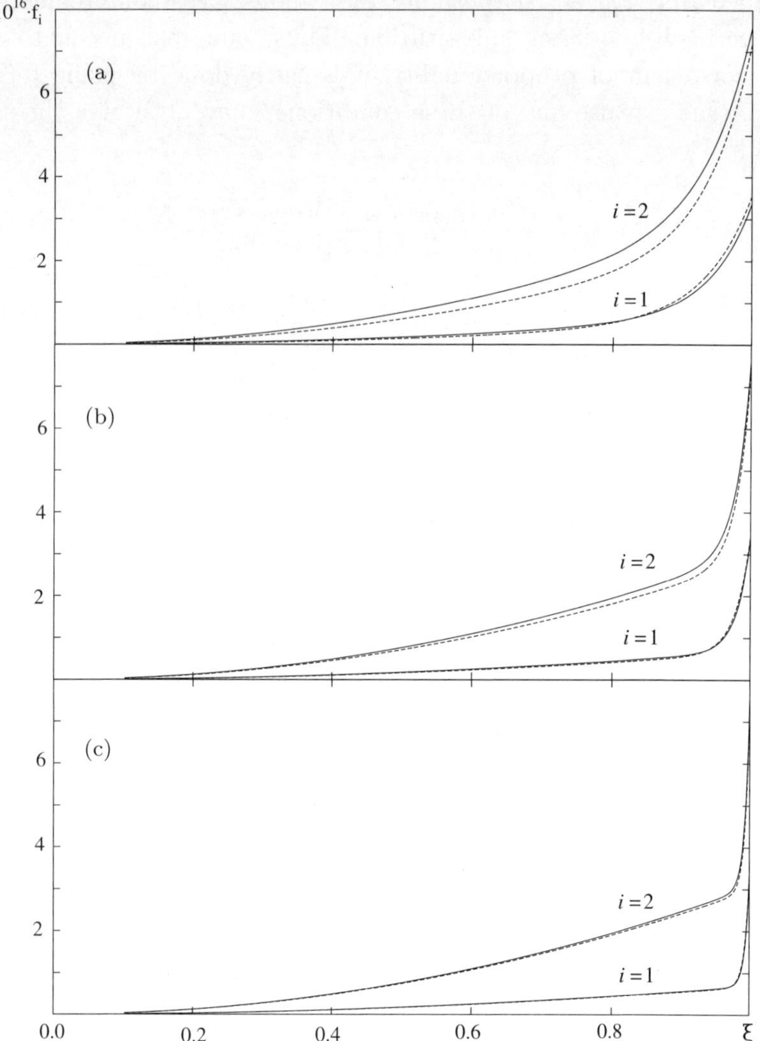

Figure 4.   Comparison of exact and approximate solutions for a ball. $R = 1 \cdot 10^{-5}$, $P = 1$, $c_{44} = 1.1 \cdot 10^{12}$, $c_{12} = 3.4 \cdot 10^{12}$, $f_{44} = f_{12} = 1 \cdot 10^{-3}$. (a) for $v_1 = 2.0 \cdot 10^{-1}$, $v_2 = 1.0 \cdot 10^{-1}$; (b) for $v_1 = 2.0 \cdot 10^{-2}$, $v_2 = 1.0 \cdot 10^{-2}$; (c) for $v_1 = 2.0 \cdot 10^{-3}$, $v_2 = 1.0 \cdot 10^{-3}$. Solid line is approximate solution, dashed line is exact solution.

6. Flexoelectric Bending of Homogeneously Polarized Circular Rod

In this section, we apply the above-described method to find the bending of a homogeneously polarized circular rod of an isotropic material.[14] Such a rod may be polarized in different directions. However, due to the fact that the flexoelectric effect is linear, we can restrict ourselves to the case of longitudinal and transverse polarization only. The response of the rod to the polarization of an arbitrary direction obviously is a superposition of responses to the longitudinal and transverse polarization. Besides, axial symmetry does not allow us to distinguish different directions of the transverse polarization.

As stated above, the classical part of strain is determined by the formal forces on the body surface. The surface of the rod consists of a cylindrical surface and the end surfaces. First we consider the cylindrical surface of the rod. Here, it is natural to use a cylindrical coordinate system.

Let cylinder be located along the longitudinal axis corresponding to coordinate x. This coordinate is also denoted as x^1. The other two curvilinear coordinates are the angular $x^2 = \phi$ and radial $x^3 = r$. Angle ϕ is measured in respect to the direction of the Cartesian z-axis. Equation of the surface is $x^3 = R$ where R is rod radius. Thus, this coordinate system belongs to the class that is needed in accordance with the method of calculations.

For such coordinates simple geometrical reasoning lead to the following equations defining the relationship between Cartesian and curvilinear coordinates:

$$\begin{cases} x = x^1 \\ y = r \sin \phi = x^3 \sin x^2 \\ z = r \cos \phi = x^3 \cos x^2 \end{cases} . \tag{63}$$

By means of simple differentiation it is easy to find the metric tensor, its determinants and the Christoffel symbols:

$$\begin{cases} g_{11} = g_{xx} = 1, \quad g_{22} = g_{\phi\phi} = r^2, \quad g_{33} = g_{rr} = 1, \quad g = r^2 \\ \Gamma^r_{\phi\phi} = -r, \quad \Gamma^\phi_{r\phi} = \Gamma^\phi_{\phi r} = \dfrac{1}{r} \end{cases} . \quad (64)$$

The remaining components are zero. Since the metric tensor $g_{\alpha\beta}$ is diagonal to find the inverse metric tensor $g^{\alpha\beta}$ is easy: one just needs to take the reciprocals of the diagonal components. Note also that the vector normal to surface has the only component $n_r = 1$.

By making perfectly straightforward calculations similar to the case of a ball described above, the following expression can be obtained for the non-classical part of the elastic displacement $\hat{u}_\alpha$ near the cylindrical surface. If the polarization P is longitudinal along the x-axis then the expression is as follows:

$$\begin{cases} \hat{u}_x = \hat{u}_1 = P\dfrac{f_{44}}{2c_{44}} e^{\lambda_x R(\xi-1)} \\ \\ \hat{u}_\phi = \hat{u}_2 = 0. \\ \hat{u}_r = \hat{u}_3 = 0 \end{cases} \quad (65)$$

For transverse polarization along z-axis $\hat{u}_\alpha$ is determined by the following equations:

$$\begin{cases} \hat{u}_x = 0 \\ \\ \hat{u}_\phi = -PR\dfrac{f_{44}}{2c_{44}} \sin\phi\, e^{\lambda_\phi R(\xi-1)} \\ \\ \hat{u}_r = P\dfrac{f_{12} + 2f_{44}}{2(c_{12} + 2c_{44})} \cos\phi\, e^{\lambda_r R(\xi-1)} \end{cases} . \quad (66)$$

Here $\xi = r/R$ is a dimensionless radial coordinate,

$$\lambda_x = \lambda_\phi = \sqrt{\dfrac{c_{44}}{v_1 + 2v_2}}, \quad (67)$$

$$\lambda_r = \sqrt{\dfrac{c_{12} + 2c_{44}}{9v_1 + 6v_2}}. \quad (68)$$

Direct application of the equations of Section 4 gives the following boundary conditions for the classical part of the displacement. For transverse polarization they are

$$
\begin{cases}
\tilde{\sigma}^{13} n_3 \big|_S = 0 \\[2mm]
\tilde{\sigma}^{23} n_3 \big|_S = \dfrac{P \sin \phi}{R^2} \cdot \dfrac{c_{44} f_{12} - c_{12} f_{44}}{c_{12} + 2c_{44}} \cdot \\[3mm]
\tilde{\sigma}^{33} n_3 \big|_S = \dfrac{P \cos \phi}{R} \cdot \dfrac{c_{44} f_{12} - c_{12} f_{44}}{c_{12} + 2c_{44}}
\end{cases}
\tag{69}
$$

If polarization is longitudinal then all $\tilde{\sigma}^{\alpha 3}$ are zero.

Boundary conditions (69) can be also expressed in terms of the Cartesian tensor component. Elementary transformations yield:

$$
\begin{cases}
\tilde{\sigma}_{xi} n_i \big|_S = 0 \\[2mm]
\tilde{\sigma}_{yi} n_i \big|_S = \dfrac{P(c_{44} f_{12} - c_{12} f_{44})}{R(c_{12} + 2c_{44})} \cdot \sin 2\phi. \\[3mm]
\tilde{\sigma}_{zi} n_i \big|_S = \dfrac{P(c_{44} f_{12} - c_{12} f_{44})}{R(c_{12} + 2c_{44})} \cdot \cos 2\phi
\end{cases}
\tag{70}
$$

From this form of boundary conditions it is immediately clear that the total force acting on the cylindrical surface is equal to zero, the integration over the angle turns these expressions to zero. It also can be easily calculated that the total bending moment of these forces vanishes while integrating over the angle ϕ.

Thus, not only the sum of formal surface forces, but also the sum of the bending moments of these forces, acting on a small part of the cylindrical surface length, are equal to zero. This means that the bending moment does not change along the homogeneously polarized rod. We emphasize that even so, the bending moment appears, but entirely by the boundary effects at the ends of the rod. Thus, in the case of homogeneously polarized rod the problem is reduced to the standard problem of the classical theory of elasticity,[15] i.e. to the determination of rod bending under the bending moments applied to its ends.

Certainly, the surface forces (69) slightly deform cross-section of the rod. However, since the main effect is the bending of the rod, we do not discuss this deformation. Note only that qualitatively it is similar to the deformation of the ball in the meridian cross-section.

To calculate the bending moment that appears at the ends of the rod, we need to specify the shape of these ends. We emphasize that we cannot limit the rod by planes because this gives sharp edges while the theory requires smooth surface. We should smooth out these sharp edges, say by a quarter of toroidal surface, or assume that the rod terminates by halves of the ball. Calculations were made for both these variants, and it obtained the same result. Omitting the details of these rather simple calculations, we note only that in the case of spherical ends one can use the results of the previous section. For edges slightly smoothed by the toroidal surface one can approximately describe a small part of this toroidal surface as cylindrical one and use the equations given above in this section. Eventually it turns out that the bending moment is

$$M_y = \frac{P\pi R^2 (c_{44} f_{12} - c_{12} f_{44})}{c_{12} + 2c_{44}}. \tag{71}$$

The other components are equal to zero. It is implied here that the rod is transversely polarized along z-axis and the equation is written for the end in the positive direction of x-axis. For the other end of the rod M_y has the opposite sign.

As is already clear from symmetry, for the longitudinal polarized rod there is no bending moment at all, this can be also proved by direct calculations. Such calculations also show that the total force acting on each end of the rod is zero both for longitudinal and for transverse polarization.

Equation (71) actually solves the problem of flexoelectric bending of the homogeneously polarized rod. One should only substitute this bending moment into standard equations from textbooks (see Ref. 15 for instance).

7. Flexoelectric Bending of Homogeneously Polarized Circular Plate

In this section, we apply the above-described method of approximate calculation to find the flexoelectric bending of a thin circular plate with radius R and thickness h, uniformly polarized normal to its plane.[14] Material of the plate is assumed isotropic. For definiteness, we assume that the average surface of the plate lies in the coordinate plane OXY.

Calculations for the plate are generally similar to those made in the previous section for the rod. However, most part of the plate surface is flat, so that one can immediately conclude, without any calculations, that the formal forces on this part of the surface are zero. As in the case of the rod, bending of the plate is determined by edge effects which are discussed below.

Thus, for most part of the plate, except for the edges, the classical part of the plate deformation is determined by standard equations without surface loading. It is well known[15] that under such conditions the displacements of the average surface of the plate $\zeta(x, y)$ obey two-dimensional differential equation

$$\Delta\Delta\zeta = 0, \tag{72}$$

where Δ is two-dimensional Laplace operator. The components of the strain tensor can be expressed in terms of ζ as follows:

$$u_{xx} = -z\zeta_{,x,x}; \quad u_{yy} = -z\zeta_{,y,y}; \quad u_{xy} = -z\zeta_{,x,y};$$

$$u_{xz} = u_{yz} = 0; \quad u_{zz} = z\,\frac{c_{12}}{c_{12} + 2c_{44}}\,(\zeta_{,x,x} + \zeta_{,y,y}). \tag{73}$$

Using polar coordinates it is easy to find an axially symmetric, regular at origin solution of the equation (72):

$$\zeta(x, y) = -\frac{Gr^2}{2} = -\frac{G}{2}(x^2 + y^2). \tag{74}$$

This solution contains a single integration constant G which is nothing but the curvature of the plate.

To find integration constant G one needs to first calculate the components of the strain tensor using (73) and then calculate the components of the strain tensor by means of standard equations $\sigma_{ij} = c_{ijkl}u_{ij}$. Thereafter, it becomes obvious that there is the linear density of the bending moment M on the boundary of the plate. For instance, at the point of intersection of the plate boundary with the coordinate axis OX this bending moment density is

$$M = \int_{-h/2}^{+h/2} \sigma_{xx}(z)z\,dz = \frac{h^3}{6} G \frac{c_{44}(3c_{12} + 2c_{44})}{c_{12} + 2c_{44}}. \tag{75}$$

Due to axial symmetry M is the same at all other points of boundary.

On the other hand M can be calculated in terms of formal forces arising due to flexoelectricity on curved surfaces on the plate boundary. As in the case of the rod, it is necessary to smooth out sharp edges using one-quarter of toroidal surface (on each sharp edge) or one-half of the toroidal surface for the whole boundary. A small part of toroidal surface can be considered as cylindrical one, so that we can apply the equations (70) (coordinate system should be rotated). For both variants of sharp edges smoothing it turns out

$$M = \frac{Ph(c_{44}f_{12} - c_{12}f_{44})}{c_{12} + 2c_{44}}. \tag{76}$$

It only remains to equate the two expressions for M and express the curvature of the plate in terms of polarization P:

$$G = \frac{6P(c_{44}f_{12} - c_{12}f_{44})}{h^2 c_{44}(3c_{12} + 2c_{44})}. \tag{77}$$

It is interesting to compare the equation (77) with the result obtained in Ref. 4 by direct minimization of the plate energy. Elementary transformations of the equations from Ref. 4 yield the equation different from (77) only in that there is 12 instead of 6 in the numerator. However, it should be kept in mind that in Ref. 4 it is considered the case when in the thin layer near the plate surface polarization falls to zero. If, in accordance with calculations presented here, we modify the calculations[4] to the case, when the polarization is strictly homogeneous, then it appears exactly equation (77). A similar dependence on the boundary conditions imposed on the polarization also holds for direct flexoelectric effect.[16]

8. Conclusion

Above we have considered the continuum theory of the flexoelectric effect in the finite-size bodies. In the part of description of the converse flexoelectric effect, this theory turns out to be quite complicated. This is due to the fact that the independent variables are the elastic displacement. Flexoelectric part of the thermodynamic potential depends on the second spatial derivatives of the elastic displacement. Moreover, in the case of flexoelectricity the elastic energy should be considered taking into account its dependence on the second spatial derivatives of elastic displacement, or the theory is self-contradictory in the general case. All of this leads to several theoretical problems.

The first theoretical problem is that derivation of the boundary conditions for the differential equations of elastic equilibrium becomes non-trivial in such a case. While the volume integration by parts, which is necessary to express the variation of the thermodynamic potential in terms of independent variates, it appears the surface integrals containing the gradients of these independent variates. Needed transformation of these surface integrals requires special mathematical tools. For this purpose it was offered to use additional surface curvilinear coordinate system or to solve the problem in curvilinear coordinates of the general form from the beginning. From a practical point of view the second approach is more convenient although it has been proved that these two approaches are mathematically equivalent. It should also be emphasized that it appears that the curvature of the boundary surface plays an important role in the boundary conditions obtained in both approaches.

Solving the problem of elastic boundary conditions derivation, we get the opportunity to solve, in principle, any boundary problems needed to describe flexoelectricity in finite-size bodies. However, the corresponding boundary problems are extremely complicated and difficult to solve. Such a boundary problem has been solved for a homogeneously polarized ball but for more complex geometries to make a similar is unrealistic. Even for the ball the solution is extremely cumbersome. Therefore, the development of approximate methods for solving such boundary problems is desired. This is the second theoretical problem.

The problems of developing a method for the approximate solution of the corresponding boundary problems were also solved and the solution is described above. It turns out that elastic displacements can be approximated as the sum of two parts. The first part was called non-classical, it is concentrated near the surface of the body and decays exponentially inside the body. To find this part it is necessary to solve fairly simple one-dimensional equations. The second, classical part is determined by the equations of the classical theory of elasticity. It turns out that the boundary conditions for these equations have a standard classical form of the boundary conditions for the body under the external forces on the surface. Thus, the finding of the classical part is reduced to the standard classical problem which does not require a separate discussion. It should be noted that these forces on the surface are formal in nature. Physically there is no force on surface, but formal forces describe the interaction between classical and non-classical part of the elastic displacement. These formal forces are expressed in terms of non-classical part of the displacement. The corresponding equations are given.

It is important to note that for homogeneous polarization the formal forces, describing the influence of the non-classical part of displacement to the classical one, appear only on the curved parts of the surface of the body. Thus, the above statement that curvature of the surface plays an important role gets a clear physical meaning. Note that the body can be, say, a polyhedron. In this case the formal forces appear only on the edges which can be treated as a limiting case of a curved surface. While calculating one should slightly smooth these edges and tend to zero the radius of smoothing at the end of the calculations. The similar approach is applied above for the particular case of the rod and thin plate which also has the sharp edges.

Thus, in framework of the theory of continuum the description flexoelectricity in finite-size bodies actually requires special theory. This theory is described above in detail. The application of this theory is illustrated in particular problems of calculation of the bending of homogeneously polarized rod and thin plate.

Acknowledgment

A. K. Tagantsev is acknowledged for reading the manuscript.

Appendix A. A Simplified Description of the Converse Flexoelectric Effect in Finite-size Bodies

By means of (15) equations (12) can be represent in following form

$$c^{\alpha\beta\gamma\delta}u_{\gamma;\delta;\beta} + f^{\gamma\delta\alpha\beta}P_{\gamma;\delta;\beta} - v^{\alpha\beta\gamma\delta\varepsilon\zeta}u_{\gamma;\delta;\zeta;\varepsilon;\beta} = 0. \tag{A.1}$$

Boundary conditions for these differential equations are used here in the form different from (13) and (14). Particularly they are

$$\Theta^{\alpha33} = 0, \tag{A.2}$$

$$\sigma^{\alpha3} - \Theta^{\alpha\beta3}{}_{;\beta} + \Theta^{\alpha33}{}_{,3} + \Theta^{\alpha(\beta\gamma)}\Gamma^{3}_{(\beta\gamma)} = 0. \tag{A.3}$$

Here the indices enclosed in parentheses run only the values 1 and 2, if there are multiple indexes in one pair of parentheses then these indices are not equal to 3 simultaneously. Equivalence of this form of boundary conditions to (13) and (14) is proved in Ref. [12]. Certainly it is assumed that a coordinate system where equation of the body surface has a form $x^3 = x_S^3$, x_S^3 is a constant.

As discussed in the main part of the chapter the solution of the equation (A.1) is presented in the form $u_\gamma = \tilde{u}_\gamma + \hat{u}_\gamma$, where $\tilde{u}_\gamma$ obey the classical equations:

$$c^{\alpha\beta\gamma\delta}\tilde{u}_{\gamma;\delta;\beta} + f^{\gamma\delta\alpha\beta}P_{\gamma;\delta;\beta} = 0. \tag{A.4}$$

From (A.4) and (A.1) it follows that $\hat{u}_\gamma$ obey differential equations:

$$c^{\alpha\beta\gamma\delta}\hat{u}_{\gamma;\delta;\beta} - v^{\alpha\beta\gamma\delta\varepsilon\zeta}\hat{u}_{\gamma;\delta;\zeta;\varepsilon;\beta} = v^{\alpha\beta\gamma\delta\varepsilon\zeta}\tilde{u}_{\gamma;\delta;\zeta;\varepsilon;\beta}. \tag{A.5}$$

Substituting $u_\gamma = \tilde{u}_\gamma + \hat{u}_\gamma$ and some transformations also yield that the boundary conditions (A.2) and (A.3) take the form:

$$v^{\alpha 3\varepsilon\delta 3\zeta}\hat{u}_{\varepsilon;\delta;\zeta} = \frac{1}{2}f^{\delta 3\alpha 3}P_\delta - v^{\alpha 3\varepsilon\delta 3\zeta}\tilde{u}_{\varepsilon;\delta;\zeta}, \tag{A.6}$$

$$c^{\alpha 3\gamma\delta}\tilde{u}_{\gamma;\delta} + f^{\gamma\delta\alpha 3}P_{\gamma;\delta} - v^{\alpha 3\gamma\delta\varepsilon\zeta}\tilde{u}_{\gamma;\delta;\zeta;\varepsilon} - v^{\alpha(\beta)\varepsilon\delta 3\zeta}\tilde{u}_{\varepsilon;\delta;\zeta;(\beta)}$$

$$+\frac{1}{2}f^{\delta 3\alpha(\beta)}P_{\delta;(\beta)} + v^{\alpha(\beta)\varepsilon\delta(\gamma)\zeta}\tilde{u}_{\varepsilon;\delta;\zeta}\Gamma^3_{(\beta)(\gamma)}$$

$$-\frac{1}{2}f^{\delta(\gamma)\alpha(\beta)}P_\delta\Gamma^3_{(\beta)(\gamma)} + c^{\alpha 3\gamma\delta}\hat{u}_{\gamma;\delta} - v^{\alpha 3\gamma\delta\varepsilon\zeta}\hat{u}_{\gamma;\delta;\zeta;\varepsilon}$$

$$- v^{\alpha(\beta)\varepsilon\delta 3\zeta}\hat{u}_{\varepsilon;\delta;\zeta;(\beta)} + v^{\alpha(\beta)\varepsilon\delta(\gamma)\zeta}\hat{u}_{\varepsilon;\delta;\zeta}\Gamma^3_{(\beta)(\gamma)} = 0. \tag{A.7}$$

Equations above are exact. In order to turn into approximation one should note that real values of $v^{\alpha\beta\gamma\delta\varepsilon\zeta}$ are small. It is convenient to assume that all of them are proportional to a scalar $v \to 0$. It is assumed also that all $c^{\alpha\beta\gamma\delta}$ is proportional to a scalar c. If $v \to 0$ then the right-hand side of (A.5) and the second term on the right-hand side of (A.6) can be neglected. Indeed, since $\tilde{u}_\gamma$ obey the classical equation (A.4), in the case of sufficiently smooth surface and in the absence of polarization gradients tend to infinity in the limit $v \to 0$, the classical part cannot have large gradients which can compensate smallness of $v^{\alpha\beta\gamma\delta\varepsilon\zeta}$. As for the non-classical part $\hat{u}_\gamma$, the situation is different: here such compensation is possible, but only if the derivatives are in x^3. Moreover, in a thin layer near the surface covariant derivatives in x^3 can be replaced by the usual derivatives, Γ-terms give only small corrections here. Thus, (A.5) and (A.6) can be approximately replaced by

$$c^{\alpha 3\gamma 3}\hat{u}_{\gamma,3,3} - v^{\alpha 3\gamma 333}\hat{u}_{\gamma,3,3,3,3} = 0, \tag{A.8}$$

$$v^{\alpha 3\varepsilon 333}\hat{u}_{\varepsilon,3,3} = \frac{1}{2}f^{\delta 3\alpha 3}P_\delta. \tag{A.9}$$

Naturally decomposition $u_\gamma = \tilde{u}_\gamma + \hat{u}_\gamma$ is not unique. By requiring that $\tilde{u}_\gamma$ obey (45), this non-uniqueness is restricted, but is not eliminated completely. It is clear that $\tilde{u}_\gamma$ does not contain a non-classical part, but it does not mean that $\hat{u}_\gamma$ does not contain a classical part. To eliminate remaining non-uniqueness one should require that $\hat{u}_\gamma$ exponentially decay inside the body. Under such condition $\hat{u}_\gamma$ obey not only (A.8) but also

$$c^{\alpha 3\gamma 3}\hat{u}_\gamma - v^{\alpha 3\gamma 333}\hat{u}_{\gamma,3,3} = 0. \tag{A.10}$$

Equations (A.10) are a system of three ordinal differential equations, the dependence on x^1 and x^2 is parametric here. Moreover, in a thin layer near the body surface one can assume that the coefficients do not depend on x^3, they are approximately equal to the surface values. Solution of such a system is easy. In a standard way one should find a fundamental basis set of solutions in the form $\hat{u}_\gamma = \bar{u}_\gamma e^{\lambda(x^3-x_S^3)}$. Obviously equations for amplitudes $\bar{u}_\gamma$ are

$$c^{\alpha 3\gamma 3}\bar{u}_\gamma = \lambda^2 v^{\alpha 3\gamma 333}\bar{u}_\gamma. \tag{A.11}$$

Equation (A.11) is a standard generalized eigenvalue problem for symmetric positive definite 3×3 matrices. Therefore, λ_n and $\bar{u}_\gamma^n$ are calculable by standard way. If the body bulk corresponds to $x^3 \leq x_S^3$ for definiteness, then λ_n equals the positive square root of the nth eigenvalue, and general representation of non-classical part is

$$\hat{u}_\gamma = \sum_{n=1}^{3} a_n \bar{u}_\gamma^n e^{\lambda_n(x^3-x_S^3)}, \tag{A.12}$$

where only the coefficients a_n are unknown. The latter can be found easily by (A.9) which yields a simple system of linear algebraic equations:

$$\sum_{n=1}^{3} \lambda_n^2 v^{\alpha 3\gamma 333}\bar{u}_\gamma^n a_n = \frac{1}{2}f^{\delta 3\alpha 3}P_\delta. \tag{A.13}$$

Note that the smallness of $v^{\alpha 3\gamma 333}$ in (A.13) is compensated by λ_n^2 and using (A.11) this system of equations can be rewritten as follows:

$$\sum_{n=1}^{3} c^{\alpha 3\gamma 3}\bar{u}_\gamma^n a_n = \frac{1}{2}f^{\delta 3\alpha 3}P_\delta. \tag{A.14}$$

Thus, the non-classical part of elastic displacements is found completely in explicit form.

As it follows from above, to find completely the non-classical part of elastic displacements one needs only the boundary conditions (A.6). The remaining boundary conditions (A.7) yield the boundary conditions for the classical equations (A.4) in this way, one should only substitute known $\hat{u}_\gamma$ to (A.7) and select the terms of the appropriate order of accuracy.

However, within this derivation of the boundary conditions for the equations (A.4) a singularity arises requiring additional analysis. This singularity is related to the fact that it appears as a subtraction of two close but large terms in which it is impossible to use the approximate expression for $\hat{u}_\gamma$ derived above. In these (and only these) terms it is necessary to consider a corrections to such $\hat{u}_\gamma$. The appropriate analysis is made below and while $\hat{u}_\gamma$ means the exact non-classical part of elastic displacements rather than approximate one defined by (A.12).

According to the above, all the terms in (A.7), where gradients of $\tilde{u}_\gamma$ are convolved with $v^{\alpha\beta\gamma\delta\varepsilon\zeta}$, should be omitted for the same reasons as in the case of the equation (A.5). Further it is convenient to introduce the notation:

$$s^\alpha = -c^{\alpha 3\gamma\delta}\hat{u}_{\gamma;\delta} + v^{\alpha 3\gamma\delta\varepsilon\zeta}\hat{u}_{\gamma;\delta;\zeta;\varepsilon} + v^{\alpha(\beta)\gamma\delta 3\zeta}\hat{u}_{\gamma;\delta;\zeta;(\beta)}$$

$$- v^{\alpha(\beta)\varepsilon\delta(\gamma)\zeta}\hat{u}_{\varepsilon;\delta;\zeta}\Gamma^3_{(\beta)(\gamma)}. \tag{A.15}$$

With this notation the boundary conditions to the classical equations can be written as follows:

$$c^{\alpha 3\gamma\delta}\tilde{u}_{\gamma;\delta} = s^\alpha - f^{\gamma\delta\alpha 3}P_{\gamma;\delta} - \frac{1}{2}f^{\delta 3\alpha(\beta)}P_{\delta;(\beta)} + \frac{1}{2}f^{\delta(\gamma)\alpha(\beta)}P_\delta\Gamma^3_{(\beta)(\gamma)}. \tag{A.16}$$

Now all the terms to be further simplified are contained in s^α. Carrying out simplification, first of all one should leave in the last term only the partial (not covariant) derivatives in x^3. The reason for this is the same as above. Next, one should express the remaining covariant derivatives in terms of partial derivatives and Γ-terms. The exact expression for the third covariant derivative of the vector is very cumbersome. But keeping in mind that these third covariant derivatives are convolved with $v^{\alpha\beta\gamma\delta\varepsilon\zeta}$, one can omit all the terms where initial vector is not differentiated at least two times. So it turns out:

$$\hat{u}_{\alpha;\beta;\gamma;\zeta} \approx \hat{u}_{\alpha,\beta,\gamma,\zeta} - \hat{u}_{\delta,\beta,\zeta}\Gamma^{\delta}_{\alpha\gamma} - \hat{u}_{\delta,\gamma,\zeta}\Gamma^{\delta}_{\alpha\beta} - \hat{u}_{\delta,\beta,\gamma}\Gamma^{\delta}_{\alpha\zeta}$$
$$- \hat{u}_{\alpha,\delta,\zeta}\Gamma^{\delta}_{\beta\gamma} - \hat{u}_{\alpha,\delta,\gamma}\Gamma^{\delta}_{\beta\zeta} - \hat{u}_{\alpha,\beta,\delta}\Gamma^{\delta}_{\gamma\zeta}. \tag{A.17}$$

With this equation and simplifications mentioned above we get:

$$s^{\alpha} = -c^{\alpha3\gamma\delta}\hat{u}_{\gamma;\delta} + v^{\alpha3\gamma\delta\varepsilon\zeta}\hat{u}_{\gamma,\delta,\zeta,\varepsilon} - v^{\alpha3\gamma\delta\varepsilon\zeta}\hat{u}_{\rho,\delta,\varepsilon}\Gamma^{\rho}_{\gamma\zeta}$$
$$- v^{\alpha3\gamma\delta\varepsilon\zeta}\hat{u}_{\rho,\zeta,\varepsilon}\Gamma^{\rho}_{\gamma\delta} - v^{\alpha3\gamma\delta\varepsilon\zeta}\hat{u}_{\rho,\delta,\zeta}\Gamma^{\rho}_{\gamma\varepsilon} - v^{\alpha3\gamma\delta\varepsilon\zeta}\hat{u}_{\gamma,\rho,\varepsilon}\Gamma^{\rho}_{\delta\zeta}$$
$$- v^{\alpha3\gamma\delta\varepsilon\zeta}\hat{u}_{\gamma,\rho,\zeta}\Gamma^{\rho}_{\delta\varepsilon} - v^{\alpha3\gamma\delta\varepsilon\zeta}\hat{u}_{\gamma,\delta,\rho}\Gamma^{\rho}_{\zeta\varepsilon} + v^{\alpha(\beta)\gamma\delta3\zeta}\hat{u}_{\gamma,\delta,\zeta,(\beta)}$$
$$- v^{\alpha(\beta)\gamma\delta3\zeta}\hat{u}_{\rho,\delta,(\beta)}\Gamma^{\rho}_{\gamma\zeta} - v^{\alpha(\beta)\gamma\delta3\zeta}\hat{u}_{\rho,\zeta,(\beta)}\Gamma^{\rho}_{\gamma\delta} - v^{\alpha(\beta)\gamma\delta3\zeta}\hat{u}_{\rho,\delta,\zeta}\Gamma^{\rho}_{\gamma(\beta)}$$
$$- v^{\alpha(\beta)\gamma\delta3\zeta}\hat{u}_{\gamma,\rho,(\beta)}\Gamma^{\rho}_{\delta\zeta} - v^{\alpha(\beta)\gamma\delta3\zeta}\hat{u}_{\gamma,\rho,\zeta}\Gamma^{\rho}_{\delta(\beta)}$$
$$- v^{\alpha(\beta)\gamma\delta3\zeta}\hat{u}_{\gamma,\delta,\rho}\Gamma^{\rho}_{\zeta(\beta)} - v^{\alpha(\beta)\varepsilon3(\gamma)3}\hat{u}_{\varepsilon,3,3}\Gamma^{3}_{(\beta)(\gamma)}. \tag{A.18}$$

Again, using the fact that to compensate smallness of $v^{\alpha\beta\gamma\delta\varepsilon\zeta}$ one needs at least two derivatives in x^3, this equation is further simplified:

$$s^{\alpha} = -c^{\alpha3\gamma\delta}\hat{u}_{\gamma;\delta} + v^{\alpha3\gamma\delta\varepsilon\zeta}\hat{u}_{\gamma,\delta,\zeta,\varepsilon} - v^{\alpha3\gamma33\zeta}\hat{u}_{\rho,3,3}\Gamma^{\rho}_{\gamma\zeta}$$
$$- v^{\alpha3\gamma\delta33}\hat{u}_{\rho,3,3}\Gamma^{\rho}_{\gamma\delta} - v^{\alpha3\gamma3\varepsilon3}\hat{u}_{\rho,3,3}\Gamma^{\rho}_{\gamma\varepsilon} - v^{\alpha3\gamma\delta3\zeta}\hat{u}_{\gamma,3,3}\Gamma^{3}_{\delta\zeta}$$
$$- v^{\alpha3\gamma\delta\varepsilon3}\hat{u}_{\gamma,3,3}\Gamma^{3}_{\delta\varepsilon} - v^{\alpha3\gamma3\varepsilon\zeta}\hat{u}_{\gamma,3,3}\Gamma^{3}_{\zeta\varepsilon} + v^{\alpha(\beta)\gamma333}\hat{u}_{\gamma,3,3,(\beta)}$$
$$- v^{\alpha(\beta)\gamma333}\hat{u}_{\rho,3,3}\Gamma^{\rho}_{\gamma(\beta)} - v^{\alpha(\beta)\gamma\delta33}\hat{u}_{\gamma,3,3}\Gamma^{3}_{\delta(\beta)}$$
$$- v^{\alpha(\beta)\gamma33\zeta}\hat{u}_{\gamma,3,3}\Gamma^{3}_{\zeta(\beta)} - v^{\alpha(\beta)\varepsilon3(\gamma)3}\hat{u}_{\varepsilon,3,3}\Gamma^{3}_{(\beta)(\gamma)}. \tag{A.19}$$

It is convenient to introduce yet another notation:

$$h^{\alpha\rho} = v^{\alpha3\gamma33\zeta}\Gamma^{\rho}_{\gamma\zeta} + v^{\alpha3\gamma\delta33}\Gamma^{\rho}_{\gamma\delta} + v^{\alpha3\gamma3\varepsilon3}\Gamma^{\rho}_{\gamma\varepsilon} + v^{\alpha3\rho\delta3\zeta}\Gamma^{3}_{\delta\zeta}$$
$$+ v^{\alpha3\rho\delta\varepsilon3}\Gamma^{3}_{\delta\varepsilon} + v^{\alpha3\rho3\varepsilon\zeta}\Gamma^{3}_{\zeta\varepsilon} + v^{\alpha(\beta)\gamma333}\Gamma^{\rho}_{\gamma(\beta)}$$
$$+ v^{\alpha(\beta)\rho\delta33}\Gamma^{3}_{\delta(\beta)} + v^{\alpha(\beta)\rho33\zeta}\Gamma^{3}_{\zeta(\beta)} + v^{\alpha(\beta)\rho3(\gamma)3}\Gamma^{3}_{(\beta)(\gamma)}. \tag{A.20}$$

Using the symmetry properties of $v^{\alpha\beta\gamma\delta\varepsilon\zeta}$ it can be rewritten as follows:

$$h^{\alpha\beta} = (v^{\alpha3\gamma3\delta3} + 2v^{\alpha3\gamma\delta33})\Gamma^{\beta}_{\gamma\delta} + (2v^{\alpha3\beta\gamma\delta3} + v^{\alpha3\beta\gamma3\delta})\Gamma^{3}_{\gamma\delta}$$

$$+ 2v^{\alpha3\beta\gamma(\delta)3}\Gamma^{3}_{\gamma(\delta)} + v^{\alpha3\gamma3(\delta)3}\Gamma^{\beta}_{\gamma(\delta)} + v^{\alpha(\gamma)\beta3(\delta)3}\Gamma^{3}_{(\gamma)(\delta)}. \quad (A.21)$$

In this notation the expression for s^{α} can be written as follows:

$$s^{\alpha} = -c^{\alpha3\gamma\delta}\hat{u}_{\gamma,\delta} + c^{\alpha3\gamma\delta}\hat{u}_{\varepsilon}\Gamma^{\varepsilon}_{\gamma\delta} + v^{\alpha3\gamma\delta\varepsilon\zeta}\hat{u}_{\gamma,\delta,\zeta,\varepsilon}$$

$$+ v^{\alpha(\beta)\gamma333}\hat{u}_{\gamma,3,3,(\beta)} - h^{\alpha\beta}\hat{u}_{\beta,3,3}. \quad (A.22)$$

Now one can proceed to the analysis of the singularity mentioned above. It appears in the first and third term of (A.22), when all the derivatives are in x^3. For all other terms there is no singularity and one can use the expression (A.12) for $\hat{u}_{\gamma}$. It is also clear that singular terms exactly cancel each other out if approximate expression (A.12) is used. This is due to the explicit form of this expression. One might think that these terms should simply be excluded, but in reality the situation is more complicated.

The fact is that $-c^{\alpha3\gamma3}\hat{u}_{\gamma,3,3} + v^{\alpha3\gamma333}\hat{u}_{\gamma,3,3,3,3}$ is equal to zero only for the main approximation for $\hat{u}_{\gamma}$. But the corrections to this approximation may result in the fact that this expression becomes finite. Henceforth we denote such corrections as w_{γ} keeping notation $\hat{u}_{\gamma}$ only for the main approximation. In this notation it turns out:

$$s^{\alpha} = -c^{\alpha3\gamma3}w_{\gamma,3} + v^{\alpha3\gamma333}w_{\gamma,3,3,3} - c^{\alpha3\gamma(\delta)}\hat{u}_{\gamma,(\delta)} + c^{\alpha3\gamma\delta}\hat{u}_{\varepsilon}\Gamma^{\varepsilon}_{\gamma\delta}$$

$$+ v^{\alpha3\gamma\delta\varepsilon\zeta}\hat{u}_{\gamma(,\delta,\zeta,\varepsilon)} + v^{\alpha(\beta)\gamma333}\hat{u}_{\gamma,3,3,(\beta)} - h^{\alpha\beta}\hat{u}_{\beta,3,3}. \quad (A.23)$$

To compensate the smallness of $v^{\alpha\beta\gamma\delta\varepsilon\zeta}$ in the fifth term of this equation there should be at least two differentiations in x^3. Thus, after some reductions one can rewrite (A.23) as follows:

$$s^{\alpha} = -c^{\alpha3\gamma3}w_{\gamma,3} + v^{\alpha3\gamma333}w_{\gamma,3,3,3} - c^{\alpha3\gamma(\delta)}\hat{u}_{\gamma,(\delta)} + c^{\alpha3\gamma\delta}\hat{u}_{\varepsilon}\Gamma^{\varepsilon}_{\gamma\delta}$$

$$+ 2(v^{\alpha3\gamma(\beta)33} + v^{\alpha3\gamma3(\beta)3})\hat{u}_{\gamma,3,3,(\beta)} - h^{\alpha\beta}\hat{u}_{\beta,3,3}. \quad (A.24)$$

It is clear that in (A.24) one should use only corrections w_{γ} of order $v^{1/2}$. Only such corrections yield a contribution of the same order as the other terms. To find the corresponding contributions, we should keep in the exact equation

$$c^{\alpha\beta\gamma\delta}(\hat{u}_\gamma + w_\gamma)_{;\delta;\beta} - v^{\alpha\beta\gamma\delta\varepsilon\zeta}(\hat{u}_\gamma + w_\gamma)_{;\delta;\zeta;\varepsilon;\beta} = v^{\alpha\beta\gamma\delta\varepsilon\zeta}\tilde{u}_{\gamma;\delta;\zeta;\varepsilon;\beta}$$

$$(A.25)$$

only the terms of the order $v^{-1/2}$. Indeed if $w_\gamma \sim v^{1/2}$ then the main terms $w_{\gamma,3,3}$ and $v^{\alpha3\gamma333}w_{\gamma,3,3,3,3}$ have just such an order. Note that although there are terms $c^{\alpha3\gamma3}\hat{u}_{\gamma,3,3}$ and $v^{\alpha3\gamma333}\hat{u}_{\gamma,3,3,3,3}$ in the equation, which are of the order $\sim v^{-1}$, now they cancel each other completely due to the fact that now $\hat{u}_\gamma$ denotes the main approximation. Right hand side of (A.25) is $\sim v$, so it should be omitted. The terms $c^{\alpha\beta\gamma\delta}w_{\gamma(;\delta;\beta)}$ and $v^{\alpha\beta\gamma\delta\varepsilon\zeta}w_{\gamma(;\delta;\zeta;\varepsilon;\beta)}$ should be omitted by the same reasons. Thus (A.25) is reduced to

$$-c^{\alpha3\gamma3}w_{\gamma,3,3} + v^{\alpha3\gamma333}w_{\gamma,3,3,3,3}$$

$$= c^{\alpha\beta\gamma\delta}\hat{u}_{\gamma(;\delta;\beta)} - v^{\alpha\beta\gamma\delta\varepsilon\zeta}\hat{u}_{\gamma(;\delta;\zeta;\varepsilon;\beta)}. \qquad (A.26)$$

Equation (A.26) requires further simplification: in the right-hand side we should keep only the terms $\sim v^{-1/2}$. For this in the first term there should be one partial derivative in x^3, and in the second term there should be three derivatives in x^3. This is why we simplify second and fourth covariant derivatives as follows:

$$\hat{u}_{\gamma;\delta;\beta} \approx \hat{u}_{\gamma,\delta,\beta} - \hat{u}_{\gamma,\rho}\Gamma^\rho_{\delta\beta} - \hat{u}_{\rho,\delta}\Gamma^\rho_{\gamma\beta} - \hat{u}_{\rho,\beta}\Gamma^\rho_{\gamma\delta}, \qquad (A.27)$$

$$\hat{u}_{\gamma;\delta;\zeta;\varepsilon;\beta} \approx \hat{u}_{\gamma,\delta,\zeta,\varepsilon,\beta} - \hat{u}_{\rho,\delta,\zeta,\varepsilon}\Gamma^\rho_{\gamma\beta} - \hat{u}_{\gamma,\rho,\zeta,\varepsilon}\Gamma^\rho_{\delta\beta}$$

$$- \hat{u}_{\gamma,\delta,\rho,\varepsilon}\Gamma^\rho_{\zeta\beta} - \hat{u}_{\gamma,\delta,\zeta,\rho}\Gamma^\rho_{\varepsilon\beta} - \hat{u}_{\rho,\delta,\varepsilon,\beta}\Gamma^\rho_{\gamma\zeta} - \hat{u}_{\rho,\zeta,\varepsilon,\beta}\Gamma^\rho_{\gamma\delta}$$

$$- \hat{u}_{\rho,\delta,\zeta,\beta}\Gamma^\rho_{\gamma\varepsilon} - \hat{u}_{\gamma,\rho,\varepsilon,\beta}\Gamma^\rho_{\delta\zeta} - \hat{u}_{\gamma,\rho,\zeta,\beta}\Gamma^\rho_{\delta\varepsilon} - \hat{u}_{\gamma,\delta,\rho,\beta}\Gamma^\rho_{\zeta\varepsilon}.$$

$$(A.28)$$

Eventually the simplified equation (A.26) takes the form:

$$-c^{\alpha3\gamma3}w_{\gamma,3,3} + v^{\alpha3\gamma333}w_{\gamma,3,3,3,3}$$

$$= (c^{\alpha3\gamma(\beta)} + c^{\alpha(\beta)\gamma3})\hat{u}_{\gamma,(\beta),3}$$

$$- c^{\alpha\beta\gamma\delta}\Gamma^3_{\delta\beta}\hat{u}_{\gamma,3} - (c^{\alpha\beta\gamma3} + c^{\alpha3\gamma\beta})\Gamma^\varepsilon_{\gamma\beta}\hat{u}_{\varepsilon,3} - 2(v^{\alpha3\gamma(\beta)33}$$

$$+ v^{\alpha(\beta)\gamma333})\hat{u}_{\gamma,(\beta),3,3,3} + 2(v^{\alpha\beta\gamma333}\Gamma^\rho_{\gamma\beta} + v^{\alpha3\gamma\delta33}\Gamma^\rho_{\gamma\delta})\hat{u}_{\rho,3,3,3}$$

$$+ (4v^{\alpha\beta\gamma\delta33}\Gamma^3_{\delta\beta} + v^{\alpha\beta\gamma3\delta3}\Gamma^3_{\delta\beta} + v^{\alpha3\gamma\delta3\beta}\Gamma^3_{\delta\beta})\hat{u}_{\gamma,3,3,3}.$$

$$(A.29)$$

Taking into account that near the body surface the Christoffel symbols and material tensors can be considered as constants, the equation (A.29) has a very specific form: all of its terms are differentiated with respect to x^3 at least once. Therefore, there exists a particular solution of this equation which obeys an equation with one less derivation:

$$-c^{\alpha3\gamma3}w_{\gamma,3} + v^{\alpha3\gamma333}w_{\gamma,3,3,3}$$

$$= (c^{\alpha3\gamma(\beta)} + c^{\alpha(\beta)\gamma3})\hat{u}_{\gamma,(\beta)}$$

$$- c^{\alpha\beta\gamma\delta}\Gamma^3_{\delta\beta}\hat{u}_\gamma - (c^{\alpha\beta\gamma3} + c^{\alpha3\gamma\beta})\Gamma^\varepsilon_{\gamma\beta}\hat{u}_\varepsilon - 2(v^{\alpha3\gamma(\beta)33}$$

$$+ v^{\alpha(\beta)\gamma333})\hat{u}_{\gamma,(\beta),3,3} + 2(v^{\alpha\beta\gamma333}\Gamma^\rho_{\gamma\beta} + v^{\alpha3\gamma\delta33}\Gamma^\rho_{\gamma\delta})\hat{u}_{\rho,3,3}$$

$$+ (4v^{\alpha\beta\gamma\delta33}\Gamma^3_{\delta\beta} + v^{\alpha\beta\gamma3\delta3}\Gamma^3_{\delta\beta} + v^{\alpha3\gamma\delta3\beta}\Gamma^3_{\delta\beta})\hat{u}_{\gamma,3,3}. \qquad (A.30)$$

Note that in the left-hand side of (A.30) there is exactly the expression that is needed in (A.24). The general solution of the inhomogeneous differential equation is the sum of a particular solution of the inhomogeneous equation and the general solution of the homogeneous equation. But the remarkable fact is that any solution of the homogeneous equation decaying away from the surface when substituted into $-c^{\alpha3\gamma3}w_{\gamma,3} + v^{\alpha3\gamma333}w_{\gamma,3,3,3}$ yields zero. So it is quite enough to consider only a particular solution, and one can simply replace $-c^{\alpha3\gamma3}w_{\gamma,3} + v^{\alpha3\gamma333}w_{\gamma,3,3,3}$ in (A.24) by right-hand side of (A.30). Having made the replacement, we obtain s^α in the form:

$$s^\alpha = -c^{\alpha3\gamma(\delta)}\hat{u}_{\gamma,(\delta)} + c^{\alpha3\gamma\delta}\hat{u}_\varepsilon\Gamma^\varepsilon_{\gamma\delta} + (c^{\alpha3\gamma(\beta)} + c^{\alpha(\beta)\gamma3})\hat{u}_{\gamma,(\beta)}$$

$$- c^{\alpha\beta\gamma\delta}\Gamma^3_{\delta\beta}\hat{u}_\gamma - (c^{\alpha\beta\gamma3} + c^{\alpha3\gamma\beta})\Gamma^\varepsilon_{\gamma\beta}\hat{u}_\varepsilon - 2(v^{\alpha3\gamma(\beta)33}$$

$$+ v^{\alpha(\beta)\gamma333})\hat{u}_{\gamma,(\beta),3,3} + 2v^{\alpha3\gamma(\delta)33}\hat{u}_{\gamma,3,3,(\delta)}$$

$$+ 2v^{\alpha3\gamma3(\beta)3}\hat{u}_{\gamma,3,3,(\beta)} - h^{\alpha\beta}\hat{u}_{\beta,3,3}. \qquad (A.31)$$

Elementary transformations lead (A.31) to a rather simple form:

$$s^\alpha = c^{\alpha(\beta)\gamma 3}\hat{u}_{\gamma,(\beta)} - c^{\alpha\beta\gamma\delta}\Gamma^3_{\delta\beta}\hat{u}_\gamma - c^{\alpha\beta\gamma 3}\Gamma^\varepsilon_{\gamma\beta}\hat{u}_\varepsilon - h^{\alpha\beta}\hat{u}_{\beta,3,3}, \quad \text{(A.32)}$$

where $h^{\alpha\beta}$ is redefined as

$$h^{\alpha\beta} = v^{\alpha(\gamma)\beta 3(\delta)3}\Gamma^3_{(\gamma)(\delta)} - v^{\alpha 3\gamma 333}\Gamma^\beta_{\gamma 3}$$
$$- 2v^{\alpha 3\beta\delta 33}\Gamma^3_{\delta 3} - v^{\alpha\varepsilon\beta 3\delta 3}\Gamma^3_{\delta\varepsilon}. \quad \text{(A.33)}$$

These equations together with (A.16) complete the construction of the approximate method considered here.

References

1. J. D. Axe, J. Harada, and G. Shirane. *Phys. Rev. B.* **1**, 1227 (1970).
2. P. V. Yudin, A. K. Tagantsev, E. A. Eliseev, A. N. Morozovska, and N. Setter. *Phys. Rev. B* **86**, 134102 (2012).
3. E. V. Bursian and O. I. Zaikovskii. *Sov. Phys. Sol. State* **10**, 1121 (1968).
4. A. K. Tagantsev and A. S. Yurkov. *J. Appl. Phys.* **112**(4), 044103 (2012).
5. A. S. Yurkov. *JETP Lett.* **94**(6), 455–458 (2011).
6. E. A. Eliseev, A. N. Morozovska, M. D. Glinchuk, and R. Blinc. *Phys. Rev. B* **79**, 165433 (2009).
7. L. E. Cross. *J. Mater. Sci.* **41**(1), 53 (2006).
8. B. Chu, W. Zhu, N. Li, and L. E. Cross. *J. Appl. Phys.* **106**(10), 104109 (2009).
9. A. S. Yurkov. Flexoelectric deformation of a homogeneously polarized ball. *arXiv:1304.1868 [cond-mat.mtrl-sci]* (2013).
10. A. S. Yurkov. *JETP Lett.* **99**(4), 214 (2014).
11. P. V. Yudin and A. K. Tagantsev. *Nanotechnology* **24**, 432001 (2013).
12. A. S. Yurkov. Mechanical boundary condition for a case when thermodynamic potential depends on strain gradients. In *Flexoelectricity in Solids: From Theory to Applications.* Ed. by A. K. Tagantsev and P. V. Yudin. World Scientific, Singapore (2016).
13. D. A. Varshalovich, A. N. Moskalev, and V. K. Khersonskii. *Quantum Theory of Angular Momentum.* World Scientific Pub. Co. (1987).
14. A. S. Yurkov. *Phys. Solid State* **57**(3), 460 (2015).

15. L. D. Landau and E. M. Lifshitz. *Theory of Elasticity*. Pergamon, Oxford (1975).

16. A. S. Yurkov and A. K. Tagantsev. Impact of surface phenomena on direct bulk flexoelectric effect in finite samples. *arXiv:1501.03365 [cond-mat.mtrl-sci]* (2015).

Chapter 6

Flexoelectricity and Phonon Spectra

P. V. Yudin

Ceramics Laboratory,
Swiss Federal Institute of Technology EPFL,
CH-1015 Lausanne, Switzerland

Novosibirsk State University,
630090 Novosibirsk, Pirogova st. 2, Russia

A. Kvasov

Ceramics Laboratory,
Swiss Federal Institute of Technology EPFL,
CH-1015 Lausanne, Switzerland

A. K. Tagantsev

Ceramics Laboratory,
Swiss Federal Institute of Technology EPFL,
CH-1015 Lausanne, Switzerland

Ferroics Laboratory, Ioffe Physical Technical Institute,
194021 St. Petersburg, Russia

Flexoelectric coupling manifests itself as interaction of acoustic and soft-mode optic phonon branches in solids. Here, we study this interaction using Landau theory and *ab initio* simulations. The outputs of the theory are upper bounds for static bulk flexoelectric coupling and a method for extraction of flexocoupling coefficients from phonon dispersion curves. In contrast to conventional methods for characterization of flexoelectric effect, where a finite sample is subject to inhomogeneous strain and polarization response is studied, the method based on phonon spectra is free of surface contributions and allows to directly extract information on bulk static and dynamic flexoelectric coefficients. Using example of

$SrTiO_3$ perovskite material it was demonstrated by means of *ab initio* simulations that the strength of dynamic bulk flexoelectric effect is comparable to that of the static effect.

1. Introduction

The flexoelectric effect looks promising for practical applications and helps to explain a number of phenomena, especially happening at the nanoscale. However, the available theoretical and experimental results are rather contradictory, attesting to a limited understanding of flexoelectricity. Flexoelectric effect actually represents four related phenomena: static and dynamic bulk flexoelectric effects, surface flexoelectric effect, and surface piezoelectricity.[1] The bulk flexoelectric effect can be split into static and dynamic effects. The static bulk flexoelectric response is controlled by the redistribution of the bound charge of a crystal driven by a strain gradient, which appears, for example, in a bent plate. The dynamic effect is absent in the quasi-static plate bending experiment, however this effect can be present as a sound wave in solids.

The static effect was widely addressed in terms of first principles calculations. The ionic contribution to static flexoelectricity was evaluated for several perovskite ferroelectrics by Maranganti and Sharma[2] using the framework offered by Tagantsev.[3] *Ab initio* calculations of this contribution were also performed by Hong *et al.*[4] and Ponomareva *et al.*[5] for $SrTiO_3$ (STO), $BaTiO_3$, and their solid solution. The first-principles calculations of the electronic and lattice contributions to flexoelectricity have been done for a number of crystals by Hong and Vanderbilt[6,7] and Stengel.[8] The calculations of the surface contribution were also done by Stengel.[9] On the other hand, there are no first principles calculations of the dynamic flexoelectric effect, and information about its magnitude is missing. At the same time, the dynamic flexoelectric effect is expected to be comparable with the static one.[3]

Despite an increasing interest to the flexoelectric effect both theoretical and experimental, it remains challenging to measure. Zubko *et al.*[10] employed a bending method to characterize the bulk static flexoelectric response in single crystals of STO. However, this method has a substantial drawback, which is the presence of surface contribution to bulk flexoelectric response. There exists an alternative method of determination of bulk flexoelectric coefficients based on the analysis of the phonon spectrum in solids, where the flexoelectric response contains both static and dynamic contributions.[1] In terms of phonons, the flexoelectric interaction can be interpreted as a repulsion between transverse acoustic (TA) and soft-mode transverse optic (TO) branches. This repulsion was observed in perovskite ferroelectrics by Axe *et al.*, who studied the dispersion of the phonons in $KTaO_3$[11] and $PbTiO_3$[12] by means of neutron scattering and by Hehlen *et al.*[13] The advantage of the method based on phonon dispersion analysis is that it does not suffer from presence of the surface contribution to the flexoelectric response, however, current approaches based on this method yields only total bulk flexoelectric coefficients and cannot resolve the static and dynamic contributions.

In this chapter, we first introduce the bulk flexoelectric effect in crystals using Landau theory in Section 2. In Section 3, we consider the repulsion of TA and soft-mode TO phonon branches in terms of flexoelectric coupling. Next Section 4 is devoted to upper bounds for static bulk flexocoupling coefficients, which are obtained from the requirement of the absence of incommensurate phases in the material. Then, in Section 5, we will demonstrate how the static and dynamic contributions to flexoelectric response can be resolved analyzing the phonon dispersion branches with the example of the simulated spectrum of STO cubic crystal. The developed method can be used with real experimental phonon spectrum to get information on flexoelectric coefficients. Section 6 describes an alternative method of calculation of the bulk flexocoupling tensors, which involves dynamical matrix. We will sum up the results in conclusion.

2. Landau Theory of Bulk Flexoelectric Coupling in Crystals

Following Yudin and Tagantsev,[1] we consider a phenomenological approach which provides an adequate description of the bulk flexoelectric effect. However, in contrast to the piezoelectric response, the treatment of the flexoelectric effect in the static (e.g. in a bent plate) and dynamic (in a sound wave) situations generally requires separate study.[3] Let us start with the static case.

2.1. *Static Flexoelectric Effect*

One traditionally introduces the flexoelectric effect via the constitutive equation for the electric polarization P_i[14]:

$$P_i = \chi_{ij} E_j + e_{ijk} u_{jk} + \mu_{klij} \frac{\partial u_{kl}}{\partial x_j}, \tag{1}$$

where E_i, u_{jk}, and $\frac{\partial u_{kl}}{\partial x_j}$ are the macroscopic electric field, the strain tensor, and its spatial gradient, respectively. The first two rhs terms of equation (1) describe the dielectric and piezoelectric responses with the tensor of the clamped dielectric susceptibility χ_{ij} and the piezoelectric tensor e_{ijk}, respectively. The last rhs term of equation (1) describes the linear polarization response to a strain gradient — flexoelectric effect. The strain tensor is defined as the symmetric part of the tensor $\frac{\partial U_i}{\partial x_j}$, where U_i is the displacement of point of the medium in the direction i:

$$u_{jk} = \frac{1}{2} \left(\frac{\partial U_j}{\partial x_k} + \frac{\partial U_k}{\partial x_j} \right). \tag{2}$$

The antisymmetric part of the tensor $\frac{\partial U_i}{\partial x_j}$

$$\Omega_{jk} = \frac{1}{2} \left(\frac{\partial U_j}{\partial x_k} - \frac{\partial U_k}{\partial x_j} \right), \tag{3}$$

corresponding to rotations of the sample as a whole. Homogeneous rotations evidently do not contribute to the polarization response. At the same time, the gradients of Ω_{jk} can contribute to the polarization response, however they can always be presented as a sum of the components of tensor $\frac{\partial u_{kl}}{\partial x_j}$, as it was shown by Indenbom *et al.*[15] The fourth-rank tensor μ_{klij} controlling the flexoelectric effect in equation (1) is the *flexoelectric tensor*, it is symmetric with respect to the permutation of the first two suffixes. The flexoelectric tensor is allowed in materials of any symmetry (including amorphous), in a sharp contrast to the piezoelectric tensor, e_{ijk}, which is a third-rank tensor and, therefore, allowed only in non-centrosymmetric materials. This makes the principal difference between piezoelectricity and flexoelectricity, as the latter is a general phenomenon having no symmetry limitations. Since the piezoelectric and flexoelectric tensors describe the properties of a material in the absence of a macroscopical electric field, these can also be defined as

$$e_{ijk} = \left(\frac{\partial P_i}{\partial u_{jk}} \right)_{E=0}, \tag{4}$$

$$\mu_{klij} = \left(\frac{\partial P_i}{\partial \left(\frac{\partial u_{kl}}{\partial x_j} \right)} \right)_{E=0}. \tag{5}$$

A more advanced description of piezoelectric and flexoelectric effects is based on minimizing of thermodynamic potential.[1] Such a minimization yields the bulk constitutive electromechanical equations:

$$E_i = \chi_{ij}^{-1} P_j - f_{klij} \frac{\partial u_{kl}}{\partial x_j} - g_{ijkl} \frac{\partial^2 P_k}{\partial x_j \partial x_l}, \tag{6}$$

$$\sigma_{ij} = c_{ijkl} u_{kl} + f_{ijkl} \frac{\partial P_k}{\partial x_l}, \tag{7}$$

where σ_{ij} and c_{ijkl} are stress and elastic stiffness, respectively. f_{ijkl} is called *flexocoupling tensor*. It is seen that, in the case where the strain gradient and the polarization are homogeneous, equation (6) reproduces the flexoelectric effect introduced by (1) with

$$\mu_{ijkl} = \chi_{kx} f_{ijxl}. \tag{8}$$

Equation (8) links the flexoelectric and flexocoupling tensors, suggesting that the flexoelectric response should be enhanced in materials with high dielectric constants such as ferroelectrics. It is clear from equation (6) that via the flexoelectric coupling the strain gradient works as an electric field. Equation (7) enables us to recognize the thermodynamically conjugated effect to the static bulk flexoelectric response — *converse flexoelectric effect*, which consists of the contribution to the mechanical stress, proportional to the gradient of polarization.

2.2. *Dynamic Flexoelectric Effect*

Now, we will discuss the so-called dynamic flexoelectric effect. While the static bulk flexoelectric effect can be viewed as an extension of the piezoelectric effect, the phenomenon treated below has no analogue in piezoelectricity. In the elastic wave in solids, the static effect can be viewed as a contribution to polarization proportional to the strain gradient (which is always present in the mechanical wave). In the time domain, there is another contribution which corresponds to the polarization response to accelerated motion of the medium, or P_i is proportional to $\ddot{U}_j$, and since in the acoustic wave $\ddot{U}_j \propto \frac{\partial u_{kl}}{\partial x_j}$, this contribution to the flexoelectric effect is called the dynamic flexoelectric effect.

On the phenomenological side, the dynamic flexoelectric effect can be taken into account by adding a mixed term to the density of kinetic energy

$$T_k = \frac{\rho}{2}\dot{U}_i^2 + \frac{\gamma_{ij}}{2}\dot{P}_i\dot{P}_j + M_{ij}\dot{U}_i\dot{P}_j, \tag{9}$$

where ρ is the density and γ_{ij} is a phenomenological tensor controlling the dynamics of polarization,[1] M_{ij} is a *flexodynamic tensor*. Minimizing the action

$$\iint (T - \Phi + u_i\sigma_i)dV\,dt, \tag{10}$$

$$\Phi = \frac{\chi_{ij}^{-1}}{2}P_jP_j + \frac{c_{ijkl}}{2}u_{ij}u_{kl} + \frac{g_{ijkl}}{2}\frac{\partial P_i}{\partial x_j}\frac{\partial P_k}{\partial x_l} \tag{11}$$

$$- \frac{f_{ijkl}}{2}\left(P_k\frac{\partial u_{ij}}{\partial x_l} - u_{ij}\frac{\partial P_k}{\partial x_l}\right) - P_iE_i, \tag{12}$$

with respect to P_i and U_j one obtains

$$E_i = \chi_{ij}^{-1}P_j - f_{klij}\frac{\partial u_{kl}}{\partial x_j} + M_{ij}\ddot{U}_j - g_{ijkl}\frac{\partial^2 P_k}{\partial x_j\partial x_l} + \gamma_{ij}\ddot{P}_j, \tag{13}$$

$$\rho\ddot{U}_i = c_{ijkl}\frac{\partial u_{kl}}{\partial x_j} + f_{ijkl}\frac{\partial^2 P_k}{\partial x_l\partial x_l} - M_{ji}\ddot{P}_j. \tag{14}$$

The last two rhs terms of equation (13) control the spatial and frequency dispersion of the polarization response. However, when we consider macroscopic manifestations of the flexoelectric response (e.g. in a dynamically bent sample or in ultrasonic acoustic wave), where $1/q$ (q — phonon wave-vector) is much larger than the typical microscopic scales, and sound wave frequency is much smaller than the typical optical phonon frequencies, these terms can be neglected. Thus, setting electric field to 0 in (13) and taking into account that $\ddot{U}_j \propto \frac{\partial u_{kl}}{\partial x_j}$, we can see that the $M_{ij}\ddot{U}_j$ term, corresponding to the dynamic flexoelectric effect, indeed provides a contribution to polarization along with f_{ijkl}.

It is instructive to eliminate $\ddot{U}_i$ between equations (13) and (14) to find a relationship controlling the total flexoelectric response in the dynamic case omitting the terms containing the higher-order second derivatives of polarization $\frac{\partial^2 P_k}{\partial x_j \partial x_l}$ and $\ddot{P}_i$:

$$E_i = \chi_{ij}^{-1} P_j - \left(f_{klij} - \frac{1}{\rho} M_{ix} c_{xjkl} \right) \frac{\partial u_{kl}}{\partial x_j}. \tag{15}$$

From this equation we can see that in view of the dynamic flexoelectric effect, the role of the flexocoupling tensor, f_{klij}, is now played by the total flexocoupling tensor:

$$f_{klij}^{\text{tot}} = f_{klij} - \frac{1}{\rho} M_{ix} c_{xjkl} = f_{klij} + f_{klij}^{\text{dyn}}. \tag{16}$$

Thus, including the dynamic case, the flexoelectric response is controlled by the total flexoelectric tensor

$$\mu_{ijkl}^{\text{tot}} = \mu_{ijkl} + \mu_{ijkl}^{\text{dyn}}, \tag{17}$$

where the dynamic contribution is correspondingly defined as

$$\mu_{klij}^{\text{dyn}} = -\frac{1}{\rho} \chi_{ix} M_{xy} c_{yjkl}. \tag{18}$$

This way, the phenomenological theory suggests that, like the static contribution, the dynamic contribution should be enhanced in the materials with high dielectric constant. The dynamic contribution to the flexoelectric effect makes it qualitatively different from the piezoelectric effect.

Order-of-magnitude estimates show that the components of tensors f_{ijkl} and f_{ijkl}^{dyn} are expected to be comparable.[1] One should mention that despite this fact, the dynamic flexoelectric effect does not always provide a contribution comparable to that of the static effect. In an acoustic wave, the dynamic effect works at full strength, however, in quasi-static experiments of plate bending, i.e. where the smallest dimension of the sample is less than the acoustic wavelength corresponding to the frequency of the external perturbation, the dynamic effect is negligible (see Refs. [1] and [10] for further details).

In the next section, we describe the interaction of the TA and TO phonon branches in terms of the continuum Landau theory, this way linking this effect with the flexoelectric coupling in the material.

3. Flexoelectric Coupling and Phonon Spectra

An important manifestation of the flexoelectric coupling is related to phonon spectra in solids. In terms of phonons, the flexoelectric interaction can be interpreted as a repulsion between TA and soft-mode TO branches. This effect was documented in perovskite ferroelectrics by Axe *et al.*, who studied the dispersion of the phonons in $KTaO_3$ (KTO)[11] and $PbTiO_3$[12] by means of neutron scattering. The temperature dependence of the dispersion curves obtained for KTO is shown in Fig. 1. As it is seen from this figure, with decreasing temperature the soft optical branch moves downward closer to the acoustic one and causes a bending of the latter. The temperature-driven acoustic phonon branch bending has also been observed in STO by Hehlen *et al.*[13] by means of Brillouin scattering.

It is possible to describe the repulsion of phonon branches in terms of the continuum Landau theory.[1] Within the validity of the continuum model we consider the long-wavelength part of the spectrum. We start with equations (13) and (14), where we rewrite the strain in terms of acoustic displacement, see equation (2). To describe the phonons, we search for solutions for polarization and displacement in the form

$$P = \tilde{P}e^{i\omega t - i\vec{q}\vec{x}}, \tag{19}$$

$$U = \tilde{U}e^{i\omega t - i\vec{q}\vec{x}}. \tag{20}$$

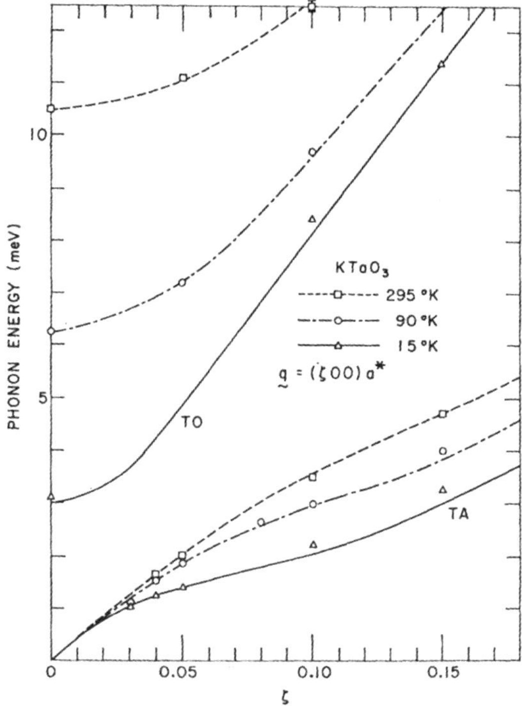

Figure 1. Temperature dependence of dispersion curves for TA and TO phonons with wave-vector $q = \frac{\pi}{a}(\xi, 0, 0)$ in KTaO$_3$, where a is the lattice constant. After Ref. [11].

Since we are interested in the transverse modes, in which electric field does not arise, we omit the term related to electric field and obtain the following set of linear homogeneous equations for amplitudes of polarization and displacement:

$$\omega^2 \gamma_{ij} \tilde{P}_j = \chi_{ij}^{-1} P_j + g_{ijkl} q_j q_l \tilde{P}_k + f_{klij} q_j q_l \tilde{U}_k - M_{ij} \tilde{U}_j \omega^2, \qquad (21)$$

$$\rho \omega^2 \tilde{U}_i = c_{ijkl} q_j q_l \tilde{U}_k + f_{ijkl} q_j q_l \tilde{P}_k - M_{ji} \tilde{P}_j \omega^2. \qquad (22)$$

Eigenfrequencies of the system (frequency ω of the transverse modes), corresponding to acoustic and optic branches, may be found from the condition of zero determinant of the set of equations (21) and (22), namely for the high symmetry [100], [110], and [111] directions of wave-vector one can write:

$$(\omega^2 - \omega_A^2)(\omega^2 - \omega_O^2) = \frac{(\omega^2 M - q^2 f_{\text{eff}})^2}{\rho\gamma}, \tag{23}$$

$$\omega_A^2 = \frac{c_{\text{eff}} q^2}{\rho}, \quad \omega_O^2 = \frac{\chi^{-1} + g_{\text{eff}} q^2}{\gamma}, \tag{24}$$

where $\vec{q}$ denotes the wave-vector, and the coefficients f_{eff}, c_{eff}, and g_{eff} are defined according to the direction:

$$[100] : c_{\text{eff}} = c_{44}, \quad f_{\text{eff}} = f_{44}, \quad g_{\text{eff}} = g_{44}, \tag{25}$$

$$[110] : \begin{cases} c_{\text{eff}} = \dfrac{1}{2}(c_{11} - c_{12}) \\[2mm] f_{\text{eff}} = \dfrac{1}{2}(f_{11} - f_{12}), \\[2mm] g_{\text{eff}} = \dfrac{1}{2}(g_{11} - g_{12}) \end{cases} \tag{26}$$

$$[111] : \begin{cases} c_{\text{eff}} = \dfrac{1}{3}(c_{11} - c_{12} + c_{44}) \\[2mm] f_{\text{eff}} = \dfrac{1}{3}(f_{11} - f_{12} + f_{44}) \\[2mm] g_{\text{eff}} = \dfrac{1}{3}(g_{11} - g_{12} + g_{44}) \end{cases} \tag{27}$$

and γ can be found from the frequency of the lowest TO mode at Γ-point and the susceptibility, namely[16]

$$\gamma = \frac{1}{\omega_O^2(0)\chi}. \tag{28}$$

The roots of equation (23) with coefficients defined by equations (25) and (27) yield the frequencies of the corresponding double degenerate TA and TO modes. The roots of equation (23) with coefficients defined by equations (26) yields the frequencies of the TA and TO modes that are polarized along the $[\bar{1}10]$ direction.

One can perform the analysis of the dispersion of the acoustic branch by using equation (23) and get information on the parameters f_{eff} and M. In the case of weak interaction between the branches (lowest order in q), the relative shift of the acoustic branch can be found as

$$\omega^2 - \omega_A^2 = -\frac{\left(q^2 f_{\text{eff}}^{\text{tot}}\right)^2}{\rho\gamma(\omega_O^2 - \omega_A^2)}, \tag{29}$$

where $f_{\text{eff}}^{\text{tot}}$ is defined as

$$f_{\text{eff}}^{\text{tot}} = f_{\text{eff}} - M\frac{c_{\text{eff}}}{\rho}. \tag{30}$$

As is clear from equation (29), the deviation of the TA branch from the linear dispersion corresponds to the difference $\omega^2 - \omega_A^2$ proportional to $(f_{\text{eff}}^{\text{tot}} q^2)^2$. The repulsive character of the interaction between the TA and TO branches (or lowering of the acoustic branch) is seen from the sign of rhs in expression (29). As for its magnitude, it is mainly controlled by the total flexocoupling coefficient (30), which has both dynamic and static contributions. One should note that, attempting to evaluate the flexocoupling tensor, such fit was already done in several papers,[13,16–18] however neglecting dynamic flexoelectricity. Thus, the lowest order in the q^2 analysis of $|\omega^2 - \omega_A^2|$ can only provide information about the total flexocoupling coefficient, not distinguishing between the static and dynamic contributions. The distinction between f and f^{tot} can be obtained by expanding the solution to equation (23) up to the next power in q^2:

$$\omega^2 - \omega_A^2 = -\frac{\left(q^2 f_{\text{eff}}^{\text{tot}} - q^4 M\frac{\chi\left(f_{\text{eff}}^{\text{tot}}\right)^2}{\rho}\right)^2}{\rho\gamma(\omega_O^2 - \omega_A^2)}. \tag{31}$$

A form of (31) suitable for analysis reads

$$\Psi(q^2) \equiv \frac{\sqrt{\rho\gamma(\omega_A^2 - \omega^2)(\omega_O^2 - \omega_A^2)}}{q^2} = \left| f_{\text{eff}}^{\text{tot}} - q^2 M\frac{\chi\left(f_{\text{eff}}^{\text{tot}}\right)^2}{\rho} \right|. \tag{32}$$

It is clear that by fitting $\Psi(q^2)$ to a linear function of q^2 one can get information on M and $f_{\text{eff}}^{\text{tot}}$. However, the resulting information on M and the components of the flexocoupling tensor is limited. Only the absolute values of M, $f_{11} - f_{12}$, and f_{44} can be provided from such analysis. We have checked that the analysis of the phonon spectrum for other directions than those addressed above unfortunately does not provide any additional information on the components of the flexocoupling tensor. Actually, all the information can be obtained from the dispersion curves for any two of the [100], [110], and [111] wave-vector directions, while the third direction may serve as a crosscheck and an evaluation of the inaccuracy of the method.

The trend of the branch repulsion described above holds when interaction between them is strong. As controlled by the difference of frequency squares in the denominator in the rhs of equation (29), the effect is expected to be strong, when the optical branch approaches the acoustic one (e.g. with decreasing temperature). As noticed by Axe,[19] if the strength of flexoelectric effect exceeds some threshold, the acoustic branch will touch the x-axis in Fig. 1, and there will be a phase transition into an incommensurate phase. We would like to discuss this situation in the next section.

4. Upper Bounds for the Static Bulk Flexocoupling Coefficients

Bounds for flexoelectric coefficients in a ferroelectric can be obtained from the analysis of the parameters of its phonon spectrum. It was shown in Section 2 that flexoelectric effect leads to a bending of the acoustic phonon branch (see Fig. 1). The acoustic branch may reach the level $\omega = 0$ at some critical wave-vector $q_c \neq 0$. This may happen if the flexoelectric coupling strength exceeds some threshold. The existence of the critical wave-vector $q_c \neq 0$ means that the system becomes unstable with respect to a spatial modulation corresponding to this wave-vector, and hence the material undergoes a phase transition into an incommensurate phase. On the other hand, for materials without incommensurate phase one can get constraints for its flexocoupling coefficients by requiring the absence of such critical wave-vectors. Below, using equations (21) and (22), we derive such constraints on the flexocoupling coefficients for the case of cubic (in the paraelectric phase) ferroelectric perovskites.

To get the constraints for the flexocoupling coefficients, in general, one can require the absence of critical vectors of any direction. Being interested in the upper limits for the coefficients, let us require the absence of critical wave-vectors only along highly symmetric axes of the crystal. In the materials addressed, such axes are the 4-fold, 3-fold, and 2-fold axes. The requirement corresponding to each of this axes will produce a constraint.

The constraint corresponding to the 4-fold axis may be derived from equation (23). Suppose there exists a wave-vector $q_c \neq 0$ with eigenfrequency $\omega = 0$. Then its magnitude must satisfy the following equation (obtained by setting $\omega = 0$ in equation (23)):

$$c_{44}\chi^{-1} + (c_{44}g_{44} - f_{44}^2)q_c^2 = 0. \tag{33}$$

Because the first term in equation (33) is positive, this equation will have no real solution if

$$f_{44}^2 < c_{44}g_{44}, \tag{34}$$

which is the sought constraint. One readily checks that if $f_{44}^2 > c_{44}g_{44}$ then a critical wave-vector necessarily appear in the limit case $\chi^{-1} \to 0$.

Analogical constraints may be derived, for the wave-vectors parallel to the 2-fold and 3-fold axes, by setting $q = \frac{1}{\sqrt{2}}(q_c, q_c, 0)$ and $q = \frac{1}{\sqrt{3}}(q_c, q_c, q_c)$ in equations (21) and (22). In the case of the 2-fold axis, there appear two pairs of coupled modes. The stability condition for the first one is identical to constraint (34), while the stability condition for the second one reads:

$$(f_{11} - f_{12})^2 < (c_{11} - c_{12})(g_{11} - g_{12}). \tag{35}$$

In the case of the wave-vector directed along a 3-fold axis the normal modes are 2-fold degenerate, and one obtains the following condition:

$$(f_{44} + f_{11} - f_{12})^2 < (c_{44} + c_{11} - c_{12})(g_{44} + g_{11} - g_{12}). \tag{36}$$

As one can check, inequality (36) follows from conditions (34) and (35) in view of the classic inequality $\frac{a+b}{2} \geq \sqrt{ab}$. Thus, equations (34) and (35) form the sought set of constraints for the flexocoupling coefficients in perovskite ferroelectrics.

For the typical ferroelectrics the upper bounds for the flexocoupling coefficients given by (34) and (35) are of the order of a few Volts. In particular for $BaTiO_3$[a]

$$|f_{44}| < 3.3V, \quad |f_{11} - f_{12}| < 7V, \tag{37}$$

and for $SrTiO_3$.[b]

$$|f_{44}| < 2.4V, \quad |f_{11} - f_{12}| < 10V. \tag{38}$$

Based on atomic order-of-magnitude estimates similar limitations can be obtained for other displacive ferroelectrics.

It is instructive to note that the above reasoning may be reformulated in terms of the ferroelectric domain wall energies. One may obtain the same upper bounds for the flexoelectric coupling coefficients by posing the requirement that a domain wall of any orientation must have positive energy. In particular, for the case of a domain wall with a normal parallel to a 4-fold axis of the perovskite crystal in tetragonal phase, the upper bound identical to (34) may be obtained as follows. Flexoelectric coupling results in renormallization of correlation energy terms in the expression for domain wall energy.[22] For this kind of wall, the flexoelectric coupling leads to the renormalization of the gradient term

$$g_{44}^{\text{eff}} = g_{44} - \frac{f_{44}^2}{c_{44}}. \tag{39}$$

For the domain wall energy to be positive, this term must be positive as well, implying $g_{44}^{\text{eff}} > 0$, which is equivalent to (34).

Concerning the derivation of bounds (34)–(36) two remarks are to be made. First, only the lowest in q terms are used in the analysis. However, it gives a correct criterion of the instability.[c] Second, the

---

[a] Calculated using parameters taken from Ref. [20].
[b] Calculated using parameters taken from Refs. [16] and [21].
[c] See work by Axe *et al.*[19] for a more detailed analysis of the problem.

constraints obtained are strict only for the case of second order phase transitions, whereas in the case $\chi^{-1} \rightarrow +0$ the material still stays in the paraelectric phase. However, equations (34)–(36) still give reasonable approximation for the upper bounds of the flexoelectric coupling coefficients in first order phase transition (like $BaTiO_3$) and incipient (like STO) ferroelectrics, where χ^{-1} approaches zero but does not reach it.

The upper bounds obtained are useful for the interpretation of experimental data on the flexocoupling coefficients. If the measured flexocoupling coefficients are essentially inconsistent with these constraints, this fact indicates that the explanation of the response characterized is beyond the static bulk flexoelectric effect.

5. Calculations of Flexoelectric Coefficients using Phonon Dispersion

We have applied the method described in Section 3, following Ref. [23] to evaluate the flexocoupling and flexodynamic tensors directly from the simulated phonon spectra. The calculations of phonon spectra were done for cubic STO exploiting the Quantum ESPRESSO (QE)[24] *ab initio* package within the GGA PBE exchange–correlation functional with ultrasoft pseudopotentials.[25] We have used an automatically generated uniform $16 \times 16 \times 16$ grid of k-points, the kinetic energy cutoff for wavefunctions was 80 Ry. The calculations of phonon spectrum are done using Density Functional Perturbation Theory (DFPT)[26] as implemented in the Phonon code. The energy threshold for self-consistency was chosen to be equal to 10^{-20} Ry with the help of convergence tests for Γ-point phonons. The phonon spectra were obtained for the [100], [110], and [111] wave-vector directions on $10 \times 10 \times 10$ automatic Monkhorst–Pack grid of q-points.

Since, within zero Kelvin DFT calculations, the cubic structure of STO is unstable (negative squares of phonon frequencies for the soft-mode phonons) we performed our calculations under hydrostatic pressures of 78–135 kBar to stabilize it, corresponding to approximately 1–2% elastic strain. Since the tensors M_{xm} and

f_{ijxl} are not critically small, in view of the Landau theory arguments and atomic estimates, such strain is expected to induce some 1–2% modifications of these tensors. On the same lines, the result of zero-Kelvin calculations should give a reasonable approximation of the finite-temperature values of these tensors. Such an approach is supported by the experience of zero-Kelvin calculations of elastic constants. This way, we believe that the estimates obtained here are valid for realistic finite temperature and pressure.

First, we would like to qualitatively reproduce the behavior observed by Axe for temperature-driven lowering of TO branch and increasing repulsion between TA and TO modes conditioned by flexoelectric effect using first principles calculations to justify our framework. We performed *ab intio* calculations of dispersion curves of STO under different hydrostatic pressures, Fig. 2. The solid lines correspond to the lower pressure ($p = 78$ kBar) when the flexoelectric interaction between the TO and TA branches is relatively strong. Here, one observes lowering of the TA conditioned by flexoelectric coupling. The short dashed lines correspond to the high pressure ($p = 135$ kBar) when the interaction is relatively weak in view of the larger distance between the TA and TO branches.

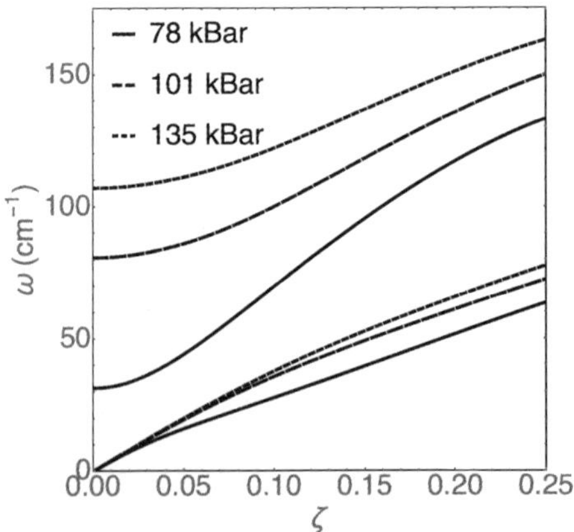

Figure 2.   Calculated phonon dispersion of TA and soft-mode TO branches with wave-vector $q = \frac{\pi}{a}(\zeta,0,0)$, where a is the lattice constant, for cubic SrTiO$_3$ stabilized by pressure. After Ref. [23].

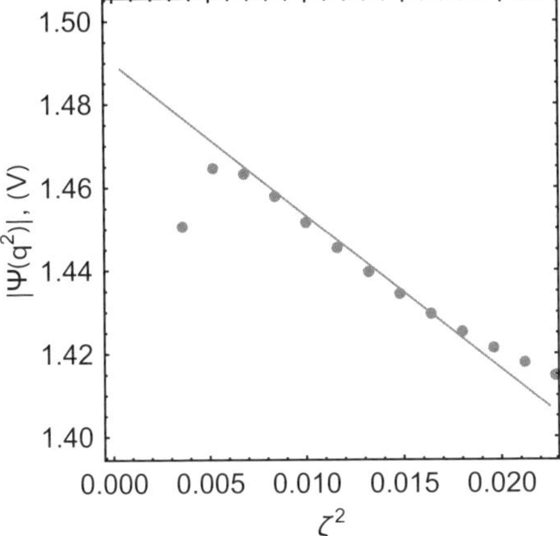

Figure 3. Absolute value of $\Psi(q^2)$ calculated using equation (32) (dots) and fitting line, which was used to obtain $f_{\text{eff}}^{\text{tot}}$ and M. To evaluate $\Psi(q^2)$, the frequencies of the TA and soft-mode TO branches entering in equation (32) were calculated for the [100] direction of wave-vector, $q = \frac{2\pi}{a}(\zeta, 0, 0)$, for cubic SrTiO$_3$ stabilized by pressure. After Ref. [23].

To get information on the static and dynamic flexocoupling tensors, the long-wavelength part of the spectra was calculated under a pressure of 78 kBar where the inter-mode coupling is a maximum. As a first step, $|M|$ and $|f_{\text{eff}}^{\text{tot}}|$ were obtained from a linear fit for $\Psi(q^2)$ as a function of q^2 for the [100] and [110] directions. An example of such a fit is shown in Fig. 3. Regarding the fit, it is worth mentioning that the reliable evaluation of the flexoelectric parameters directly from the spectrum is a challenging computational task. The problem is that, on one hand, one is interested in the long-wavelength limit of the spectrum, on the other hand, close to the Γ-point, the violation of the acoustic sum rules corrupts the simulated spectrum.[26] Thus, for this evaluation, only a limited interval of the eigenvectors should be used for the fit, as it was done in Fig. 3. Here, we used the simulated TA and TO frequencies for ω^2 and ω_O^2, respectively, $\rho = 5174 \frac{\text{kg}}{\text{m}^3}$. ω_A^2 and γ were found using equations (24) and (28) with the susceptibility ($\chi/\epsilon_0 = 2420$) and elastic constants (Table 1) obtained from the tangent of dispersion curves near the Γ-point. Then, based on the results of the fit, using (30) we evaluate the values of $|M|$, $|f_{11} - f_{12}|$, and $|f_{44}|$, which are given in Table 2.

Table 1. Stiffness tensor for cubic $SrTiO_3$ under a pressure of 78 kBar. (a) Calculated from dynamical matrix. (b) Calculated from the tangent of dispersion curves. (c) Experimental data.[21]

$(\times 10^{11}\,\frac{N}{m^2})$	(a) Dyn. mat.	(b) Phonon disp.	(c) Experiment
c_{11}	2.77 ± 0.05	3.1 ± 0.1	3.16
c_{12}	1.06 ± 0.05	0.9 ± 0.1	1.03
c_{44}	0.92 ± 0.05	1.0 ± 0.1	1.22

Table 2. Total $(f_{\text{eff}}^{\text{tot}} = f_{\text{eff}} - M\frac{c_{\text{eff}}}{\rho})$, static (f), and dynamic (M) flexocoupling coefficients for cubic $SrTiO_3$ under a pressure of 78 kBar obtained from phonon dispersion.

$\left\|f_{44}^{\text{tot}}\right\|$ (V)	1.5 ± 0.1
$\left\|f_{11}^{\text{tot}} - f_{12}^{\text{tot}}\right\|$ (V)	< 0.4
$\left\|f_{44}\right\|$ (V)	0.3 ± 0.1
$\left\|f_{11} - f_{12}\right\|$ (V)	2.4 ± 0.2
$\left\|M\right\|(\frac{Vs^2}{m^2})$	$(6 \pm 1) \times 10^{-8}$
$\left\|M\right\|\frac{c_{44}}{\rho}$ (V)	1.2 ± 0.2

Analyzing Table 2 one observes that the values of total and static flexocoupling coefficients can drastically differ, which attests to the importance of dynamic effect.

The spectrum for the [111] direction was found in a reasonable agreement with results obtained for those for the [100] and [110] directions, enabling us to evaluate the inaccuracy of our calculations given in this table. The suggested method allows to find some combinations of flexoelectric coefficients within the sign ambiguity, however the relative sign is known, see rhs in equation (32).

The proposed method of evaluating the components of flexo-dynamic and flexocoupling tensors from the simulated spectra can also be applied for the determination of these components from the experimental phonon dispersion curves of materials exhibiting a ferroelectric soft mode. Thus, having experimental information about phonon dispersion, dielectric susceptibility, and density of the material, one can exploit the above described method to find the flexodymanic tensor and certain combinations of flexocoupling coefficients experimentally. It is of importance to mention that, formally, this method can be applied to the phonon spectrum of any material, however in the absence of low-energy TO mode, it will not give reliable information on the sought tensors. The point is that in such a situation the self-dispersion of the TA mode (which is always present and related to other effects) is expected to be of the same order of magnitude as the dispersion caused by flexoelectricity.

6. Calculations of Flexoelectric Coefficients using Dynamical Matrix

In this section, we are interested in the evaluation of the dynamic bulk flexoelectric effect and its comparison with the static effect for an STO crystal using microscopic approach in the Born-charge approximation. To do this, the tensors M_{xm} and f_{ijxl} were obtained using the expansion of the dynamical matrix $A_{ip,i'p'}(\vec{q})$ calculated with the long-range contribution of the macroscopic electric field being excluded.[3] Here, the suffixes i and i' enumerate the Cartesian components of the atomic displacements while p and p' — atoms in an elementary unit cell of the crystal. Specifically, one needs the coefficients of the following long-wavelength expansion of the dynamical matrix

$$A_{ip,i'p'}(\vec{q}) = A^{(0)}_{ip,i'p'} + A^{(1)j}_{ip,i'p'} q_j + \frac{1}{2} A^{(2)jl}_{ip,i'p'} q_j q_l + \cdots , \qquad (40)$$

and the matrix $\Gamma_{ip,i'p'}$, that is the inverse of the singular matrix $A^{(0)}_{ip,i'p'}$ defined in a special way.[27] The results of Tagantsev's approach,[3] written for a cubic crystal of perovskite structure in the Born charge approximation, can be summarized as follows

$$M_{ij} = \frac{\chi^{-1}}{v} Q_{ix,p} \Gamma_{xp,jp'} \left(m_{p'} - \frac{1}{s} \sum_{p''} m_{p''} \right), \qquad (41)$$

$$f_{ijkl} = \frac{\chi^{-1}}{v} Q_{kx,p} \left(N^{ijl}_{x,p} + N^{jil}_{x,p} - N^{lij}_{x,p} \right), \qquad (42)$$

$$N_{i,p}^{jkl} = \frac{1}{2} \sum_{p''} \Gamma_{ip,i'p'} \left(A_{i'p',jp''}^{(2)kl} - \frac{\delta_{p'p''}}{s} \sum_{q'q''} A_{i'q',jq''}^{(2)kl} \right), \tag{43}$$

where all summations are done over all s atoms of the unite cell. Here v is the volume of the unit cell; $Q_{ij,p}$ and m_p are the matrix of the Born charges and the mass of pth atom in the unit cell. The dielectric susceptibility can be found as[27]

$$\chi_{ij} = \frac{1}{v} Q_{ix,p} \Gamma_{xp,yp'} Q_{jy,p'}. \tag{44}$$

Thus to find the tensors χ_{xm}, M_{xm}, and f_{ijxl}, only the matrices $A_{ip,i'p'}^{(0)}$ and $A_{ip,i'p'}^{(2)jl}$ (available once the dynamical matrix $A_{ip,i'p'}(\vec{q})$ is calculated), the matrix of the Born charges $Q_{ix,p}$ and the volume of the unit cell are needed.

The evaluation of the contribution of dynamic flexoelectricity to the total flexocoupling tensor f_{ijxl}^{tot}, as is clear from equation (30), requires the values of the elastic constants c_{mlij} which we also evaluated using the dynamical matrix, specifically for perovskite cubic structure using the relationship[27]

$$c_{ijkl} = \frac{1}{2v} \sum_{p,p'} A_{ip,kp'}^{(2)jl}. \tag{45}$$

We found $A_{ip,i'p'}(\vec{q})$ and $Q_{ix,p}$ using our first principles calculations for STO stabilized by pressure of 78 kBar, when TA and TO branches stand close to each other, and the flexoelectric coupling is maximal. To find the independent components of tensors $\chi_{ij} = \chi \delta_{ij}$, $M_{ij} = M \delta_{ij}$, f_{ijkl}, and c_{ijkl} (χ, M, f_{11}, f_{12}, f_{44}, c_{11}, c_{12}, c_{44}), we evaluate the wave-vector dependence of the dynamical matrix $A_{ip,i'p'}(\vec{q})$ for the [100] and [110] directions of the reciprocal space, namely for [100] direction of q-vector ($\vec{q} = q_1 = q$):

$$A_{ip,i'p'}(q) = A_{ip,i'p'}^{(0)} + \frac{1}{2} A_{ip,i'p'}^{(2)11} q^2, \tag{46}$$

for [010] direction of q-vector ($\vec{q} = q_2 = q$):

$$A_{ip,i'p'}(q) = A_{ip,i'p'}^{(0)} + \frac{1}{2} A_{ip,i'p'}^{(2)22} q^2, \tag{47}$$

Table 3. Material parameters for cubic $SrTiO_3$ ($a = 3.886$ Å) under a pressure of 78 kBar obtained using the dynamical matrix and from analysis of phonon dispersion curves. f and M are flexocoupling and flexodynamic coefficients. f^{tot} is defined in equation (30). ρ is the density ($\rho = 5174 \frac{\text{kg}}{\text{m}^3}$) calculated by the mass of atoms in the unit cell divided by the cell volume. (a) Calculated from dynamical matrix. (b) Obtained from the simulated phonon dispersion curves. (c) Experimental data.[16,18,21]

	(a) Dyn. mat.	(b) Phonon disp.	(c) Experiment
f_{11} (V)	1.11 ± 0.05		
f_{12} (V)	-1.30 ± 0.05		
f_{44} (V)	-0.28 ± 0.05	± 0.3	
$M\left(\times 10^{-8} \frac{\text{Vs}^2}{\text{m}^2}\right)$	(6 ± 0.5)	$\pm(6 \pm 1)$	
$M\frac{c_{44}}{\rho}$ (V)	1.2 ± 0.1	$\pm(1.2 \pm 0.2)$	
$\left|f_{44}^{\text{tot}}\right|$ (V)	1.5 ± 0.1	1.5 ± 0.1	$1.2\text{--}2.2$
$\frac{1}{2}\left|f_{11}^{\text{tot}} - f_{12}^{\text{tot}}\right|$ (V)	< 0.2	< 0.4	$0.6\text{--}0.7$

and similarly for [110] direction ($q_1 = q_2 = q$):

$$A_{ip,i'p'}(q) = A^{(0)}_{ip,i'p'} + \frac{1}{2}\left(A^{(2)11}_{ip,i'p'} + 2A^{(2)12}_{ip,i'p'} + A^{(2)22}_{ip,i'p'} \right) q^2. \qquad (48)$$

Then, we obtain the sought matrices $A^{(0)}_{ip,i'p'}$ and $A^{(2)jl}_{ip,i'p'}$ and finalize the calculations using equations (41)–(45) to find bulk flexocoupling tensors. The results of the calculations are presented in Table 3.

Analyzing Table 3 one makes a remarkable observation on the very strong renormalization of the flexocoupling tensor by dynamic flexoelectricity. It is seen that the corresponding components of the tensors f_{ijkl} and f_{ijkl}^{tot} can drastically differ. This implies that the traditional neglect of dynamic flexoelectricity in the dynamic electromechanical simulations incorporating flexoelectricity is inadmissable.

The method based on the analysis of dynamical matrix allows to find all independent components of static flexocoupling tensor unlike the method based on phonon dispersion. It is seen from Table 3 that the obtained results yielded by the two methods are in good agreement with each other. Table 3 also gives a comparison between the obtained results and the available experimental data by Tagantsev *et al.*,[18] who determined total flexoelectric coefficients for crystalline STO based on experimental data of Brillouin scattering by Hehlen *et al.*[13] Since, in a sound wave the bulk flexoelectric effect includes static and dynamic contributions, therefore the analysis, which was done by Tagantsev with the contribution of the dynamic flexoelectric effect being omitted, gave only the total values of the flexoelectric coefficients. This way, for the moment, experimental information is only available on some components of the tensor f^{tot}_{ijkl}.

Despite, currently, all experimental studies and simulations involving flexoelectricity ignore the dynamic effect, we found these effects to be of a comparable magnitude, in agreement with earlier order-of-magnitude estimates,[3] which confirms the fact of the necessity of including the dynamic contribution, while extracting the flexocoupling coefficients from the phonon spectrum.

7. Conclusions

Concluding this section, we would like to highlight some essential moments. We demonstrated a method of extraction of information on flexodynamic and flexocoupling tensors using the simulated spectrum of STO. This method can be applied to the treatment of experimental phonon spectra, being currently the only method enabling experimental determination of the values of the bulk flexoelectric tensors, since static experiments in a finite sample yield the sum of the static bulk and surface contributions, which, in general, are comparable. We obtained the values of the upper bounds for static bulk flexocoupling coefficients from the requirement of the absence of incommensurate phase in the material. We evaluated the strength of the dynamic bulk flexoelectric effect and compared it to that of the static effect. We identified the situation of STO, where the dynamic effect can be a few times stronger than static. The agreement between our calculations and the experimental data, obtained from the Brillouin and neutron scattering experiments, is reasonable.

Acknowledgments

The authors are indebted to Dragan Damjanovic, Pavlo Zubko, and Anna Morozovska for useful discussions. The authors also gratefully acknowledge funding from the Swiss National Science Foundation, Grant No. 200020-144463/1. The Research was also supported by grant of the government of the Russian Federation No. 2012-220-03-434.

References

1. P. V. Yudin and A. K. Tagantsev. *Nanotechnology.* **24**(43), 432001 (36pp) (2013).
2. R. Maranganti and P. Sharma. *Phys. Rev. B.* **80**, 054109 (2009).
3. A. K. Tagantsev. *Phys. Rev. B.* **34**, 5883–5889 (1986).
4. J. Hong, G. Catalan, J. F. Scott, and E. Artacho. *J. Phys. Condens. Matt.* **22**(11), 112201 (2010).
5. I. Ponomareva, A. K. Tagantsev, and L. Bellaiche. *Phys. Rev. B.* **85**, 104101 (2012).
6. J. Hong and D. Vanderbilt. *Phys. Rev. B.* **84**, 180101 (2011).
7. J. Hong and D. Vanderbilt. *Phys. Rev. B.* **88**, 174107 (2013).
8. M. Stengel. *Phys. Rev. B.* **88**, 174106 (2013).
9. M. Stengel. *Phys. Rev. B.* **90**, 201112 (2014).
10. P. Zubko, G. Catalan, A. Buckley, P. R. L. Welche, and J. F. Scott. *Phys. Rev. Lett.* **99**(16), 167601 (2007).
11. J. D. Axe, J. Harada, and G. Shirane. *Phys. Rev. B.* **1**, 1227–1234 (1970).
12. G. Shirane, J. D. Axe, J. Harada, and J. P. Remeika. *Phys. Rev. B.* **2**, 155–159 (1970).
13. B. Hehlen, L. Arzel, A. K. Tagantsev, E. Courtens, Y. Inaba, A. Yamanaka, and K. Inoue. *Phys. Rev. B.* **57**, R13989–R13992 (1998).
14. S. M. Kogan. *Sov. Phys.-Sol. State.* **5**(10), 2069–2070 (1964).
15. V. L. Indenbom, E. B. Loginov, and M. A. Osipov. *Kristalografija.* **26**, 1157 (1998).
16. V. G. Vaks. *Introduction into Microscopic Theory of Ferroelectrics.* Nauka, Moscow, (1973).
17. E. Farhi, A. K. Tagantsev, R. Currat, B. Hehlen, E. Courtens, and L. A. Boatner. *Eur. Phys. J. B.* **15**(14), 615–623 (2000).
18. A. K. Tagantsev, E. Courtens, and L. Arzel. *Phys. Rev. B.* **64**, 224107 (2001).

19. J. D. Axe, J. Harada, and G. Shirane. *Phys. Rev. B.* **1**, 1227–1234 (1970).

20. A. K. Tagantsev, L. E. Cross, and J. Fousek. *Domains in Ferroic Crystals and Thin Films.* Springer, New York (2010).

21. M. Adachi, Y. Akishige, T. Asahi, K. Deguchi, K. Gesi, K. Hasebe, T. Hikita, T. Ikeda, and Y. Iwata. *Landolt–Bornstein, Numerical Data and Functional Relationships in Science and Technology, New Series,* Vol. III/36, Ferroelectrics and Related Substances: Oxides. Springer-Verlag, Berlin Heidelberg New-York (2001).

22. E. A. Eliseev, A. N. Morozovska, M. D. Glinchuk, and R. Blinc. *Phys. Rev. B.* **79**, 165433 (2009).

23. A. Kvasov and A. K. Tagantsev. *Phys. Rev. B.* **92**, 054104 (2015).

24. P. Giannozzi, S. Baroni, N. Bonini, M. Calandra, R. Car, C. Cavazzoni, D. Ceresoli, G. L. Chiarotti, M. Cococcioni, I. Dabo, A. D. Corso, S. de Gironcoli, S. Fabris, G. Fratesi, R. Gebauer, U. Gerstmann, C. Gougoussis, A. Kokalj, M. Lazzeri, L. Martin-Samos, N. Marzari, F. Mauri, R. Mazzarello, S. Paolini, A. Pasquarello, L. Paulatto, C. Sbraccia, S. Scandolo, G. Sclauzero, A. P. Seitsonen, A. Smogunov, P. Umari, and R. M. Wentzcovitch. *J. Phys. Condens. Matt.* **21**(39), 395502–395516 (2009).

25. Andrea Dal Corso. *Computational Material Science* **95**, 337 (2014).

26. P. Giannozzi and S. Baroni. Density-functional perturbation theory. In S. Yip, (ed.), *Handbook of Materials Modeling.* Springer Netherlands (2005), pp. 195–214.

27. M. Born and K. Huang. *Dynamical Theory of Crystal Lattices.* Oxford University Press, Oxford (1962).

Chapter 7

Impact of Flexoelectric Effect on Electro-mechanics of Moderate Conductors

A. N. Morozovska and O. V. Varenyk

Institute of Physics, National Academy of Sciences of Ukraine, 46, pr. Nauki, 03028 Kyiv, Ukraine

S. V. Kalinin

Center for Nanophase Materials Science, Oak Ridge National Laboratory, Oak Ridge, TN, 37831

Strong coupling between electrochemical potentials, concentrations of electrons, ions, and strains mediated by flexoelectric effect is a ubiquitous feature of moderate conductors, in particular mixed ionic–electronic moderate conductors (MIECs), the materials of choice in devices ranging from electroresistive and memristive elements to ion batteries and fuel cells. Corresponding mechanisms that govern bias-concentration-strain changes (Vegard expansion, elastic dipole, deformation potential, and flexoelectric coupling) are analyzed.

Proposed theoretical analysis can be readily extended towards the bias-induced strains in the planar capacitor and scanning probe microscopy-like geometries and proved that the impact of flexoelectric effect can be critically important for understanding their local electromechanical response.

1. Introduction

Mixed ionic–electronic moderate conductors (MIECs), such as solid electrolytes with rechargeable ions or vacancies, which also can be mobile, free electrons, and/or holes, can display a reversible dynamics of the space charge layers that leads to a pronounced resistive switching between meta-stable states with high and low resistance[1-6] and unique dynamic properties (including hysteretic) electromechanical response.[7-14] Though MIECs are promising candidates for the non-volatile memory devices, the physical principles of the phenomena are far not clear. In general memristive systems cannot store energy, but they "remember" the total charge transfer due to the metastable changes of their conductance.[15] Strukov *et al.*[16,17] demonstrated that memristive behavior[18] can be inherent to thin semiconductor films, when the drift-diffusion kinetic equations for electrons, holes and mobile donors/acceptors are strongly coupled, at that the memory resistance depends on the thickness ratio of the doped and pure regions of semiconductor. Notably the space charge dynamics in MIEC thin films were studied theoretically mostly in the framework of linear drift-diffusion Poisson–Planck–Nernst theory and diluted species approximation.[16,17,19,20]

One can expect a strong correlation between the electrochemical and electromechanical response in MIECs.[2] Actually, the dynamic redistribution of mobile species (ions, vacancies, and electrons) concentration caused by electromigration (electric field-driven for charged species) and diffusion (concentration gradient-driven for both charged and neutral species) mechanisms can change the lattice molar volume. The changes in the volume result in local electrochemical stresses, so called "Vegard stress" or "chemical pressure".[21,22] The Vegard mechanism plays a decisive role in the origin and evolution of local strains caused by the point defect kinetics in solids.[23,24]

Overall, there are several different mechanisms contributing to the local Vegard-type stresses:

Contribution I. Spatial redistribution of mobile atoms, ions, protons, or vacancies leads to the changes of the surrounding force matrix causing local stresses $\sigma_{ij}(\mathbf{r}, t)$. These stresses are not directly related with the charge of mobile species of type m, they are primary defined with the local changes of their instant concentration $N_m(\mathbf{r}, t)$ in comparison with its equilibrium value N_{m0}. Within linear theory

$$\sigma_{ij}^m(\mathbf{r}, t) \propto \beta_{ij}^m (N_m(\mathbf{r}, t) - N_{m0}), \tag{1}$$

where β_{ij}^m is a *Vegard strain* tensor of the material conjugated with the mobile species of type m.[22] Note that $N_m(\mathbf{r}, t)$ cannot be arbitrarily high, because the maximal volume allowed per one mobile point defect (ion, proton, or vacancy) is limited,[10,25,26] since the defects have their physical size (up to several unit cells) due to the strong changes of the force matrix appeared under the defect inclusion. The concentration limit of mobile defects in the spatial region was named as "steric" effect.[25,26]

Contribution II. Impurity atoms (or different vacancies) *ionization* leads to the changes of their ionic radii that in turn leads to the local stress. In the case of linear response impurity ionization can cause a local stress proportional to the difference of ionized defects concentration from its equilibrium value, i.e.

$$\sigma_{ij}^{ed}(\mathbf{r}) \propto w_{ij}^d\big(N_d^+(\mathbf{r}) - N_{d0}^+\big) + w_{ij}^a\big(N_a^-(\mathbf{r}) - N_{a0}^-\big), \qquad (2)$$

where $w_{kl}^{a,d}$ is called an *elastic dipole* or *chemical strain* tensor of the material.[27] The structure of the tensor is controlled by the symmetry (crystalline or Curie group symmetry) of the material; for isotropic or cubic media it is diagonal, $w_{jk}^{a,d} = w^{a,d}\delta_{jk}$ (hereinafter δ_{jk} is the Kronecker-delta symbol). $N_d^+(\mathbf{r})$ and $N_a^-(\mathbf{r})$ are inhomogeneous concentration of ionized donors and acceptors, N_{d0}^+ and N_{d0}^+ are their equilibrium concentrations. The local concentrations $N_d^+(\mathbf{r})$ and $N_a^-(\mathbf{r})$ obey Fermi–Dirac (FD) statistics. Experimental methods for $w_{kl}^{a,d}$ determination are relatively well established. For instance, one could either directly study the strain of a given sample with the changes of stoichiometry (see e.g. Ref. [22]) or consider the set of several samples with slightly different composition (solid solution).

Contribution III. Electrons and holes accumulation or depletion via deformation potential mechanism leads to the local stress (see Ref. [11] and Appendix A):

$$\sigma_{ij}^e(\mathbf{r}, t) = \xi_{ij}^C(n(\mathbf{r}, t) - n_0) + \xi_{ij}^V(p(\mathbf{r}, t) - p_0), \qquad (3)$$

where $n(\mathbf{r}, t)$ is an instant concentration of electrons in conduction band, n_0 is their equilibrium concentration; $p(\mathbf{r}, t)$ is the concentration of holes in the valence band, p_0 is their equilibrium concentration. $\xi_{ij}^{C,V}$ is a deformation potential tensor of electrons (C) and holes (V).[28–30] We are aware that equation (3) is a rough approximation, neglecting the electron wave-vector $\mathbf{k}$ dispersion and assuming that all electrons under consideration are located in the immediate vicinity of the global minimum $\mathbf{k} = 0$ of their simple conduction band. Surely, equation (3) is unable to involve complex structure of the conduction band such as several minima or spin effects. The properties of the introduced deformation potential tensor $\xi_{ij}^{C,V}$ are determined by the point symmetry of the material.

Contribution IV. Inhomogeneous electric fields, which are inevitably present in systems with inhomogeneous space charge (e.g. in the vicinity of the localized sources of external electric field), induces elastic stresses linearly proportional to the polarization gradient due to the flexoelectric coupling[31–33]:

$$\sigma_{ij}^{f} = -f_{ijkl}\frac{\partial P_k}{\partial x_l}. \tag{4}$$

Here f_{ijkl} are the components of flexocoupling tensor,[33–36] $P_k(\mathbf{r})$ are polarization components.

2. Thermodynamic Approach for the Flexo-electrochemical Effect Description

Taking into account the flexoelectric effect and the Vegard law of chemical expansion,[37] the local stresses σ_{ij} and strains u_{ij} induced by the charged defects (ionized donors or acceptors) are related as the modified Hook's law, whose form follows from equations (1)–(4):

$$\sigma_{ij} = \left(\begin{array}{c} c_{ijkl}u_{kl} - f_{ijkl}\dfrac{\partial P_k}{\partial x_l} - q_{ijkl}P_kP_l - w_{ij}^{a}\left(N_a^{-} - N_{a0}^{-}\right) \\[2mm] -w_{ij}^{d}\left(N_d^{+} - N_{d0}^{+}\right) - \xi_{ij}^{C}(n - n_0) - \xi_{ij}^{V}(p - p_0) \end{array} \right). \tag{5a}$$

Here c_{ijkl} is the tensor of elastic stiffness (hereinafter we use the Einstein summation convention for all repeating indexes). For the sake of generality we included the contribution of electrostriction stress with coefficients q_{ijkl}.

Equation (5a) can be rewritten for the local strain as follows:

$$u_{ij} = \begin{pmatrix} s_{ijkl}\sigma_{kl} + F_{ijkl}\dfrac{\partial P_k}{\partial x_l} + Q_{ijkl}P_kP_l + \Xi_{ij}^C(n - n_0) \\[2mm] + \Xi_{ij}^V(p - p_0) + W_{ij}^a\big(N_a^- - N_{a0}^-\big) + W_{ij}^d\big(N_d^+ - N_{d0}^+\big) \end{pmatrix}.$$

(5b)

Here, s_{ijkl} is the tensor of elastic compliances, $F_{ijkl} = f_{ijpq}s_{pqkl}$ and $Q_{ijkl} = q_{ijpq}s_{pqkl}$ are flexocoupling and electrostriction stress tensors. Chemical strain tensors have the form $W_{ij}^d = w_{kl}^d s_{klij}$, $W_{ij}^a = w_{kl}^a s_{klij}$, deformation potentials $\Xi_{ij}^C = \xi_{kl}^C s_{klij}$ and $\Xi_{ij}^V = \xi_{kl}^V s_{klij}$.

The local strain caused by donors and acceptors redistribution from uniform equilibrium state leads to the shift of their chemical potential levels. So that their equilibrium concentrations in the Boltzmann–Planck–Nernst (BPN) approximation depend on the strain as:

$$N_d^+ \approx N_{de}^+ \exp\left(\frac{w_{jk}^d u_{jk} - eZ_d\varphi}{k_BT}\right),$$
$$N_a^- \approx N_{ae}^- \exp\left(\frac{w_{jk}^a u_{jk} + eZ_a\varphi}{k_BT}\right),$$

(6a)

where e is the elementary charge, Z_d and Z_a are donor (acceptors) ionization degree. Boltzmann constant $k_B = 1.3807 \times 10^{-23}$ J/K, T is the absolute temperature, $\varphi(\mathbf{r})$ is electric potential.

Allowing for the realistic FD statistics and regarding donor level infinitely thin, expressions (6a) become more complicated:

$$N_d^+ = N_d^0\big(1 - f\big(E_d + w_{jk}^d u_{jk} - eZ_d\varphi - E_F\big)\big).$$

(6b)

FD distribution function is $f(x) = (1 + \exp(x/k_BT))^{-1}$; N_d^0 is the concentration of defect atoms. E_d and E_F are the donor and Fermi level positions.

Elastic strain induces conduction (valence) band edge shift which is proportional to the strain in the linear approximation,[28,29]

$$E_C(u_{ij}(\mathbf{r})) = E_C(0) - \xi_{ij}^C u_{ij}(\mathbf{r}), \quad E_V(u_{ij}(\mathbf{r})) = E_V(0) + \xi_{ij}^V u_{ij}(\mathbf{r}),$$

where E_C and E_V are the energetic position of the bottom of conduction band and the top of the valence band respectively. Neglecting the strain-induced changes in the density of states in the energy bands, one can express the impact on the strain of the equilibrium concentration of the electrons in the conduction and holes in the valence bands[11]:

$$
\begin{aligned}
n(\mathbf{r}) &\approx n_e \exp\left(\frac{\xi_{ij}^C u_{ij}(\mathbf{r}) + e\varphi(\mathbf{r})}{k_B T}\right), \\
p(\mathbf{r}) &\approx p_e \exp\left(\frac{\xi_{ij}^V u_{ij}(\mathbf{r}) - e\varphi(\mathbf{r})}{k_B T}\right).
\end{aligned}
\tag{7}
$$

Equations (5) demonstrate that impurity concentration variation induces stress/strain, while equations (6) and (7) show that the strains produce the concentration changes.

The electric potential φ included in equations (5)–(7) obeys the Poisson-type equation:

$$\varepsilon_b \frac{\partial^2 \varphi}{\partial x_i \partial x_j} = \frac{\partial P_i}{\partial x_i} - e\left(p - n - Z_a N_a^- + Z_d N_d^+\right), \tag{8}$$

here ε_b is a background permittivity,[38] $\mathbf{P}$ is a soft mode related polarization of incipient or proper ferroelectric.

Equations (5)–(8) should be substituted by the condition of mechanical compatibility, $e_{ijk}\partial^2 u_{ij}/\partial x_j \partial x_k = 0$, mechanical equilibrium equation $\partial \sigma_{ij}/\partial x_j = 0$ and material equation relating polarization and electric field, e.g. for linear dielectrics $P_i = \varepsilon_0 \chi_{ij} E_j$, for paraelectrics or ferroelectrics it can be nonlinear Landau–Ginzburg equation.

In most cases, the changes of band structure due to the external pressure is rather weak and when calculating the space charges distributions the stress contribution can be neglected in the first approximation, the so-called decoupling approximation. Then the ionic and electrostatic field distributions are substituted in equations (5)–(7) to obtain local electromechanical response.

3. Flexo-electrochemical Effect Manifestation in Electrochemical Strain Microscopy

Development of strains is a phenomenon ubiquitous in solid-state electrochemical devices[22,39] and electroresistive and memristive electronics. On the other hand, electrochemically generated strains can be utilized to design different electromechanical devices,[40,41] or diagnostic tool.[7,42] In particular Electrochemical Strain Microscopy (ESM)[9–42] uses the periodic nanoscale electrochemical strains generated by a biased scanning probe of microscope to detect local electromechanical response at the 10–100 nanometer scale.

The local electromechanical response, defined as the surface displacement u_i divided on the voltage applied to the probe, can be measured by the ESM. The schematic of the experiment is shown in Fig. 1(a).

Figure 1(b) illustrates the difference between three-dimensional and one-dimensional geometry. Diffusion of ionized defects and drift in electric field make their distribution inhomogeneous; the latter coupled with Vegard and flexoelectric effects leads to the inhomogeneous strain and surface shift measured by SPM.

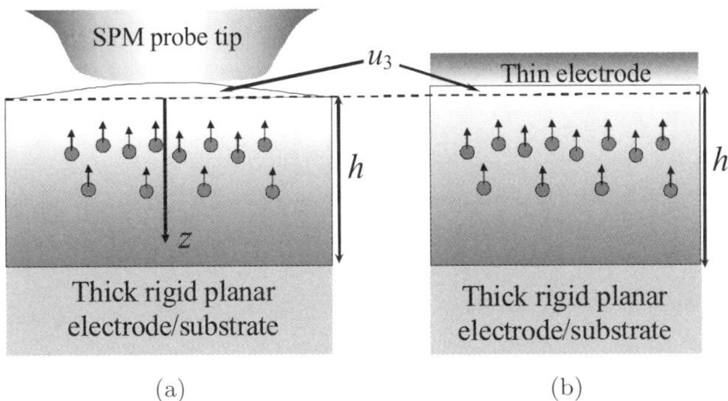

Figure 1.   Schematics of ESM measurements with a flattened scanning probe microscope (SPM) tip (a) is approximated by the (b) strain response of the one-dimensional-system, where u_3 is the surface vertical displacement for fixed back interface. Electric voltage is applied to the tip (a) or top electrode (b), h is the thickness of MIEC film. (Adapted from Ref. [11]).

In order to demonstrate how the general formalism (1)–(8) can be applied to the description of the flexo-electrochemical coupling effect let us consider a typical one-dimensional problem of MIEC film clamped to a rigid substrate and neglect the electrostriction contribution as quadratic effect. The geometry of considered problem is shown in Fig. 1(b). For cubic symmetry equation (5b) gives the relations between dilatation strains and stresses:

$$u_{ii} = (s_{11} - s_{12})\sigma_{ii} + s_{12}(\sigma_{11} + \sigma_{22} + \sigma_{33}) + F_{ii33}\frac{\partial P_3}{\partial z}$$

$$+ W_d\left(N_d^+ - N_{d0}^+\right) + W_a\left(N_a^- - N_{a0}^-\right)$$

$$+ \Xi^C(n - n_0) + \Xi^V(p - p_0). \tag{9}$$

Hereinafter the Vegard tensor is regarded diagonal, $W_{ij} = W\delta_{ij}$ (δ_{ij} is delta Kronecker symbol) and $i = 1, 2, 3$. Boundary conditions for strains in a thick single-domain film are $u_{11}(0) = u_{22}(0) = u_m$, u_m is a misfit strain, and $\sigma_{3i}(h) = 0$. Film thickness is h. Since $F_{1133} = F_{2233} \equiv F_{13}$, $u_{11}(z) = u_{22}(z)$. Mechanical equilibrium condition $\partial\sigma_{ij}/\partial x_j = 0$ gives Lame equation $\partial\sigma_{33}/\partial z = 0$, that leads to $\sigma_{33} = $ const. that allowing for the boundary conditions $\sigma_{3i}(h) = 0$ gives $\sigma_{33} = 0$. Using mechanical compatibility conditions, non-zero stresses and strains acquire the form:

$$\sigma_{22} = \sigma_{11} = \left(\begin{array}{c} \dfrac{u_m}{s_{11} + s_{12}} - \dfrac{F_{13}}{s_{11} + s_{12}}\dfrac{\partial P_3}{\partial z} \\[2mm] W_d\left(N_d^+ - N_{d0}^+\right) + W_a\left(N_a^- - N_{a0}^-\right) \\ -\dfrac{+ \Xi^C(n - n_0) + \Xi^V(p - p_0)}{s_{11} + s_{12}} \end{array}\right), \tag{10a}$$

$$u_{11} = u_{22} = u_m, \tag{10b}$$

$$u_{33} = \frac{2s_{12}u_m}{s_{11} + s_{12}} + \left(\frac{s_{11} - s_{12}}{s_{11} + s_{12}}\right)\left(W_d\left(N_d^+ - N_{d0}^+\right) + W_a\left(N_a^- - N_{a0}^-\right)\right)$$

$$+ \left(\frac{s_{11} - s_{12}}{s_{11} + s_{12}}\right)\left(\Xi^C(n - n_0) + \Xi^V(p - p_0)\right)$$

$$+ \left(F_{33} - \frac{2s_{12}F_{13}}{s_{11} + s_{12}}\right)\frac{\partial P_3}{\partial z}. \tag{10c}$$

In fact equation (10) presents the concrete form of more general bias-strain-concentration equation describing the linear relation between the concentration of diffusing species, flexoelectric and electronic effects in mixed ionic–electronic conductors.

Then using Poisson equation (8) in one-dimensional case for linear dielectrics with $P_3 \approx \varepsilon_0(\varepsilon_{33} - 1)E_3$, equation (10c) allows to calculate normal displacement of the film surface as[11]:

$$u_3(0) \approx -\int_0^h dz \left(\begin{array}{l} \lambda(\tilde{\Xi}^C, \tilde{\gamma})(n - n_0) + \lambda(\tilde{\Xi}^V, -\tilde{\gamma})(p - p_0) \\ +\mu(\tilde{\beta}^a, \tilde{\gamma})(N_a^- - N_{a0}^-) + \mu(\tilde{\beta}^d, -\tilde{\gamma})(N_d^+ - N_{d0}^+) \end{array} \right).$$

$$(11)$$

It is seen from equation (11) that the MIEC surface displacement is proportional to the total charge of each species. Thus only the injected charges control the displacement. Note, that the relation between the total charge and electrostatic potential on the semiconductor surface are well established.[43]

In equation (11), we introduced the designations for the flexoelectrochemical coupling constants as[11]:

$$\lambda(\tilde{\Xi}, \tilde{\gamma}) = \tilde{\Xi}_{33} - \frac{2s_{12}\tilde{\Xi}_{11}}{s_{11} + s_{12}} + \left(\tilde{\gamma}_{33} - \frac{2s_{12}\tilde{\gamma}_{12}}{s_{11} + s_{12}} \right) \frac{e}{\varepsilon_0\varepsilon_{33}}, \qquad (12a)$$

$$\mu(\tilde{\beta}, \tilde{\gamma}) = -\tilde{\beta}_{33} + \frac{2s_{12}\tilde{\beta}_{11}}{s_{11} + s_{12}} + \left(\tilde{\gamma}_{33} - \frac{2s_{12}\tilde{\gamma}_{12}}{s_{11} + s_{12}} \right) \frac{e}{\varepsilon_0\varepsilon_{33}}, \qquad (12b)$$

$$\tilde{\Xi}_{ij}^{C,V} = \Xi^{C,V}(s_{11} + 2s_{12})\delta_{ij}, \qquad \tilde{\beta}_{ij}^{a,d} = W^{a,d}(s_{11} + 2s_{12})\delta_{ij},$$

$$\tilde{\gamma}_{33} = \tilde{\gamma}_{22} = \tilde{\gamma}_{11} = s_{11}f_{11} + 2s_{12}f_{12}, \qquad (12c)$$

$$\tilde{\gamma}_{12} = f_{11}s_{12} + f_{12}(s_{11} + s_{12}), \qquad \tilde{\gamma}_{44} = f_{44}s_{44},$$

where the first terms originated from the deformation potential or Vegard tensors, while the last ones originated from the flexoelectric coupling.

Flexoelectric effect contribution into the coupling constants λ and μ from equations (12) is estimated in the Table 1 of Ref. [11].

The flexoelectric contribution ranges from 0.1 to 10 eV for non-ferroelectric crystalline ionics, which is comparable or much higher than the chemical expansion and deformation potential contributions, which are within the range (0.5–5) eV for the materials. For incipient ($SrTiO_3$) and proper ($Pb(Zr,Ti)O_3$ and $BaTiO_3$) ferroelectrics the flexoelectric effect contribution is much higher than the other ones.

In dimensionless units the strain–voltage response depends on the parameter $eU/(k_B T)$, as anticipated from the diode theory for the case $h \gg R_S$, where R_S is the screening length in the film.

Note, that for the ionically blocking planar top and substrate electrodes the identities $\int_0^h dz(N_a^- - N_{a0}^-) = 0$ and $\int_0^h dz(N_d^+ - N_{d0}^+) = 0$ are valid, and so only the first two "electronic" terms give contribution to displacement per equation (11). Electronic strain–voltage response of the MIEC film with holes and ionized acceptors that is placed between ionically-blocking electrodes is shown in Fig. 2(a). The electronic strain–voltage response demonstrate strong

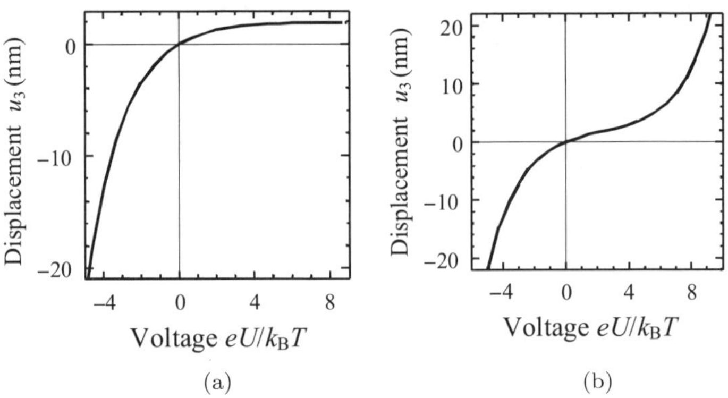

(a)   (b)

Figure 2.   Strain response of the one-dimensional-system. Electronic (a) and (b) mixed ionic–electronic strain–voltage responses $u_3(U)$ of the MIEC film sandwiched between plain ionically-blocking electrodes (a) and (b) ionically-blocking top electrode and ionically-conducting bottom electrode. Graphs are calculated for acceptor concentration $N_{a0}^- = 10^{24}\,\mathrm{m}^{-3}$, room temperature $T = 300K$, coupling constants $\lambda(\tilde{\Xi}^V, -\tilde{\gamma}) = 10^{-31}\,\mathrm{m}^3$ and $\mu(\tilde{\beta}^a, \tilde{\gamma}) = 10^{-30}\,\mathrm{m}^3$, MIEC film thickness $h = 100R_S$. Response is shown in dimensionless units. The geometry of considered problem is shown in the Fig. 1(b). (Adapted from Ref. [11]).

asymmetry under the change of electric voltage polarity: for positive $U > 0$ strong saturation occurs at very small response values, while for negative $U < 0$ the response rapidly increases linearly and reaches noticeable values at $eU/(k_B T) \sim 5$. The origin of the strong voltage asymmetry is the conservation of the full amount of mobile ionized acceptors, which are negatively charged.

The strain–voltage response of the MIEC film placed between the electrodes, one or both of which is ionically-conducting, belongs to the mixed ionic–electronic type; it is shown in Fig. 2(b). The pronounced asymmetry effect seen from the Fig. 2(b) originates from the fact that we put $\lambda(\tilde{\Xi}^V, -\tilde{\gamma}) \ll \mu(\tilde{\beta}^a, \tilde{\gamma})$, because the typical electronic contribution $\tilde{\Xi}^V_{ii} \sim 10^{-31}\,\mathrm{m}^3$ is an order of magnitude smaller than the ionic, $\tilde{\beta}^a_{ii} \sim 10^{-30}\,\mathrm{m}^3$. The voltage behavior (symmetry or weak asymmetry) of the curves for ionic exchange with ambient follows from the fact that the problem is actually identical to that of the charge accumulation at the interface of an intrinsic semiconductor.[43]

Quasi-static hysteresis loops of the strain–voltage response are presented in the Fig. 3. They correspond to the surface deformation appearing in MIEC with mobile donors and electrons under the inhomogeneous ESM probe electrical field, changing with time in a sinusoidal manner (geometry is shown in the Fig. 1(a)).

For demonstration we use the Gaussian form of the potential drop at the film surface, namely the voltage is applied between the probe and bottom electrode $U(r,t) = U_0 \exp(-r^2/r_0^2) \sin(\omega t)$, and regard that the tip lateral size r_0 is much smaller that the lateral size of the computation cell. Tip and bottom electrodes were assumed to be blocking for donors and electrons. Vertical displacement of MIEC surface is shown directly under the probe. Different color of the curves corresponds to the different frequencies of the external field. Simulations are performed in the framework of decoupling approximation, which means that redistribution of the charge carries leads to appearing of the local strain and displacement, while the strain as itself does not impact on charge carrier distribution.

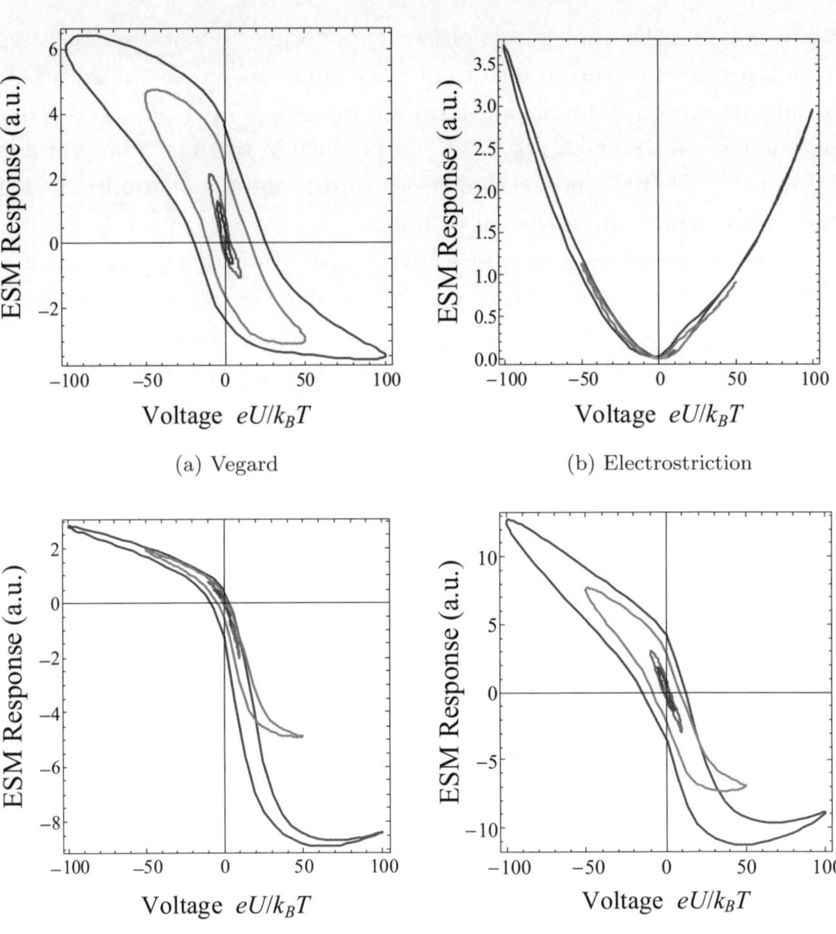

(a) Vegard

(b) Electrostriction

(c) Flexocoupling

(d) Total

Figure 3. Local ESM response of MIEC. Local ESM response $\tilde{u}_3$ calculated for five different voltage amplitudes $eU/k_BT = 1, 5, 10, 50, 100$ (corresponding to the ends of the loops) and dimensionless frequency $\omega t_e = 10^{-3}$ for the following model cases. Electrodes are blocking for the both types of carriers. The geometry of considered problem is shown in the Fig. 1(a). Other parameters are listed in the Table B1 of Appendix B.[44] Part (a) is calculated with the Vegard contribution only, without electrostriction and flexocoupling contributions. Part (b) — with the electrostriction only, without Vegard and flexocoupling contributions. Part (c) — with the flexocoupling only, without Vegard and electrostriction contributions. Part (d) is the total ESM response including Vegard, electrostriction, and flexocoupling contributions.

We regarded that the electron obeys Boltzmann statistics that does not saturate with the voltage increase, while ionized donor amount saturate with the voltage increase for FD statistics. The response calculated using FD statistics for donors demonstrates ultra-strong asymmetry under the change of electric voltage polarity, because it becomes purely electronic at negative voltage and almost ionic at positive voltages.

Part (a) in the Fig. 3 is calculated with the Vegard contribution only, without electrostriction and flexocoupling contributions. Part (b) — with the electrostriction only, without Vegard and flexocoupling contributions. Part (c) — with the flexocoupling only, without Vegard and electrostriction contributions. Part (d) is the total ESM response including Vegard, electrostriction, and flexocoupling contributions. As one can see from the figure parts, all three contributions becomes comparable at voltages $|eU/k_BT| < 100$ (at higher voltages the linear relation between the polarization and electric field, $P_i = \varepsilon_0 \chi_{ij} E_j$, may become invalid for the electrostriction contribution calculations) and all of them contain inherent nonlinearity. The loop opening originated from the Vegard strain is visible even at very low frequency $\omega t_e < 0.001$, meanwhile the electrostriction response has quasi-parabolic shape, and the flexoelectric coupling caused a rather slim strongly asymmetric ESM loop, and the asymmetry origin is the same as in the one-dimensional case considered above. At small voltages the Vegard and flexoelectric strain–voltage response is linear in the static limit and becomes elliptic with the frequency increase.

4. Summary Remarks

Strong coupling between electrochemical potentials, concentrations of electrons, ions, and strains mediated by flexoelectric effect is a ubiquitous feature of moderate conductors, in particular MIECs, the materials of choice in devices ranging from electroresistive and memristive elements to ion batteries and fuel cells. Corresponding mechanisms that govern bias-concentration-strain changes are the Vegard expansion (other name elastic dipole), deformation potential, electrostriction and flexoelectric coupling. Notably, that the contribution of the flexoelectric coupling to the local surface displacement of the moderate conductors is dominant under the space charge blocking boundary conditions.

Proposed analysis can be readily extended towards the bias-induced strains in the planar capacitor and scanning probe microscopy-like geometries and proved that the impact of flexoelectric effect can be critically important for understanding their local electromechanical response.

Acknowledgments

Authors are very grateful to Prof. A. K. Tagantsev for multiple stimulating discussions and extremely fruitful ideas, which essentially improved the chapter content and results presented, and to Dr. E. A. Eliseev for his critical remarks.

Appendix A

Let us demonstrate the validity of equations (5) and (6) from the chapter for the electrons in the conductive band (energy E_C), and for ionized donor (energy E_d) obeying quantum FD statistics.

We start from the expression for the free energy density of electrons in conductive band:[45]

$$\frac{F}{V} = \frac{1}{V}(E + U - TS). \tag{A.1}$$

Kinetic + electrostatic energy of electrons in conductive band in the classical Boltzmann and parabolic approximations is

$$E_{el}^{k} + U_{el} = \int_{E_C}^{\infty} \varepsilon g_C(\varepsilon) f(\varepsilon - E_g - e\varphi) d\varepsilon - e\varphi n_C$$

$$= \frac{3}{2} N_C k_B T \cdot F_{3/2}\left(\frac{E_g + e\varphi}{k_B T}\right) - e\varphi n_C$$

$$\approx \left(\frac{3}{2} k_B T - e\varphi\right) n_C. \tag{A.2}$$

Here $g_C(\varepsilon) \approx \frac{\sqrt{2m_n^3\varepsilon}}{2\pi^2\hbar^3}$ in parabolic (effective mass) approximation, $f(x) = (1+\exp(x/k_BT))^{-1}$ is the FD distribution function. Electron concentration in conductive band is $n_C = \int_0^\infty d\varepsilon \cdot g_C(\varepsilon)f(\varepsilon - E_g - e\varphi)$, that gives $n_C = N_C\exp\left(\frac{E_g+e\varphi}{k_BT}\right)$ in Boltzmann approximation, while $n_C = N_C F_{1/2}\left(\frac{E_g+e\varphi}{k_BT}\right)$ in parabolic approximation. The band gap is $E_g = E_F - E_C$, E_C is the conductive band energy, E_F is a Fermi energy. Electrostatic potential is φ.

The entropy density for Fermi gas of electrons is

$$S_{el} = -k_B\sum_\alpha(f_\alpha\ln f_\alpha + (1-f_\alpha)\ln(1-f_\alpha)). \tag{A.3a}$$

The probability of the occupation of the αth state by an electron is $f_\alpha = f\left(\frac{-E_g+\varepsilon_\alpha-e\varphi}{k_BT}\right)$. The summation in equation (A.3a) is performed over the states on conduction band denoted by the summation index "α". In Boltzmann approximation

$$f_\alpha \approx \exp\left(-\frac{-E_g + \varepsilon_\alpha - e\varphi}{k_BT}\right), \quad f_\alpha = \frac{n_C}{N_C}\exp\left(-\frac{\varepsilon_\alpha}{k_BT}\right),$$

and so the entropy acquires the form[11]:

$$S_{el} \approx -k_B\sum_\alpha(f_\alpha\ln f_\alpha - f_\alpha) = k_B N_C\left(\frac{n_C}{N_C}\ln\left(\frac{n_C}{N_C}\right) - \frac{n_C}{N_C}\right). \tag{A.3b}$$

Kinetic + electrostatic energy of the simple (infinitely thin and non-degenerated) donor impurity level with energy E_d is

$$E_{don}^k + U_{don} = (e\varphi - E_d)N_d^0(1 - f_d), \tag{A.4a}$$

$$f_d = \left(1 + \exp\left(\frac{E_d - E_F - e\varphi}{k_BT}\right)\right)^{-1},$$

$$1 - f_d = \left(1 + \exp\left(\frac{E_F - E_d + e\varphi}{k_BT}\right)\right)^{-1}. \tag{A.4b}$$

The total concentration of donor atoms is N_k^0. Immediately concentration of the ionized donors N_k^+ is

$$\frac{N_d^+}{N_d^0} = (1 - f_d) \equiv \left(1 + \exp\left(\frac{-E_d + E_F + e\varphi}{k_B T}\right)\right)^{-1}. \qquad \text{(A.4c)}$$

So that $E_{don}^d + U_{don} = (e\varphi - E_d)N_d^+$.

The entropy density for electrons on the infinitely thin donor level is[45]:

$$S_d = -k_B(f_d \ln f_d + (1 - f_d)\ln(1 - f_d))$$

$$\equiv -k_B\left(\frac{N_d^+}{N_d^0}\ln\left(\frac{N_d^+}{N_d^0}\right) + \left(1 - \frac{N_d^+}{N_d^0}\right)\ln\left(1 - \frac{N_d^+}{N_d^0}\right)\right). \qquad \text{(A.5)}$$

Using equations (A.2), (A.3b), (A.4), and (A.5), the free energy density (A.1) can be expressed in term of its independent variables strain u_{ij}, electron and ionized donors concentrations n_C, and N_k^+, and temperature T as:

$$\frac{F}{V} = e\varphi\left(N_d^+ - n_C\right) + n_C E_g(u_{ij})$$

$$+ k_B T N_C\left(\frac{n_C}{N_C}\ln\left(\frac{n_C}{N_C}\right) - \frac{n_C}{N_C}\right) - N_d^+ E_d$$

$$+ k_B T N_d^0\left(\frac{N_d^+}{N_d^0}\ln\left(\frac{N_d^+}{N_d^0}\right) + \left(1 - \frac{N_d^+}{N_d^0}\right)\ln\left(1 - \frac{N_d^+}{N_d^0}\right)\right).$$

$$\text{(A.6)}$$

Then the stress tensor can be found from its definition

$$\sigma_{ij} = \frac{\partial}{\partial u_{ij}}\left(\frac{F}{V}\right)\bigg|_{T, n_C, N_d^+} = n_C\left(\frac{\partial E_g}{\partial u_{ij}} - \frac{k_B T}{N_C}\frac{\partial N_C}{\partial u_{ij}}\right) + N_d^+\frac{\partial E_d}{\partial u_{ij}}$$

$$\approx n_C \Xi_{ij}^C + N_d^+ w_{ij}^d. \qquad \text{(A.7)}$$

Linear approximation on strain was used in equation (A.7), namely:

$$E_g(u_{ij}) = E_g(0) + \Xi_{ij}^C u_{ij} + \ldots, \qquad E_d(u_{ij}) = E_d(0) + w_{ij}^d u_{ij} + \ldots. \qquad \text{(A.8)}$$

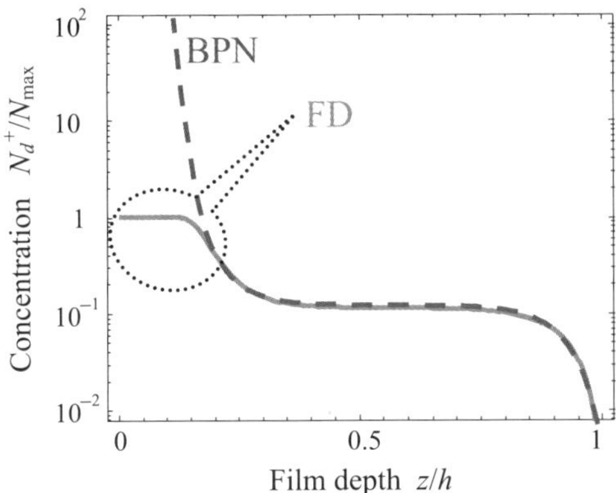

Figure A1. Typical difference in the spatial distribution of ionized defects calculated in the BPN approximation (dashed curve) and allowing for the FD statistics (solid curve). Negatively charged probe electrode is located at $z = 0$, positively charged (or grounded) electrode is located at $z = h$. Capacitor geometry is shown in the Fig. 1(b). (Adapted with permission from Ref. 10)

Here, we neglect the strain dependence of the density of states N_C and regarded N_d^0 as strain-independent. Here Ξ_{ij}^C is a deformation potential tensor, w_{ij}^d is a Vegard stress tensor.

The variative derivative of F on the donor concentration gives the contribution to the electrochemical potential

$$\xi_d = -\frac{\partial F}{\partial N_d^+} = -e\varphi - w_{ij}^d \sigma_{ij} + k_B T \ln\left(\frac{N_d^+}{N_d^0 - N_d^+}\right). \qquad (A.9)$$

Actually, from equation (A.9), donor concentration is $N_d^+ = N_d^0(1 - f(e\varphi + w_{ij}^d u_{ij} + \xi_d))$, where $f(x) = (1 + \exp(x/k_B T))^{-1}$ is the FD distribution function.

Figure (A1) shows typical difference in the spatial distribution of ionized defects calculated in the BPN approximation and allowing for the FD statistics.

References

1. A. Sawa. *Mater. Today.* **11**, 28 (2008).
2. B. Magyari-Kope and M. Tendu. *Nanotechnology.* **22**, 254029 (2011).
3. D.-S. Shang, L. Shi, J.-R. Sun, and B.-G. Shen. *Nanotechnology.* **22**, 254008 (2011).

4. Y. Kim, S. Kelly, A. Morozovska, E. K. Rahani, E. Strelcov, E. Eliseev, S. Jesse, M. Biegalski, N. Balke, N. Benedek, D. Strukov, J. Aarts, I. Hwang, S. Oh, J. S. Choi, T. Choi, B. H. Park, V. Shenoy, P. Maksymovych, and S. Kalinin. *Sergei. Nano Lett.* **13**, 4068–4074 (2013).

5. R. Waser, and M. Aono. *Nat. Mater.* **6**, 833 (2007).

6. K. Szot, M. Rogala, W. Speier, Z. Klusek, A. Besmehn, and R. Waser. *Nanotechnology.* **22**, 254001 (2011).

7. A. N. Morozovska, E. A. Eliseev, and S. V. Kalinin. *Appl. Phys. Lett.* **96**, 222906 (2010).

8. A. N. Morozovska, E. A. Eliseev, N. Balke, and S. V. Kalinin. *J. Appl. Phys.* **108**, 053712 (2010).

9. N. Balke, S. Jesse, A. N. Morozovska, E. Eliseev, D. W. Chung, Y. Kim, L. Adamczyk, R. E. Garcia, N. Dudney, and S. V. Kalinin. *Nat. Nanotechnol.* **5**, 749–754 (2010).

10. A. N. Morozovska, E. A. Eliseev, O. V. Varenyk, Y. Kim, E. Strelcov, A. Tselev, N. V. Morozovsky, and S. V. Kalinin. *J. Appl. Phys.* **116**, 066808 (2014).

11. A. N. Morozovska, E. A. Eliseev, A. K. Tagantsev, S. L. Bravina, L.-Q. Chen, and S. V. Kalinin. *Phys. Rev. B.* **83**, 195313 (2011).

12. A. N. Morozovska, E. A. Eliseev, G. S. Svechnikov, and S. V. Kalinin. *Phys. Rev. B.* **84**, 045402 (2011).

13. A. N. Morozovska, E. A. Eliseev, and S. V. Kalinin. *J. Appl. Phys.* **111**, 014114 (2012).

14. A. N. Morozovska, E. A. Eliseev, S. L. Bravina, F. Ciucci, G. S. Svechnikov, L.-Q. Chen, and S. V. Kalinin. *J. Appl. Phys.* **111**, 014107 (2012).

15. L. O. Chua, and S. M. Kang. *Proc. IEEE.* **64**, 209 (1976).

16. D. B. Strukov, G. S. Snider, D. R. Stewart, and R. S. Williams. *Nature.* **453**, 80 (2008).

17. D. B. Strukov, J. L. Borghetti, and R. S. Williams. *Small.* **5**(9), 1058 (2009).

18. L. O. Chua. *IEEE T. Circuits Syst.* **18**, 507 (1971).

19. Y. Gil, O. M. Umurhan, and I. Riess. *Solid State Ionics.* **178**, 1–12 (2007).

20. Y. Gil, O. M. Umurhan, and I. Riess. *J. Appl. Phys.* **104**, 084504 (2008).

21. S. M. Allen, and J. W. Cahn. *Acta Metall. Mater.* **27**(6), 1085 (1979).

22. X. Zhang, W. Shyy, and A. M. Sastry. *J. Electrochem. Soc.* **154**, A910 (2007).

23. M. Tang, H. Y. Huang, N. Meethong, Y. H. Kao, W. C. Carter, and Y. M. Chiang. *Chem. Mater.* **21**(8), 1557 (2009).

24. M. Tang, W. C. Carter, J. F. Belak, and Y.-M. Chiang. *Electrochim. Acta.* **56**(2), 969 (2010).

25. M. S. Kilic, M. Z. Bazant, and A. Ajdari. *Phys. Rev. E.* **75**, 021502 (2007).

26. M. S. Kilic, M. Z. Bazant, and A. Ajdari. *Phys. Rev. E.* **75**, 021503 (2007).

27. D. A. Freedman, D. Roundy, and T. A. Arias. *Phys. Rev. B.* **80**, 064108 (2009).

28. N. W. Ashcroft, and N. D. Mermin. *Solid State Physics.* Holt, Rinehart, and Winston, New York (1976), 826 pages.

29. A. I. Anselm. *Introduction to Semiconductor Theory.* Mir, Moscow, Prentice-Hall, Englewood Cliffs, NJ (1981).

30. G. L. Bir, and G. E. Pikus. Symmetry and deformation effects in semiconductors. Moscow (1972).

31. V. S. Mashkevich, and K. B. Tolpygo. *Sov. Phys. JETP.* **4**, 455 (1957).

32. Sh. M. Kogan. *Solid State Phys.* **5**(10), 2829 (1963).

33. A. K. Tagantsev. *Phys. Rev. B.* **34**, 5883 (1986).

34. G. Catalan, L. J. Sinnamon, and J. M. Gregg. *J. Phys. Condens. Mat.* **16**, 2253 (2004).

35. E. A. Eliseev, A. N. Morozovska, M. D. Glinchuk, and R. Blinc. *Phys. Rev. B.* **79**(16), 165433-1-10 (2009).

36. P. V. Yudin, and A. K. Tagantsev. *Nanotechnology.* **24**, 432001 (2013).

37. S. Golmon, K. Maute, S. H. Lee, and M. L. Dunn. *Appl. Phys. Lett.* **97**, 033111 (2010).

38. A. K. Tagantsev, G. Gerra, and N. Setter. *Phys. Rev. B.* **77**, 174111 (2008).

39. Y. T. Cheng and M. W. Verbrugge. *J. Appl. Phys.* **104**, 083521 (2008).

40. T. Mirfakhrai, J. D. W. Madden, and R. H. Baughman. *Mater. Today* **10**, 30 (2007).

41. T. E. Chin, U. Rhyner, Y. Koyama, S. R. Hall, and Y. M. Chiang. *Electrochem. Solid St.* **9**, A134 (2006).

42. A. K. Pannikkat, and R. Raj. *Acta Mat.* **47**, 3423 (1999).

43. S. M. Sze. *Physics of Semiconductor Devices*, 2nd edn. Wiley-Interscience, New York (1981).

44. O. V. Varenyk, M. V Silibin, D. A Kiselev, E. A. Eliseev, S. V. Kalinin, and A. N. Morozovska. http://arxiv.org/abs/1411.0966 (Accepted to *JAP*).

45. M. Y. Gureev, A. K. Tagantsev, and N. Setter. *Phys. Rev. B.* **83**, 184104 (2011).

Chapter 8

Role of Flexoelectricity in Multidomain Ferroelectrics

R. Ahluwalia

Institute of High Performance Computing, Singapore, 138632

A. K. Tagantsev

Ceramic Laboratory, Swiss Federal Institute of Technology EPFL,
CH-1015 Lausanne, Switzerland

Ferroics Laboratory, Ioffe Physical Technical Institute,
St. Petersburg 194021, Russia

P. Yudin

Ceramic Laboratory, Swiss Federal Institute of Technology EPFL,
CH-1015 Lausanne, Switzerland

Novosibirsk State University, 2 Pirogova street,
Novosibirsk 630090, Russia

N. Setter

Ceramic Laboratory, Swiss Federal Institute of Technology EPFL,
CH-1015 Lausanne, Switzerland

N. Ng

Institute of High Performance Computing, Singapore, 138632

D. J. Srolovitz

Department of Materials Science and Engineering,
Department of Mechanical Engineering, University of Pennsylvania,
Philadelphia, PA 19104, USA

We have studied the influence of flexoelectric coupling on the domain patterns in ferroelectrics using phase field modeling. The phase field simulations predict that a high strength of the flexoelectric coupling leads to formation of a fine structure in domain patterns in ferroelectrics. The fine structure forms when the coupling strength exceeds a critical value and is related to local transition into an incommensurate phase. Depending on the parameters, a structure with stripe patterns with antiparallel polarizations or a pattern with two-dimensional arrays of alternating vortices is found. A single mode analytical theory is used to study the stability of incommensurate phases. It is shown that above a critical strength of the flexoelectric coupling, an incommensurate phase must necessarily exist, either as a meta-stable state and at high enough strength, as the only stable state. Even when the flexoelectric coupling is not strong enough to form an incommensurate phase, the domain walls and the overall domain patterns are still influenced by flexoelectricity. When an external field is applied to the multidomain state, flexoelectricity leads to a pinning of domain walls. This shows that flexoelectricity can also influence the effective electro-mechanical properties of multidomain ferroelectrics.

1. Introduction

There is a growing body of evidence which suggests that flexoelectric coupling can play a significant role in ferroelectric materials.[1-13] In a series of papers, Cross and co-workers showed that a giant flexoelectric response can be obtained in some ferroelectric materials.[1-5] It is not clear if this large response is an intrinsic unit cell level property or is arising from extrinsic contributions such as surfaces and domain walls. Thus, the first step to understand this high response should be to understand how the flexoelectric coupling influences ferroelectricity, even if no external strain gradient is applied. Typically, ferroelectric materials have heterogeneities such as domain walls which can have large polarization and strain gradients. Since flexoelectricity is a coupling between electrical polarizations and strain gradients or, conversely, between strain and polarization gradients, it is reasonable to assume that the domain walls and consequently the overall domain structure will be influenced by this effect. In this chapter, we explore the impact of this coupling on multidomain ferroelectrics.

When a ferroelectric material is cooled to a temperature below the curie temperature, it spontaneously forms domains of the equivalent variants of the ferroelectric state.[14] For example in context of tetragonal ferroelectrics, there are broadly two kinds of domains: $180°$ domains which have domain walls that separate regions with antiparallel polarizations and $90°$ domains which separate regions which have polarizations perpendicular to each other. While the $180°$ domain walls are characterized by gradients in polarizations only, $90°$ domains have gradients in both strains and polarizations due to the fact that domain wall separates tetragonal variants with different orientations. We first examine the nature of flexoelectric coupling associated with these domain walls. Flexoelectric coupling in the bulk is characterized by flexocoupling constants that enter the electromechanical constitutive equations[15]

$$E_i = \chi_{ij}^{-1} P_j - f_{klij}\frac{\partial \varepsilon_{kl}}{\partial x_j},$$

$$\sigma_{ij} = C_{ijkl}\varepsilon_{kl} + f_{ijkl}\frac{\partial P_k}{\partial x_l}, \tag{1}$$

where the dummy suffix summation rule is accepted. Here E_i and σ_{ij} represent the electric and stress fields, ε_{ij} is the strain tensor, P_i is the polarization, C_{ijkl} is the elastic constant tensor, and χ_{ij}^{-1} are the components of the inverse dielectric susceptibility tensor. The flexocoupling coefficients f_{ijkl} are measured in volts and, according to estimates based on atomic considerations, are typically in the 1–10 V range.[16] For tetragonal ferroelectrics, the appropriate flexocoupling coefficients are f_{11}, f_{12}, and f_{44} (in Voigt notation). For a $180°$ wall, f_{44} is the relevant coefficient. It is easy to see that non-zero f_{44} implies that a shear stress will be generated at the domain wall due to polarization gradient. For $90°$ domains not only f_{44} is relavant, but also flexocoupling coefficients f_{11} and f_{12} which lead to uniaxial stresses at the domain boundaries. This suggests that flexoelectric coupling will influence the thermodynamics of multidomain ferroelectrics and can influence the overall domain patterns.

In general, flexoelectric coupling biases the system towards creating polarization and strain gradients. On the other hand, there is always an energy penalty associated with creating domain walls which implies that creation of a gradient in polarizations costs extra energy. Thus, we are dealing with competing interactions. Competing interactions can lead to formation of "incommensurate" or modulated structures in solids.[17] In 1970,[18] Axe *et al.* pointed out that the phonon dispersion curves for the ferroelectric $KTaO_3$ suggest that mode coupling (which is, as presently known, due to flexoelectrity) can lead to the formation of the incommensurate state in perovskite ferroelectrics. In fact, a large number of ferroic materials are known to exhibit modulated phases.[19–21] These ideas have prompted us to revisit the theory of ferroelectric domains, incorporating the effects of flexoelectric coupling. An understanding of the influence of flexoelectricity of ferroelectric domains is crucial to interpret the experimental data which predicts large values of the flexocoupling constants. If the large flexoelectric response is a fundamental property of the lattice, then its influence of the underlying domain structure must be significant, considering domain walls have strains and polarization gradients, even in the absence of external strain gradients. Since domain walls can migrate under external electric fields, a related issue is the influence of flexoelectric coupling on the effective polarization-electric field (P-E) response of a ferroelectric. Motivated by these considerations, we used the phase field approach to make a parameteric study of how domain switching behavior changes as the strength of the flexoelectric is increased.

We perform phase-field simulations within the Ginzburg–Landau framework, where the polarization components are the order parameters. For simplicity, we restrict our analysis to a two-dimensional model that describes domain patterns having continuous translational symmetry along the third dimension (z-axis). In this approximation, a ferroelectric undergoing a transition from a cubic to tetragonal phase can be viewed as a two-dimensional ferroelectric undergoing a square to rectangular phase transition (see e.g. Ref. [22]). The new feature of the present work is the incorporation of flexoelectric coupling in the Ginzburg–Landau framework (in addition to electrostrictive and electrostatic interactions). To understand the role played by each flexoelectric tensor component, we consider the cases: (i) $f_{44} \neq 0, f_{11} = f_{12} = 0$, (ii) $f_{11} \neq 0, f_{12} = f_{44} = 0$, and (iii) $f_{12} \neq 0, f_{11} = f_{44} = 0$, where f_{ij} are the flexocoupling coefficients (in Voigt notation) for a system undergoing a cubic to tetragonal transition.

The chapter is organized as follows. Section 2 describes the Ginzburg–Landau model with flexoelectric coupling. In Section 3, we present numerical simulation results for domain patterns for different flexoelectric couplings. A single mode thermodynamic analysis to understand the simulation results is presented in Section 4. Section 5 discusses the effect of flexoelectricity on electric field induced domain switching. Finally, the chapter ends with a summary and conclusions.

2. Model

We consider a two-dimensional model to describe domain patterns in a system with continuous translational symmetry along the third dimension (z-axis). We assume that the component of the polarization along the translation axis is negligible ($P_z = 0$). In such cases, a perovskite ferroelectric undergoing a cubic to tetragonal phase transition can be viewed as an auxiliary two-dimensional ferroelectric with a square to rectangular phase transition. The parameters for this two-dimensional ferroelectric may be calculated from the properties of the three-dimensional ferroelectric.

The total free energy of the auxiliary two-dimensional ferroelectric with flexoelectric coupling can be expressed as

$$F = \int d\vec{r}[f_{Landau} + f_{elastic} + f_{flex} + f_{estat}], \qquad (2)$$

where $d\vec{r} = dxdy$. f_{Landau} is expressed in terms of the polarization $\vec{P}$ components and their gradients as

$$f_{Landau} = \alpha_1\left(P_x^2 + P_y^2\right) + \alpha_{11}\left(P_x^4 + P_y^4\right) + \alpha_{12}P_x^2 P_y^2$$
$$+ \alpha_{111}\left(P_x^6 + P_y^6\right) + \alpha_{112}\left(P_x^4 P_y^2 + P_x^2 P_y^4\right) \qquad (3)$$

$$+ \sum_i \frac{\lambda_1}{2} \left[\left(\frac{\partial P_i}{\partial x} \right)^2 + \left(\frac{\partial P_i}{\partial y} \right)^2 \right]$$

$$+ \sum_i \frac{\lambda_2}{2} \left[\left(\frac{\partial^2 P_i}{\partial x^2} \right) + \left(\frac{\partial^2 P_i}{\partial y^2} \right) \right]^2 ,$$

where $i = x, y$. In this phenomenological model, we describe the low-order gradient energy (quadratic in the first spatial derivatives of the polarization) choosing only one coefficient λ_1. For perovskite materials, in terms of the conventional fourth-rank tensor[23] g_{ijkl} for the correlation energy density with 3 independent coefficients g_{1111}, g_{1122}, and g_{1212}, such choice implies $g_{1111} = -g_{1122} = g_{1212} = \lambda_1$. These components satisfy the isotropic medium condition $g_{1111} - g_{1122} = 2g_{1212}$. Thus, the gradient energy used in our model corresponds to a specific type of isotropic medium. A higher-order gradient energy (quadratic in the second spatial derivatives of the polarization) is also described by only one coefficient λ_2. These simplifications are done in order to isolate the effect of flexoelectricity on the domain wall orientations from that of the anisotropy of the gradient energy. Note that the higher-order correlation energy is often omitted in the analysis of ferroelectric domain patterns. However, it is important for the description of modulated structures.[24]

The appropriate elastic energy, incorporating electrostrictive coupling, is

$$f_{elastic} = \frac{C_{11}}{2} \left(\left(u_{xx} - u_{xx}^0 \right)^2 + \left(u_{yy} - u_{yy}^0 \right)^2 \right)$$

$$+ C_{12} \left(u_{xx} - u_{xx}^0 \right) \left(u_{yy} - u_{yy}^0 \right) + \frac{C_{44}}{2} \left(u_{xy} - u_{xy}^0 \right)^2 , \quad (4)$$

where the linearized strain components are expressed in terms of the displacements u_i as $u_{xx} = (\partial u_x / \partial x)$, $u_{yy} = (\partial u_y / \partial y)$, and $u_{xy} \equiv (\partial u_x / \partial y + \partial u_y / \partial x)$. Here u_{ij}^0 represents the electrostrictive strain tensor,[a]

$$u^0_{xx} = Q_{11}P^2_x + Q_{12}P^2_y, \quad u^0_{yy} = Q_{11}P^2_y + Q_{12}P^2_x, \quad u^0_{xy} = Q_{44}P_xP_y,$$

$$(5)$$

where Q_{ij} are the electrostrictive constants which couple strains and squares of polarizations. The contribution from the flexoelectric coupling can be expressed as

$$f_{flex} = -\frac{f_{11}}{2}\left(P_x\frac{\partial u_{xx}}{\partial x} + P_y\frac{\partial u_{yy}}{\partial y} - u_{xx}\frac{\partial P_x}{\partial x} - u_{yy}\frac{\partial P_y}{\partial y}\right)$$

$$-\frac{f_{12}}{2}\left(P_x\frac{\partial u_{yy}}{\partial x} + P_y\frac{\partial u_{xx}}{\partial y} - u_{xx}\frac{\partial P_y}{\partial y} - u_{yy}\frac{\partial P_x}{\partial x}\right)$$

$$-\frac{f_{44}}{2}\left(P_x\frac{\partial u_{xy}}{\partial y} + P_y\frac{\partial u_{xy}}{\partial x} - u_{xy}\left(\frac{\partial P_y}{\partial x} + \frac{\partial P_x}{\partial y}\right)\right). \quad (6)$$

The total stresses, including the flexoelectricity contributions, are

$$\sigma_{xx} = \sigma^e_{xx} + \left[f_{11}\left(\frac{\partial P_x}{\partial x}\right) + f_{12}\left(\frac{\partial P_y}{\partial y}\right)\right],$$

$$\sigma_{yy} = \sigma^e_{yy} + \left[f_{11}\left(\frac{\partial P_y}{\partial y}\right) + f_{12}\left(\frac{\partial P_x}{\partial x}\right)\right], \quad (7)$$

$$\sigma_{xy} = \sigma^e_{xy} + f_{44}\left[\left(\frac{\partial P_y}{\partial x}\right) + \left(\frac{\partial P_x}{\partial y}\right)\right],$$

where σ^e_{ij} represents the electrostrictive part of the stress tensor

$$\sigma^e_{xx} = C_{11}(u_{xx} - u^0_{xx}) + C_{12}(u_{yy} - u^0_{yy}),$$

$$\sigma^e_{yy} = C_{11}(u_{yy} - u^0_{yy}) + C_{12}(u_{xx} - u^0_{xx}), \quad (8)$$

$$\sigma^e_{xy} = C_{44}(u_{xy} - u^0_{xy}).$$

The electrostatic contribution to the free energy is given as

$$f_{estat} = \left(\varepsilon_b\frac{\vec{E} \bullet \vec{E}}{2}\right), \quad (9)$$

---

[a] There exist a confusion in the literature concerning the Voight notation for Q_{44}. Here, we define $Q_{44} = 4Q_{2323}$.

where $\vec{E} = -\vec{\nabla}\varphi$ is the electric field, and φ and ε_b are the electrostatic potential and background dielectric permittivity of the material. The electrostatic potential can be calculated from Gauss's law $\vec{\nabla} \bullet \vec{D} = \rho(\vec{r})$, where $\vec{D} = \varepsilon_b \vec{E} + \vec{P}$ and ρ is the free charge density. Here we consider a system without free charge, for which Gauss's law leads to the following constraint

$$\vec{\nabla} \bullet \left(-\varepsilon_b \vec{\nabla}\varphi + \vec{P} \right) = 0. \tag{10}$$

Polarization evolution kinetics are governed by the time-dependent Ginzburg–Landau equations

$$\frac{\partial P_i}{\partial t} = -\Gamma \left(\frac{\delta F}{\delta P_i} - E_i \right), \quad i = x, y, \tag{11}$$

where E_i is the electric field, Γ is a kinetic coefficient, and $\delta F/\delta P_i$ is the variational derivative of the free energy with respect to the polarization.

The full equations of motion can be expressed as

$$-\frac{1}{\Gamma}\frac{\partial P_x}{\partial t} = 2\alpha_1 P_x + 4\alpha_{11}P_x^3 + 2\alpha_{12}P_x P_y^2 + 6\alpha_{111}P_x^5$$
$$+ \alpha_{112}(4P_x^3 P_y^2 + 2P_x P_y^4)$$
$$- 2P_x(Q_{11}\sigma_{xx}^e + Q_{12}\sigma_{yy}^e) - Q_{44}\sigma_{xy}^e P_y$$
$$- f_{11}\left(\frac{\partial u_{xx}}{\partial x}\right) - f_{12}\left(\frac{\partial u_{yy}}{\partial x}\right) - f_{44}\left(\frac{\partial u_{xy}}{\partial y}\right)$$
$$- \lambda_1 \nabla^2 P_x + \lambda_2 \nabla^2(\nabla^2 P_x) + \frac{\partial \varphi}{\partial x},$$

$$-\frac{1}{\Gamma}\frac{\partial P_y}{\partial t} = 2\alpha_1 P_y + 4\alpha_{11}P_y^3 + 2\alpha_{12}P_y P_x^2 + 6\alpha_{111}P_y^5$$
$$+ \alpha_{112}(4P_y^3 P_x^2 + 2P_y P_x^4)$$
$$- 2P_y(Q_{11}\sigma_{yy}^e + Q_{12}\sigma_{xx}^e) - Q_{44}\sigma_{xy}^e P_x$$

$$- f_{11}\left(\frac{\partial u_{yy}}{\partial y}\right) - f_{12}\left(\frac{\partial u_{xx}}{\partial y}\right) - f_{44}\left(\frac{\partial u_{xy}}{\partial x}\right)$$

$$- \lambda_1 \nabla^2 P_y + \lambda_2 \nabla^2(\nabla^2 P_y) + \frac{\partial \varphi}{\partial y}, \tag{12}$$

where $\nabla^2_{=}(\partial^2/\partial x^2) + (\partial^2/\partial y^2)$ is the Laplacian operator. The displacement field dynamics are given by the dissipative force balance equations

$$\rho\frac{\partial^2 u_x}{\partial t^2} - \eta\nabla^2\frac{\partial u_x}{\partial t} = \frac{\partial \sigma_{xx}}{\partial x} + \frac{\partial \sigma_{xy}}{\partial y}$$

$$\rho\frac{\partial^2 u_y}{\partial t^2} - \eta\nabla^2\frac{\partial u_y}{\partial t} = \frac{\partial \sigma_{xy}}{\partial x} + \frac{\partial \sigma_{yy}}{\partial y}, \tag{13}$$

where η is a viscosity that is used to drive the system toward mechanical equilibrium $\partial \sigma_{ij}/\partial x_j = 0$. Equations (12) and (13) subject to the constraint (10) are used to study the dynamics of the polarizations.

3. Simulations of Bulk Domain Patterns

To simulate the domain patterns, we solve equations (10), (12), and (13), using a finite difference method. The computational domain is a square. To mimic a bulk system, we apply periodic boundary conditions for the polarizations and the electrostatic potential. For computational purposes, dimensionless length and time scales are introduced as $\vec{r'} = \vec{r}/\delta$ and $t^* = (t/\delta)\sqrt{C_{11}/\rho}$. The polarizations and displacements are scaled as $P_i = P_0 P_i^*$ and $u_i = Q_{11}P_0^2\delta u_i^*$ respectively. Here P_0 is a characteristic polarization and δ is the minimal spatial scale for our numerical simulations. The free energy parameteres are scaled as $\alpha_1^* = \alpha_1/|\alpha_1|$, $\alpha_{11}^* = \alpha_{11}P_0^2/|\alpha_1|$, $\alpha_{12}^* = \alpha_{12}P_0^2/|\alpha_1|$, $\alpha_{111}^* = \alpha_{111}P_0^4/|\alpha_1|$ $\alpha_{112}^* = \alpha_{112}P_0^4/|\alpha_1|$. The flexocoupling coefficients are scaled as $f_{ij}^* = (f_{ij}Q_{11}P_0/\delta|\alpha_1|)$. The rescaled gradient coefficients are expressed as $\lambda_1^* = \lambda_1/(\delta^2|\alpha_1|)$ and $\lambda_2^* = \lambda_2/(\delta^4|\alpha_1|)$. The rescaled viscosity and the kineatic coefficient are expressed as $\eta^* = \eta\sqrt{C_{11}/\delta^2\rho}$ and $\Gamma^* = \Gamma|\alpha_1|\delta\sqrt{/\rho/C_{11}}$, respectively. The numerical simulations were performed using the dimensionless variables t^*, P^*, u^*, and r'.

We initialize the system with small, random fluctuations in the polarization around zero. Physically, this corresponds to quenching from the paraelectric state into the ferroelectric one. The simulations reported in this chapter are performed for the mechanically constrained case with $\langle u_{xx} \rangle = \langle u_{yy} \rangle = \langle u_{xy} \rangle = 0$. Here symbol $\langle \rangle$ denotes averaging over the computational domain. We present the mechanically constrained case here since it can describe the ferroelastic as well as ferroelectric domains. For the mechanically unconstrained case, the reader is refered to Ref. [12].

Although the results presented here can be applied to any ferroelectric, for our simulations we have used parameters close to those of $Pb(Zr,Ti)O_3$ (PZT) with aTi content of 80%.[25]

Since λ_1 and λ_2 are not known for PZT, we have chosen a value of λ_1 of the order of the coefficients for the gradient terms in $PbTiO_3$ and $BaTiO_3$.[14] We choose λ_2 such that the terms controlled by λ_1 and λ_2 to be of the same order of magnitude when the polarization changes by its "atomic" value (e/a^2 where e and a are at a typical interatomic distance) at the distance a (as expected from order-of-magnitude estimates). Note that this set of the model parameters corresponds to a tetragonal homogeneous ferroelectric (i.e. the homogeneous ferroelectric state will correspond to a tetragonal system).

Using the parameters from Table 1, we simulate bulk domain patterns for three different cases: $f_{11} = f_{44} = 0$, $f_{12} \neq 0$, $f_{12} = f_{44} = 0$, $f_{11} \neq 0$ and $f_{11} = f_{12} = 0$, $f_{44} \neq 0$. We first need

Table 1.　Material parameters used in the simulations.

$\alpha_1 = -14.8 \times 10^7\,\mathrm{Vm/C}$	$\alpha_{11} = -3.1 \times 10^7\,\mathrm{Vm^5/C^3}$	$\alpha_{12} = 6.3 \times 10^8\,\mathrm{Vm^5/C^3}$
$\alpha_{111} = 2.5 \times 10^8\,\mathrm{Vm^9/C^5}$	$\alpha_{112} = 9.7 \times 10^8\,\mathrm{Vm^9/C^5}$	$\lambda_1 = 1.5 \times 10^{-10}\,\mathrm{Vm^3/C}$
$\lambda_2 = 5.9 \times 10^{-28}\,\mathrm{Vm^5/C}$	$C_{11} = 1.7 \times 10^{11}\,\mathrm{N/m^2}$	$C_{12} = 7.9 \times 10^{10}\,\mathrm{N/m^2}$
$C_{44} = 1.1 \times 10^{11}\,\mathrm{N/m^2}$	$Q_{11} = 8.1 \times 10^{-2}\,\mathrm{m^4/C^2}$	$Q_{12} = -2.4 \times 10^{-2}\,\mathrm{m^4/C^2}$
$Q_{44} = 6.4 \times 10^{-2}\,\mathrm{m^4/C^2}$	$P_0 = 0.69\,\mathrm{C/m^2}$	

to specify the kinetic parameters in equations ((12) and (13)). In absence of reliable data for the viscosity and the mobility, we take $\Gamma^* = 1$ and $\eta^* = 10$. This choice of the kinetic coefficients ensures that mechanical equilibrium is established faster than polarization relaxation. For each case, we initiate the structure in a random paraelectric condition, i.e. the polarizations, displacements and the potentials are initialized with small fluctuations around zero. For convenience of comparison between the different cases, the same initial random seed is used in all cases. The simulations were evolved until the domain pattern was stabilized; in the present case this was achieved within $t^* = 2 \times 10^5$ time steps. The minimal spatial scale of our numerical simulations δ was set equal to 1 nm. We have found this value to be small enough to describe the delicate features of the polarization modulations, e.g. ripples at the domain wall as seen in Figs. 1(e) and (f). The numerical results presented in the paper are carried with $128 \times 128\,\mathrm{nm}^2$ unit cell. To test if our results are cell-size sensitive, we have performed simulations at few sizes and find no significant change in the domain patterns.

3.1. *Coupling of Polarization with Shear Strain Gradients: f_{44}*

Consider the case where the polarization is coupled only to shear strain gradients and the coupling to gradients of uniaxial strains is negligible. Simulations are performed for several values of the flexo-electric coefficient f_{44} along with the corresponding non-flexoelectric case $f_{44} = f_{12} = f_{11} = 0$. Figure 1 shows the patterns for the case $\langle \varepsilon_{xx} \rangle = \langle \varepsilon_{yy} \rangle = \langle \varepsilon_{xy} \rangle = 0$. For the non-flexoelectric case, a domain pattern with two ferroelastic variants and with 90° uncharged (head to tail) domain walls is observed (Figs. 1(a) and (b)). This pattern persists upon increasing f_{44} until $f_{44} = 5\,\mathrm{V}$. For $f_{44} > 5\,\mathrm{V}$, we obtain states where there are antiparallel domains within the ferroelastic domains. Figures 1(c) and (d) show this four domain state pattern for $f_{44} = 7\mathrm{V}$. To illustrate the nature of the domain walls, in Fig. 2(a) we plot the profile of the polarization along the double arrow in Fig. 1(d). Note that at this value of f_{44}, although

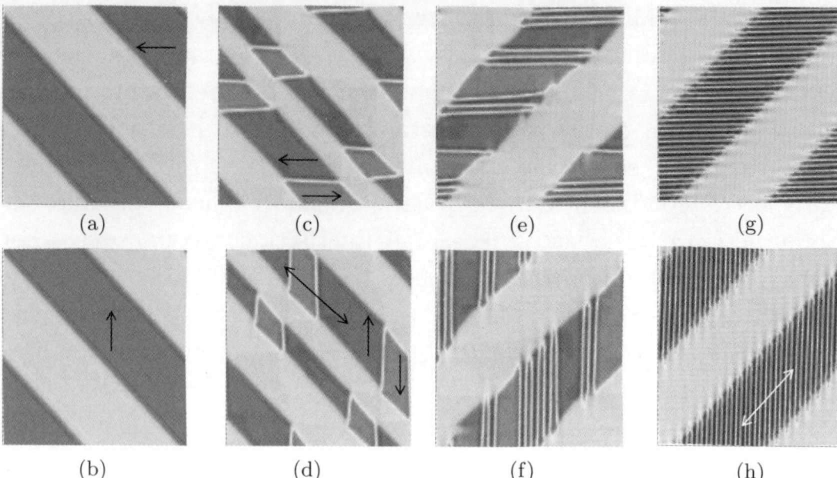

Figure 1.   Domain pattern for the mechanically-constrained case for different flexoelectric coupling strengths (a, b) $f_{44} = 0$. (c, d) $f_{44} = 7\,\text{V}$. (e, f) $f_{44} = 9\,\text{V}$, and (g, h) $f_{44} = 12\,\text{V}$ in the $128 \times 128\,\text{nm}^2$ simulation cell. In the top row of images (a, c, e, g) polarization *x-component* is shown, the darkness being minimal for zero polarization and maximal for polarization left. In the bottom row of images (b, d, f, h) polarization *y-component* is shown, the darkness being minimal for zero polarization and maximal for polarization up. Single-headed arrows indicate polarization direction, while double-headed arrows indicate cross-sections shown in Fig. 2.

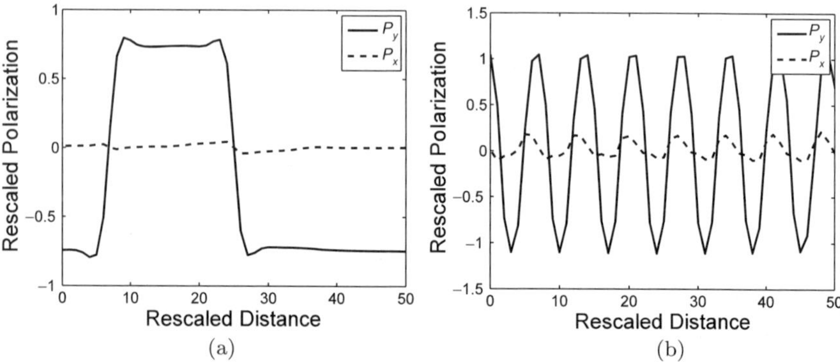

Figure 2.   Polarization profiles corresponding to the black double arrow in Fig. 1(d) is shown in (a) and that corresponding to white double arrow in Fig. 1(h) is shown in (b).

the polarization is still homogeneous within the domains, "spikes" are observed at the domain walls. The spikes are clearly a consequence of the flexoelectric coupling. For $f_{44} = 12V$, a clear "herring bone" pattern with fine antiparallel domains in each ferroelastically distinct variant is observed (Figs. 1(g) and (h)). Note that such patterns were observed in the early work of Merz, although at a different length scales.[26] In Fig. 2(b), we plot the polarization profiles along the white double arrow in Fig. 1(h). It can be seen that the polarization with in the domains is non-uniform, almost varying sinusoidally in space. This is clear evidence that an "incommensurate" phase is formed due to flexoelectricity.

3.2. *Coupling of Polarization with Uniaxial Strain Gradients: f_{11}*

Here, we consider the case of flexoelectric coupling with gradient of the uniaxial strain along the polarization direction $f_{12} = f_{44} = 0$, $f_{11} \neq 0$. Figure 3 shows the domain patterns for the mechanically constrained case for this situation. Figure 4 shows a magnified view of the polarization vectors in representative regions. The domain patterns remain similar to the non-flexoelectric case shown in Figs. 2(a) and (b) until $f_{11} = 8\,V$. When $f_{11} = 8\,V$, one observes formation of an additional narrow ferroelastic domain (Figs. 3(a) and (b)). For $f_{11} > 8\,V$, the density of the 90° domain walls increases with increasing f_{11}. The increasing density of domain walls may be understood in terms of the domain wall energies. It has been shown that the flexoelectric coupling renormalizes the gradient coefficient in the free energy, effectively decreasing the domain wall energy.[12,13] The decreased domain wall energy explains the increase in number of domain walls as strength of the flexoelectric coupling is increased.

The domain pattern drastically changes for $f_{11} = 10\,V$ (Figs. 3(c) and (d)). All four variants are present in this case and a very fine 90° domain structure appears. This fine domain pattern co-exists with domains having homogeneous polarizations. A magnified view of the region in the box in that figure is shown in Fig. 4(a). By plotting polarization profiles, similar to those in Fig. 2, we have confirmed

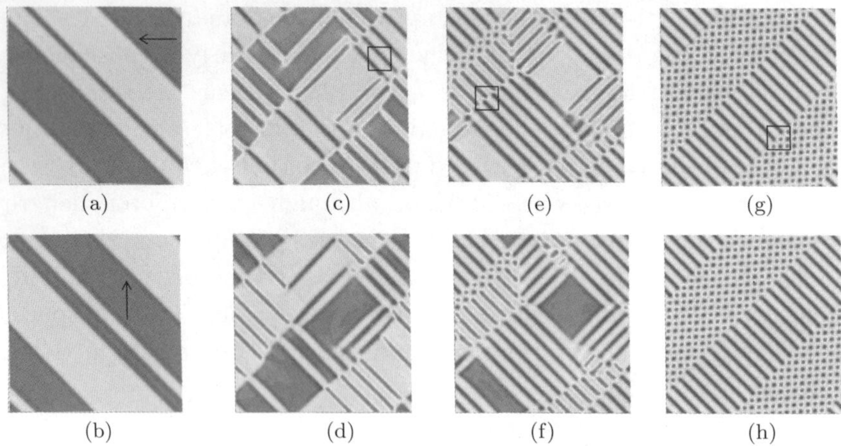

(a)   (c)   (e)   (g)

(b)   (d)   (f)   (h)

Figure 3.   Domain patterns for the mechanically-constrained case when $f_{11} \neq 0$ for different flexoelectric coupling strengths. The patterns correspond to (a, b) $f_{44} = 8\,\text{V}$. (c, d) $f_{44} = 10\,\text{V}$. (e, f) $f_{44} = 11\,\text{V}$, and (g, h) $f_{44} = 12\,\text{V}$. In the top row of images (a, c, e, g) polarization *x-component* is shown, the darkness being minimal for zero polarization and maximal for polarization left. In the bottom row of images (b, d, f, h) polarization *y-component* is shown, the darkness being minimal for zero polarization and maximal for polarization up. Arrows indicate polarization direction, while squares indicate areas for which the polarization distribution is shown in Fig. 4. The size of the simulation cell is $128 \times 128\,\text{nm}^2$.

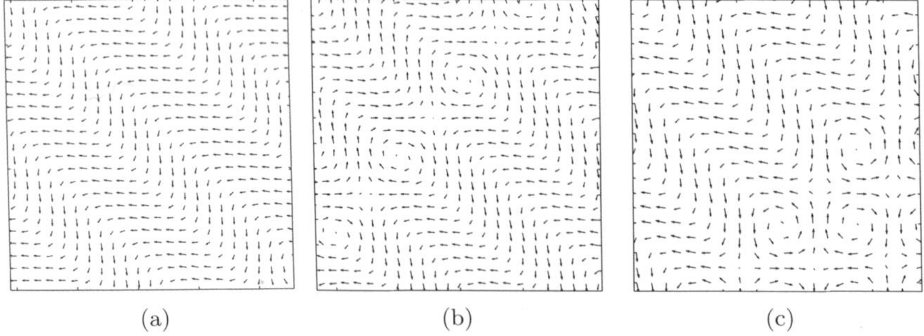

(a)   (b)   (c)

Figure 4.   Magnified view of the polarization distributions for the marked regions in Fig. 3. The polarization vector fields plotted in (a, b, c) correspond to areas indicated with squares in Fig. 3 (c, e, g) respectively and to flexocoupling coefficients f_{44} of the magnitudes $10\,\text{V}$, $11\,\text{V}$, and $12\,\text{V}$ respectively.

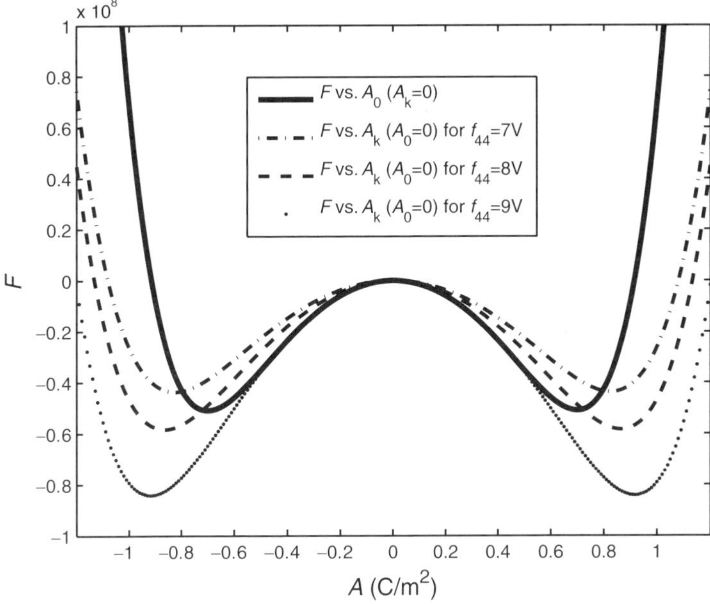

Figure 5. Free energy as a function of the modulation amplitude for different values of f_{44}. The free energy of the homogeneous phase is also shown.

that the fine-stripe domains in Fig. 5 are modulated with non-uniform polarizations within the domains. Thus, we observe a one-dimensional modulated phase. Unlike the one-dimensional in Fig. 1 where the modulation was between antiparallel domains, here the modulation is similar to 90° domains. For $f_{11} = 11\text{V}$ (Figs. 3(e) and (f)), we observe four different types of domains of this one-dimensional modulated phase: the initial one, the one rotated by 90°, and two additional ones that are shifted by a half-period. These domains co-exist with the homogeneous polarization domains. Further increase of f_{11} leads to the appearance of a new two-dimensional checkerboard modulated structure and the homogeneous polarization domains completely disappear. For $f_{11} = 12\,\text{V}$, the two-dimensional checkerboard pattern co-exists with the one-dimensional modulated pattern (Figs. 3(g) and (h)). Figure 4(c) shows the local polarization distribution for a typical region at an interface of the co-existing patterns. This clearly shows that the checkerboard pattern is actually an array of polarization vortices where the alternating vortices have opposite sense. It can be shown that this two-dimensional modulation may be described by the modulation

$$P_x = A \cos \left(\frac{k_{\min} x}{\sqrt{2}} \right) \cos \frac{\left(\frac{k_{\min} y}{\sqrt{2}} \right)}{\sqrt{2}},$$

$$P_y = A \sin \left(\frac{k_{\min} x}{\sqrt{2}} \right) \sin \frac{\left(\frac{k_{\min} y}{\sqrt{2}} \right)}{\sqrt{2}},$$

$$(14)$$

where $k_{\min}$ is the magnitude of the modulating wave-vector.[12]

An interesting issue is the transition from one-dimensional modulated structure to the two-dimensional checkerboard pattern. The numerical simulations reveal that this transition starts at the boundaries of two different domains of the one-dimensional modulated structure. Such "domain walls" are interesting and unusual; see, e.g. the region inside the box in the image for $f_{11} = 11\,\mathrm{V}$ for which the polarization vectors are shown in Fig. 4(b). We can observe a row of polarization vortices at the domain wall. These vortices are precursors to the formation of a two-dimensional modulated phase. We should remark that such domain walls have been observed in recent experiments on free standing PZT nanostructures.[27] The region indicated with the arrow in Fig. 3(e) is the evidence of the growth of the two-dimensional modulated structure from the domain wall between two shifted one-dimensional-modulations of the same orientation.

Above we have considered the case $f_{12} = f_{44} = 0$, $f_{11} \neq 0$. We also studied the $f_{11} = f_{44} = 0$, $f_{12} \neq 0$ case and obtained qualitatively similar patterns, that is why the above results can be qualitatively applied for the both couplings of the polarization with uniaxial strains.

4. Single Mode Analysis

The simulations presented in the previous section show that modulated/incommensurate phases can form above a critical value of the flexoelectric constant. The existence of such phases has been explained using linear stability analysis.[12,13] To develop an understanding of the thermodynamic stability of the predicted modulated patterns, we have analyzed a simplified one-dimensional model for the full nonlinear free energy for the case $f_{11} = f_{12} = 0$, $f_{44} \neq 0$ (note that while the calculation is performed for the $f_{44} \neq 0$ case, the main conclusions are generic and valid for other flexoelectric couplings as well). For simplicity, we assume only one component of the polarization P_y and use the model corresponding to the stress free case. We further make the simplification that the polarization can vary only along a direction transverse to the polarization, i.e. $(\partial P_y/\partial y) = 0$. Using these simplifications (neglecting electrostrictive effects), the total free energy can be expressed as

$$F = \int dx \left(\alpha_1 P_y^2 + \alpha_{11} P_y^4 + \alpha_{111} P_y^6 + \frac{\lambda_1}{2} \left(\frac{\partial P_y}{\partial x} \right)^2 + \frac{\lambda_2}{2} \left(\frac{\partial^2 P_y}{\partial x^2} \right)^2 \right)$$
$$+ \int dx \left(\frac{C_{44}}{2} \varepsilon_{xy}^2 + \frac{f_{44}}{2} \left(P_y \left(\frac{\partial \varepsilon_{xy}}{\partial x} \right) - \varepsilon_{xy} \left(\frac{\partial P_y}{\partial x} \right) \right) \right). \qquad (15)$$

Assuming a stress free system $\sigma_{xy} = \delta F/\delta \varepsilon_{xy} = 0$, the strain ε_{xy} is expressed as

$$\varepsilon_{xy} = \frac{f_{44}}{C_{44}} \frac{\partial P_y}{\partial x}. \qquad (16)$$

Substituting (16) in (15), we get

$$F = \int dx \left(\alpha_1 P_y^2 + \alpha_{11} P_y^4 + \alpha_{111} P_y^6 + \frac{\lambda_1 - G_2}{2} \left(\frac{\partial P_y}{\partial x} \right)^2 \right.$$
$$\left. + \frac{\lambda_2}{2} \left(\frac{\partial^2 P_y}{\partial x^2} \right)^2 \right), \qquad (17)$$

where $G_2 = f_{44}^2/C_{44}$. An important conclusion one can draw from (17) is that flexoelectric coupling leads to a reduction in the domain wall energy. Indeed, because $G_2 > 0$, the effective correlation term $\lambda_1^* = \lambda_1 - G_2$ becomes smaller than without flexoelectric effect. This explains the increase in the domain wall density with increasing strength of the flexoelectric coupling that has been observed in the simulations. To analyze this free energy, we use a single mode approximation of the form

$$P_y = A_0 + A_k \sin(kx). \tag{18}$$

Here, A_0 is the homogeneous polarization and A_k is the amplitude of the modulated phase. Substituting (18) into (17), the free energy density (averaged over one wavelength) is calculated as

$$f(A_0, A_k, k) = \alpha_1 A_0^2 + \alpha_{11} A_0^4 + \alpha_{111} A_0^6 + \frac{1}{2}\alpha_1 A_k^2 + \frac{3}{8}\alpha_{11} A_k^4$$

$$+ \frac{5}{16}\alpha_{111} A_k^6 + 3\alpha_{11} A_0^2 A_k^2$$

$$+ \alpha_{111}\left(\frac{15}{2} A_0^4 A_k^2 + \frac{45}{8} A_0^2 A_k^4\right)$$

$$+ \frac{(\lambda_1 - G_2)k^2 + \lambda_2 k^4}{4} A_k^2. \tag{19}$$

The critical wave-vectors (corresponding to modulation which will appear first) can be derived from minimization of (19) with respect to wave-vector k. The condition for the extremum yields

$$\frac{\delta f(A_0, A_k, k)}{\delta k} = 0 \Rightarrow 2(\lambda_1 - G_2)k + 4\lambda_2 k^3 = 0. \tag{20}$$

Equation (20) has only one solution $k = 0$ if $\lambda_1 - G_2 > 0$ while for $\lambda_1 - G_2 < 0$ there are 3 solutions:

$$\begin{aligned}
k &= 0 & G_2 &< \lambda_1, \\
k &= 0, \ k = \pm\sqrt{\frac{G_2 - \lambda_1}{2\lambda_2}} & G_2 &> \lambda_1.
\end{aligned} \tag{21}$$

In the latter case, $k = 0$ corresponds to the maximum in the energy, while $k = \pm\sqrt{\frac{G_2 - \lambda_1}{2\lambda_2}}$ is the expression for the sought critical wave-vectors.

In the regime $G_2 > \lambda_1$, by substitution of $k = \pm\sqrt{\frac{G_2-\lambda_1}{2\lambda_2}}$ into equation (19) we obtain

$$f(A_0, A_k) = \alpha_1 A_0^2 + \alpha_{11} A_0^4 + \alpha_{111} A_0^6 + \frac{1}{2}\left(\alpha_1 - \frac{(G_2 - \lambda_1)^2}{8\lambda_2}\right) A_k^2$$

$$+ \frac{3}{8}\alpha_{11} A_k^4 + \frac{5}{16}\alpha_{111} A_k^6 + 3\alpha_{11} A_0^2 A_k^2$$

$$+ \alpha_{111}\left(\frac{15}{2} A_0^4 A_k^2 + \frac{45}{8} A_0^2 A_k^4\right). \tag{22}$$

This free energy expression can be used to analyze the stability of the different phases as f_{44} increases. Particularly we will be interested in competition of homogeneous ferroelectric state ($A_k = 0$, $A_0 \neq 0$ in (22)) and modulated parent phase ($A_0 = 0$, $A_k \neq 0$ in (22)).

The free energy of the homogeneous state is given as

$$f(A_0, A_k = 0) = \alpha_1 A_0^2 + \alpha_{11} A_0^4 + \alpha_{111} A_0^6, \tag{23}$$

and the free energy of the modulated phase is given as

$$f(A_0 = 0, A_k) = \frac{1}{2}\left(\alpha_1 - \frac{(G_2 - \lambda_1)^2}{8\lambda_2}\right) A_k^2 + \frac{3}{8}\alpha_{11} A_k^4 + \frac{5}{16}\alpha_{111} A_k^6. \tag{24}$$

In Fig. 5, we show the free energy of the modulated state as a function of the modulation amplitude A_k, for different values of f_{44}. The homogeneous free energy (which is independent of f_{44}) is also shown as a function of A_0 in the same plot. From comparison of the depth of the minima in this figure one concludes that the homogeneous ferroelectric state becomes meta stable above a certain value of f_{44}. This implies that a coexistence of the homogeneous phase is possible in a range of values for the flexoelectric coefficients, which has indeed been observed in the simulation of domain patterns of Figs. 1 and 3.

Next, we want to test for the stability of the homogeneous state. In Fig. 6, we plot the quantity $f(A_0 = P_0, A_k)$ as a function of the

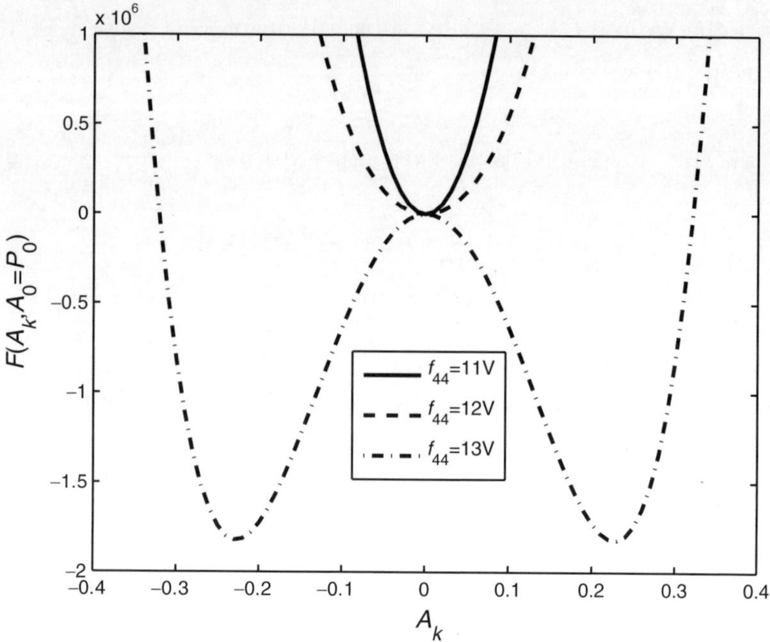

Figure 6.　$f(A_0 = P_0, A_k)$ as a function of amplitude A_k for different values of f_{44}.

amplitude A_k. Here P_0 is the minima of the homogeneous free energy given as

$$P_0^2 = \frac{-\alpha_{11} + \sqrt{\alpha_{11}^2 - 3\alpha_1\alpha_{111}}}{3\alpha_{111}}. \tag{25}$$

This figure clearly shows that for the present set of parameters, the homogeneous single domain state $A_0 = P_0$ and $A_k = 0$ becomes unstable to form a modulated state for $f_{44} \geq 15\,\text{V}$.

The above analysis has clearly demonstrated incommensurate phase must exist above a critical strength of the flexoelectric coupling, either as a metastable or a stable state. In fact, at very high value of the flexoelectric coupling, the homogeneous state becomes completely unstable and the incommensurate state is the only stable state.

5. Influence of Flexoelectricity on Domain Switching

Having established that flexoelectricity can significantly influence the domain patterns, the next question that arises is that what is the influence of flexoelectricity on effective properties which are controlled by motion of domain walls. For example, when an electric field is applied to a multidomain ferroelectric, the field induced motion of domain walls can significantly influence the polarization vs. electric-field response. Properties such as susceptibility and piezoelectric behavior can also be influenced by domain walls. Thus, it is also important to understand the impact of flexoelectricity on such properties. In this section, we investigate the impact of flexoelectricity on the polarization vs. electric-field response of a multidomain ferroelectric state.

To illustrate the basic idea, we focus on the mechanically unconstrained (stress-free) case. Since the system is stress free, no ferroelastic domains are observed for this case. In absence of an applied electric field, the domain pattern obtained after cooling from paraelectric state is a single ferroelastic variant which may have antiparallel 180° domain walls. The top row in Fig. 7 shows the pattern for different strengths of the flexoelectric couplings. As seen earlier in Fig. 1, the density of domain walls increases with increasing strength of the flexoelectric coupling. To understand how these domain states respond to an externally applied field, we performed additional simulations where we applied a gradually increasing electric field along $+x$-direction to the domain states shown in top row of Fig. 7. The external field can be introduced by adding an additional term to the free energy in equation (2). The contribution of the external field is expressed as $f_{ext} = -E_x^{ext} P_x$.

With the configurations in top row of Fig. 7 as initial conditions, the electric field along the $+x$ direction is continuously increased from zero to $E_x^{ext} = 50\,\mathrm{MV/m}$ in $t^* = 2 \times 10^4$ time steps. The bottom row of Fig. 7 shows the final patterns. It is clear that complete domain switching is achieved for the non-flexoelectric case, where as only partial switching is observed for the cases when flexoelectricity is incorporated. In Fig. 8, we show the fraction of switched domains vs.

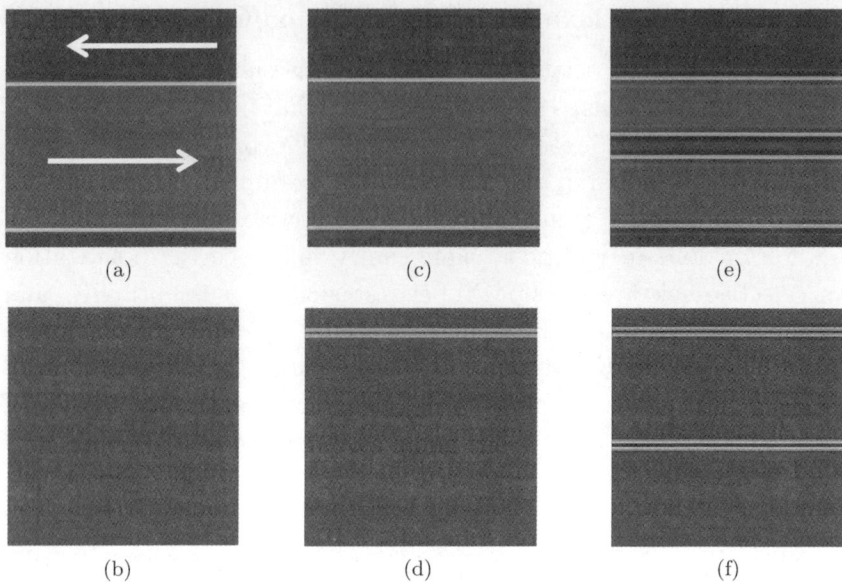

Figure 7.   Domain patterns for the stress-free case when $f_{44} \neq 0$ with no external field (top row) and externally applied field of $E_x^{ext} = 50\,\mathrm{MV/m}$ (bottom row). The *x-component* of the polarization is shown for each case. The patterns correspond to (a, b) $f_{44} = 0$, (c, d) $f_{44} = 5\,\mathrm{V}$, (c, d) and (e, f) $f_{44} = 8\,\mathrm{V}$. The white arrows indicate polarization directions. The size of the simulation cell is $128 \times 128\,\mathrm{nm}^2$.

the applied field for different strengths of the flexoelectric coefficient. Here, the fraction of switched domains is calculated as the ratio of area of domains with $P_x > 0$ to that of the total area of the simulation cell. This figure shows that domain switching initiates at increasingly higher value of the electric field as the flexoelectric coupling becomes stronger. This may be understood in terms of domain wall mobility. With increase of flexoelectric coupling domain walls become narrower, as controlled by the decrease of the effective correlation term. For narrower domain walls numerical "pinning" due to finite mesh size becomes stronger. Due to this, a higher electric field needs to be applied to move the domain walls for larger coupling strengths. Similar effect may be observed in real systems where pinning forces both due to pierls barrier and those caused by defects are known to increase drastically with the decrease of domain wall width.[28]

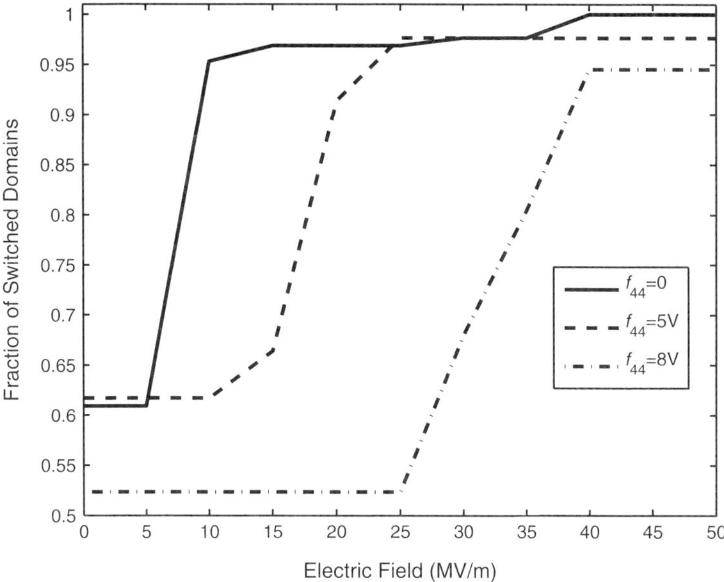

Figure 8. Fraction of switched domains as a function of the applied electric field for the situation shown in Fig. 7.

6. Summary and Conclusions

We have studied the nature of ferroelectric domains, taking into account the flexoelectric coupling using both phase field simulations and analytical theory. We simulated bulk domain patterns in cubic to tetragonal ferroelectrics with a two-dimensional-model. To understand the effect of each component of the flexoelectric tensor, we have separately analyzed the $f_{12} = f_{44} = 0$, $f_{11} \neq 0$, and $f_{12} = f_{11} = 0$, $f_{44} \neq 0$ cases. In order to understand the role of the strength of the flexoelectric interaction, we have tried a range of values for the coefficients. We find that ferroelectric domain patterns change dramatically as the strength of the appropriate flexoelectric coupling is increased. These patterns are very sensitive to the value of the flexoelectric coupling constants. For example, in Figs. 1–3, the domain patterns change significantly by changing the coupling constants by just 1 V. This clearly highlights the need for accurate determination of the flexoelectric coefficients. We also find that, there is a critical value of f_{ij} above which modulated phases form with both polarization and strain gradients. The inhomogeneous distribution of the polarization depends on which component of f_{ij} is non-zero. For example, when the polarization is coupled to shear strain gradients ($f_{44} \neq 0$), a stripe modulation with antiparallel polarization is observed. For the mechanically-constrained case, a "herring bone" domain pattern with antiparallel polarization domains within each ferroelastic variant is obtained. When the polarizations are coupled to gradients in the uniaxial strains ($f_{11} \neq 0$ or $f_{12} \neq 0$), a unique two-dimensional modulation of the polarization is obtained. This modulation corresponds to an array of localized polarization vortices, where neighboring vortices have opposite senses. Polarization vortices are usually believed to arise due to depolarization fields; however, we have shown that flexoelectricity can also lead to localized polarization vortices.

We also analytically studied the influence of flexoelectrity on ferroelectric behavior using a single mode approach for a simple one-dimensional model. Our calculations confirm that flexoelectric coupling renormalizes the domain wall energy of the ferroelectrics. We also show that above a critical value of the flexoelectric coupling strength, an incommensurate state must necessarily exist, either as a meta-stable state and at high enough coupling strength, as the only stable state.

Finally, we found that flexoelectric effects can strongly influence ferroelectric materials even if modulated phase does not form. For example, flexoelectric coupling also influences the domain wall width, thereby influencing their mobility and switching behavior induced by electric field. This underscores the need to carefully measure the flexoelectric coupling strengths in order to interpret recent experiments where flexoelectric effects have been shown to be important.

References

1. W. Ma and L. E. Cross. *Appl. Phys. Lett.* **3440** (2002).
2. W. Ma and L. E. Cross. *Appl. Phys. Lett.* **072905** (2005).
3. W. Ma and L. E. Cross. *Appl. Phys. Lett.* **232902** (2006).

4. W. Zhu, J. Y. Fu, N. Li, and L. E. Cross. *Appl. Phys. Lett.* **192904** (2006).

5. A. Gruverman, B. J. Rodriguez, A. I. Kingon, R. J. Nemanich, A. K. Tagantsev, J. S. Cross, and M. Tsukuda. *Appl. Phys. Lett.* **728** (2003).

6. H. Lu, C.-W. Bark, D. Esque de los Ojos, J. Alcala, C. B. Eom, G. Catalan, and A. Gruverman. *Science,* **59** (2012).

7. D. Lee, A. Yoon, S. Y. Jang, J.-G. Yoon, J.-S. Chung, M. Kim, J. F. Scott, and T. W. Noh. *Phys. Rev. Lett.* **057602** (2011).

8. G. Catalan, A. Lubk, A. H. G. Vlooswijk, E. Snoeck, C. Magen, A. Jannssens, G. Risspens, G. Rijnders, D. H. A. Blank, and B. Noheda. *Nat. Mater.* **966** (2011).

9. E. A. Eliseev, A. N. Morozovska, M. D. Glinchuk, and R. Blinc. *Phys. Rev. B.* **165433** (2009).

10. P. V. Yudin, A. K. Tagantsev, E. A. Eliseev, A. N. Morozovska, and N. Setter. *Phys. Rev. B.* **134102** (2012).

11. Y. Gu, M. Li, A. N. Morozovska, Y. Wang, E. A. Eliseev, V. Gopalan, and L. Q. Chen. *Phys. Rev. B.* **174111** (2014).

12. R. Ahluwalia, A. K. Tagantsev, P. V. Yudin, N. Setter, N. Ng, and D. J. Srolovitz. *Phys. Rev. B.* **174105** (2014).

13. P. V. Yudin, R. Ahluwalia, and A. K. Tagantsev. *Appl. Phys. Lett.* **082913** (2014).

14. A. K. Tagantsev, L. E. Cross, and J. Fousek. *Domains in Ferroic Crystals and Thin Films,* 1st Edn, Springer, Berlin (2010).

15. V. L. Indenbom, E. B. Loginov, and M. A. Osipov. *Kristalografija.* **1957** (1981).

16. Sh. M. Kogan, *Sov. Phys. Sol. State.* **2069** (1963).

17. A. D. Bruce and R. A. Cowley. *Structural Phase Transitions.* Taylor & Francis, London (1981).

18. J. D. Axe, J. Harada, and G. Shirane. *Phys. Rev. B.* **1227** (1970).

19. R. Blinc and A. P. Levanyuk, Incommensurate phase in dielectrics. In: Agranovich V. M. and Maradudin A. A. (eds.). *Modern Problems in Condensed Matter.* Amsterdam, North-Holland Physics Publishing, a division of Elsevier Science Publishers B.V. (1986), p. 410.

20. J. Petzelt. *Phase Transit.* **155** (1981).

21. I. Maclaren, R. Villaurrutia, and A. Pelaiz-Barranco. *J. Appl. Phys.* **034109** (2010).

22. K. Dayal and K. Bhattacharya. *Acta Mater.* **1907** (2007).

23. J. Hlinka and P. Marton. *Phys. Rev. B.* **104104** (2006).

24. B. Houchmanzadeh, J. Lajzerowicz, and E. Salje. *Phase Transit.* **77** (1992).

25. M. J. Haun, Z. Q. Zhuang, E. Furman, S. J. Jang, and L. E. Cross. *Ferroelectrics.* **45** (1989).

26. W. J. Merz. *Phys. Rev.* **690** (1954).

27. L. W. Chang, V. Nagarajan, J. F. Scott, and J. M. Gregg. *Nano Lett.* **2553** (2013).

28. I. Suzuki and Y. Ishibashi. *Ferroelectrics.* **64**, 181 (1985).

Chapter 9

Flexoelectricity Impact on the Domain Wall Structure and Polar Properties

A. N. Morozovska

Institute of Physics, National Academy of Sciences of Ukraine,
46, pr. Nauki, 03028 Kiev, Ukraine

S. V. Kalinin

Center for Nanophase Materials Science,
Oak Ridge National Laboratory,
Oak Ridge, TN, 37831

E. A. Eliseev

Institute for Problems of Materials Science,
National Academy of Sciences of Ukraine,
3, Krjijanovskogo, 03142 Kiev, Ukraine

The impact of the flexoelectric coupling on the polarization structure and carriers accumulation in domain walls (DW) is studied self-consistently within the framework of Landau–Ginzburg–Devonshire (LGD) theory. We obtained that the flexoelectric coupling induces the polarization component that is perpendicular to the DW plane (so-called Neel component), that in turn leads to the depolarization field and electric potential variation across the nominally uncharged DW. The potential variation leads to the carrier's accumulation and thus increases the wall static conductivity. The origin of Neel-type polarization component presents a fundamental interest, but its value is typically very small raising the question about the effect of experimental manifestation. Since the electron density at the wall becomes at least several times higher than in the single-domain region for known values of flexoelectric coupling coefficients, the contrast can be detected by the electric current atomic force microscopy (c-AFM).

1. Overview

1.1. *Polarization Structure and Electronic Properties at the Domain Wall in Ferroics*

For multiaxial ferroics with multicomponent order parameters, analysis of polarization structure in the domain wall (DW) necessitates taking into account the coupling between order parameter components and their gradients mediated by stress accommodation.[1] For instance, the biquadratic coupling between polarization and structural order parameter was shown to be important.[1-4] The coupling can lead to the appearance of polarization in the structural domains (twins), however the conditions of such manifestations are usually very strict.[1] Situation is similar for multiferroics (namely ferromagnetics–ferroelectrics), where a local magnetic moment can appear in the ferroelectric DW due to either biquadratic[5] or inhomogeneous coupling.[6,7]

Polarization vector inside a "nominally" uncharged DW in multiaxial ferroelectrics can have all three components.[8,9] Hereinafter the term "nominally" uncharged means that polarization vectors in domains are parallel to the wall plane, but the situation can be different in the DW, where the perpendicular component can be non-zero. Hereafter, we use the following notations (see Fig. 1): the

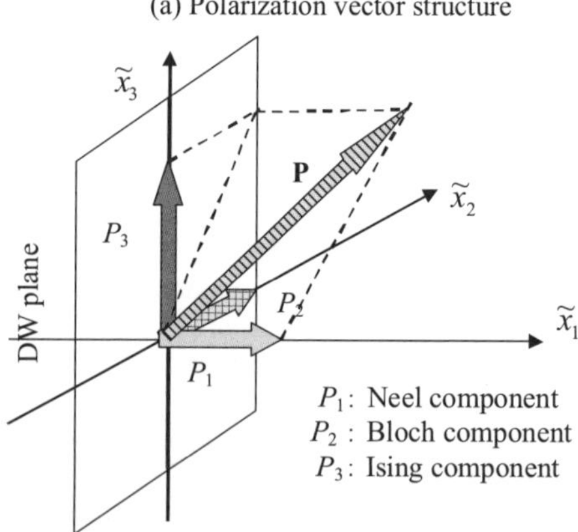

Figure 1.   Polarization vector structure showing Ising, Bloch and Neel components orientation. (Adapted from Ref. [40].)

polarization component P_3, parallel to the spontaneous polarization $\pm P_S$ in the domains, is regarded as the Ising-type; the component P_2, parallel to the wall plane, but perpendicular to Ising-type component, which vanishes far from the wall, is regarded as Bloch-type component; and component P_1, normal to the wall, is regarded as Néel-type component. Note that the Néel-type component P_1 is associated with the non-zero divergence of polarization vector and hence should be considered jointly with associated depolarization fields.

Despite attempts to describe polarization behavior in multicomponent ferroics were made since the early days of ferroelectricity,[10–12] the progress with understanding of their DW structure appeared limited. For instance Fedosov and Darinskii[12] found that the polarization component normal to DW demonstrates a weak deviation from a constant distribution in early 1971. This leads to the internal electric field appearance and hence to the potential step in DW plane, which is consistent with *ab initio* calculations.[13] Relatively recently Hlinka and Márton[14] calculated numerically the structure of twin walls in tetragonal perovskite crystal $BaTiO_3$ in the framework of the phenomenological Landau–Ginzburg–Devonshire (LGD) theory. Then Maryam Taherinejad *et al.* considered ferroelectric twin domains and 180° DWs of [110] orientation with the help of *ab initio* calculations[15] and showed the possibility of Bloch components of polarization to appear. Ferroelectric DW resembling Neel walls in ferromagnetics were predicted in thin ferroelectric $LiNbO_3$ films.[8] Polarization can appear in elastic twins and antiphase boundaries of incipient ferroelectrics–ferroelastic like $SrTiO_3$ and $CaTiO_3$ due to the biquadratic coupling between their structural order and polarization in the antiferrodistortive phase.[1,16] Bloch-type DWs can originate in the rhombohedral $BaTiO_3$ at low temperature[17] and tetragonal $PbTiO_3$.[9]

Existence of charged DWs in ferroelectric semiconductors should result in accumulation or depletion of the space charge on adjacent sides of the DW, and should thus lead to enhanced conductivity at these sites.[18] Modern experiments approved by theoretical modelling revealed the strong enhancement, e.g. Sluka *et al.* reported about 9-order contrast of static conductivity in charged DWs in tetragonal $BaTiO_3$.[19] Seidel *et al.* using scanning probe microscopy (SPM) methods revealed room-temperature metallic conductivity of nominally uncharged 180° and 109° DWs in $BiFeO_3$.[20,21] Farokhipoor *et al.* have shown that nominally uncharged as-grown 71° DWs in $BiFeO_3$ can also exhibit enhanced conductivity.[22] Nominally uncharged vortex structures fabricated in $BiFeO_3$ show an order of magnitude increase in conductivity over single domain regions.[23] DW conductance was reported in $ErMnO_3$ by Meyer *et al.*[24] DWs are conducting in $LiNbO_3$.[25,26] Guyonnet *et al.* observed conductive DWs in $Pb(Zr,Ti)O_3$.[27] Then Maksymovych *et al.* report about metallic conductivity of 180° DWs in $Pb(Zr,Ti)O_3$.[28] The variability of the conductivity response in ferroelectric DWs was observed by combined Current-Atomic Force Microscopy (c-AFM) and Piezoresponse Force Microscopy (PFM).[20,23,28–30]

LGD theory is a powerful method to study DWs' conductivity in ferroelectric-semiconductors. In 1969,[31] the conductivity mechanism was described for the first time stemming from compensation of polarization charge discontinuity by mobile carriers in the material. Analytical LGD theory was further developed for charged walls in uniaxial[32,33] and multiaxial tetragonal ferroelectrics,[19,34] nominally uncharged walls in rhombohedral ferroelectrics,[35] improper ferro-electrics[36] and twin walls in incipient ferroelectrics–ferroelastics.[37] Recent advances in both experiment and theory of the relation between DWs conductivity and their structure could be found in Ref. [38].

1.2. *Flexoelectricity Impact on the DW Structure and Electronic Properties*

Since the gradients of ferroelectric polarization and elastic strains are inherent and rather strong in DWs, i.e. inside the Region of the wall intrinsic width around the wall plane, flexoelectric coupling can essentially influence both the structure of the DW's and consequently its electronic properties. Below, we briefly overview relevant works devoted to the flexoelectricity impact on the DW structure and properties.

Using LGD theory, it was shown that flexoelectric coupling induces a new polarization component in DWs of tetragonal $BaTiO_3$ with a structure qualitatively different from the classical Bloch-wall structure.[39] The wall polarization rotates in opposite directions on the two sides of the wall and passes through zero in the center, in contrast to classical Bloch-type walls, where the polarization vector draws a helix on passing from one domain to the other. The effect lowers the symmetry of 180° walls, which are oblique with respect to the cubic crystallographic axes, while {100} and {110} walls stay "untouched". Being of the Ising type in the absence of the flexoelectric interaction, the inclined DW acquires a new polarization component with a structure qualitatively different from the classical Bloch-wall structure. The flexoelectric coupling lowers the DW energy and gives rise to its additional anisotropy, which is comparable to that conditioned by elastic anisotropy. The atomic order-of-magnitude estimates show that the Bloch-type polarization component may be comparable with spontaneous polarization suggesting that, in general, it is mandatory to include the flexoelectric coupling in DW simulations in ferroelectrics. However, being in-plane of the wall, the Bloch component of polarization does not lead to appearance of the bound charge and hence is not expected to considerably affect electronic properties of the material. In contrast, Neel-type polarization component may lead to the carrier's accumulation. The appearance of Neel-type polarization component in the nominally uncharged 180° DWs in tetragonal $Pb(Zr,Ti)O_3$ was predicted in Refs. [34 and 28].

With a special attention to flexoelectric coupling Eliseev *et al.*[40] considered the structure of the 180° DW of arbitrary orientation in rhombohedral phase of $BaTiO_3$. Bloch and Neel components of polarization are shown to appear in the DW. Dependent on the wall orientation, two types of DW behaviors are identified, namely the low-energy "achiral" walls with odd polarization profile and the "chiral" walls with mixed parity polarization distribution. Due to the flexoelectric coupling one has to distinguish DWs with different signs of gradient of Ising polarization component (i.e. "negative" and "positive" DWs). Moreover, these two types of walls have different structure and energy in rhombohedral ferroelectrics. Note, that the presence of additional components of polarization in the DWs in tetragonal phase of $BaTiO_3$ were confirmed with the help of *ab initio* calculations and phase field modelling.[41]

The combined effect of flexoelectricity and rotostriction, that is the coupling between the strain tensor u_{ij} and the dyadic product of antiferrodistorsive order parameter (oxygen tilt Φ_i) components $\Phi_i\Phi_j$, can lead to a spontaneous polarization and pyroelectricity in the vicinity of antiphase boundaries, structural twin walls, surfaces, and interfaces in the octahedrally tilted phase of otherwise non-ferroelectric perovskites such as $CaTiO_3$, $SrTiO_3$, and $EuTiO_3$. In particular a spontaneous polarization in the $SrTiO_3$ antiphase and twin boundaries arises at temperatures lower than the antiferrodistortive structural phase transition temperature of $T_S \sim 105\,\mathrm{K}$ in agreement with previously unexplained experimental results.[42–44] Carriers accumulation caused by the strain-induced band structure changes originated via the deformation potential mechanism, structural order parameter gradient, rotostriction and flexoelectric coupling is explored near the twin DW-surface junctions in the ferroelastic materials.[37] Approximate analytical results show that inhomogeneous elastic strains, which exist in the junctions between the twin walls and the surface due to the rotostriction coupling, decrease the local band gap via the deformation potential and flexoelectric coupling mechanisms. Also, flexoelectric and rotostriction coupling leads to the appearance of the polarization and electric fields proportional to the structural order parameter gradient in the twin walls — surface junctions.[37] Finally flexo-antiferrodistortive coupling, defined as the coupling between the antiferroelectric polarization A_j and oxygen tilt Φ_i gradients, becomes a necessary complement to the well-known flexoelectric coupling.[45] The coupling is universal for all antiferrodistortive systems and can lead to the formation of incommensurate, spatially-modulated phases in multiferroics. It may cause a noticeable influence of the flexo-antiferrodistortive coupling on the structure and physical properties of domain boundaries in the antiferrodistortive materials. It was shown[46] that this type of coupling could lead to the formation of spatially modulated periodic structures in Sm-doped $BiFeO_3$.

Below, we present a Landau theory study of internal structure of DW driven by flexoelectric coupling. The generic theory will be applied to prototypic perovskite ferroelectric $BaTiO_3$. We will mainly focus on Neel polarization component and on DW conductivity. Also aspects of symmetry and angular anisotropy of DW structure and intrinsic energy will be covered.

2. Free Energy and Coupled Equations

Let us consider nominally uncharged $180°$ DW in the bulk of n-type ferroelectric-semiconductor $BaTiO_3$. Within LGD theory, equations of state for polarization components can be derived from the variation of the corresponding thermodynamic potential (e.g. Gibbs potential functional or Helmholz free energy). For the $m\bar{3}m$ symmetry in the crystallographic frame the expression for the Gibbs potential has the form:[34]

$$G = a_i P_i^2 + a_{ij} P_i^2 P_j^2 + a_{ijk} P_i^2 P_j^2 P_k^2 + \frac{g_{ijkl}}{2} \frac{\partial P_i}{\partial x_j} \frac{\partial P_k}{\partial x_l} - Q_{ijkl}\sigma_{ij}P_k P_l$$

$$+ \frac{F_{ijkl}}{2} \left(\sigma_{ij} \frac{\partial P_k}{\partial x_l} - P_k \frac{\partial \sigma_{ij}}{\partial x_l} \right) - \frac{s_{ijkl}}{2}\sigma_{ij}\sigma_{kl} - P_i E_i, \tag{1}$$

where E_i is electric field (including the depolarization one), u_{ij} are elastic strains, σ_{ij} are elastic stresses, P_i are polarization components related to the soft mode (in what follows we use a term polarization as a shorthand). Summation is performed over all repeated indices.

In equation (1) a_i, a_{ij}, and a_{ijk} are expansion coefficients of the second, fourth and sixth-order dielectric stiffness tensors correspondingly, gradient coefficients are g_{ijkl}, F_{ijkl} is the flexoelectric tensor; Q_{ijkl} is electrostriction tensor, s_{ij} are elastic compliances, φ is the electric potential, E_k are electric-field components, ε_b is background permittivity[47] and $\varepsilon_0 = 8.85 \times 10^{-12}\,\text{F/m}$ the universal dielectric constant. The latter term in equation (1) is the electrostatic energy of free space charges with density ρ in the electric field with potential φ related with electric field E_i by definition $E_k = -\partial\varphi/\partial x_k$.

Electrostatic potential φ should be determined self-consistently from the Poisson equation

$$\varepsilon_0\varepsilon_b\frac{\partial^2\varphi}{\partial x_i^2} = \frac{\partial P_i}{\partial x_i} - \rho. \tag{2}$$

The charge density $\rho = e(N_d^+ + p - n)$, where $e = 1.6 \times 10^{-19}\,\mathrm{C}$ is the electron charge, n is the concentration of the electrons in the conduction band; p is the concentration of holes in the valence band; N_d^+ is the concentration of ionized donors. Mass action law in n-doped material leads to the fact that holes concentration can be regarded negligibly small in comparison with the electrons, which are minority carriers.

Euler–Lagrange equations of state for polarization components and elastic stresses were derived by the variation of the functional (1a) on the polarization components P_i and stresses σ_{ij}[34,40]:

$$\frac{\partial G}{\partial P_i} - \frac{\partial}{\partial x}\left(\frac{\partial G}{\partial\left(\dfrac{\partial P_i}{\partial x}\right)}\right) = 0, \tag{3a}$$

$$\frac{\partial G}{\partial \sigma_{ij}} - \frac{\partial}{\partial x}\left(\frac{\partial G}{\partial\left(\dfrac{\partial \sigma_{ij}}{\partial x}\right)}\right) = -u_{ij}. \tag{3b}$$

Equations (3) should be solved along with mechanical equilibrium conditions

$$\frac{\partial \sigma_{ij}}{\partial x_i} = 0, \tag{4a}$$

and Saint–Venant compatibility relations

$$e_{ikl}e_{jmn}\left(\frac{\partial^2 u_{ln}}{\partial x_k \partial x_m}\right) = 0. \tag{4b}$$

Here e_{ikl} is an antisymmetric Levi–Civita tensor.

We introduce coordinate set $\{\tilde{x}_1, \tilde{x}_2, \tilde{x}_3\}$ related to the DW as it is shown in Fig. 2. Far away from the DW-surface juctions (i.e. deeply in

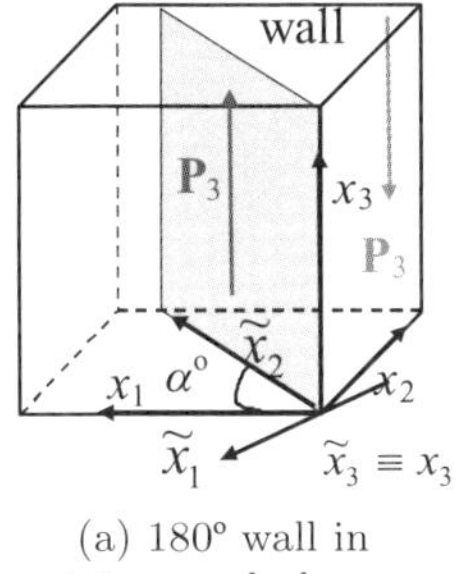

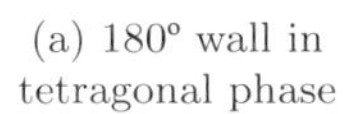

(a) 180° wall in
tetragonal phase

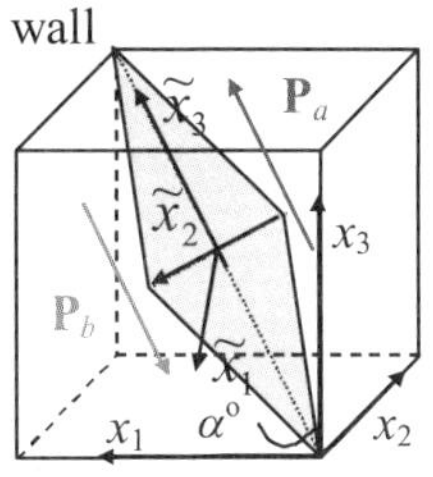

(b) 180° wall in
rhombohedral phase

Figure 2. Rotated coordinate frame $\{\tilde{x}_1, \tilde{x}_2, \tilde{x}_3\}$ choice for $180°$ nominally uncharged DWs in tetragonal (a) and rhombohedral (b) ferroelectrics; α is the wall rotation angle counted from crystallographic axis x_3. The distance from the wall is $\tilde{x}_1$ in both cases.

the bulk) we can consider one-dimensional problem in the reference frame $\{\tilde{x}_1, \tilde{x}_2, \tilde{x}_3\}$ with physical quantities depending only on the coordinate $\tilde{x}_1$ normal to the wall. The new coordinates in the wall-related reference frame can be obtained by following transformation:

$$\tilde{x}_1 = x_1 \cos\alpha - x_2 \sin\alpha, \quad \tilde{x}_2 = x_1 \sin\alpha + x_2 \cos\alpha, \quad \tilde{x}_3 = x_3, \quad (5)$$

where $\{x_1, x_2, x_3\}$ are the coordinates in the pseudo-cubic crystallographic reference frame, α is the DW tilt angle with respect to the crystallographic axis x_3 [see Fig. 2(a)].

Elastic boundary conditions are $\tilde{\sigma}_{ij}(\tilde{x}_1 \rightarrow \pm\infty) = 0$. In the case of $\tilde{x}_1$-dependent solution, compatibility relation $e_{ikl}e_{jmn}(\partial^2 \tilde{u}_{ln}/\partial\tilde{x}_k\partial\tilde{x}_m) = 0$ leads to the conditions of constant strains $\tilde{u}_2 = \text{const.}$, $\tilde{u}_3 = \text{const.}$, $\tilde{u}_4 = \text{const.}$, while general form dependences like $\tilde{u}_1 = \tilde{u}_1(\tilde{x}_1)$, $\tilde{u}_5 = \tilde{u}_5(\tilde{x}_1)$, and $\tilde{u}_6 = \tilde{u}_6(\tilde{x}_1)$ do not contradict to these relations. Mechanical equilibrium conditions $\partial\tilde{\sigma}_{ij}/\partial\tilde{x}_i = 0$ could be written as $\partial\tilde{\sigma}_1/\partial\tilde{x}_1 = 0$, $\partial\tilde{\sigma}_5/\partial\tilde{x}_1 = 0$, $\partial\tilde{\sigma}_6/\partial\tilde{x}_1 = 0$. Since $\tilde{\sigma}_{ij}(\tilde{x}_1 \rightarrow \pm\infty) = 0$, one obtains $\tilde{\sigma}_1 = \tilde{\sigma}_5 = \tilde{\sigma}_6 = 0$. Finally, elastic stresses non-zero components in rotated coordinate frame and Voight notations are:

$$\tilde{\sigma}_2 = \frac{s_{11}U_2 - s_{12}U_3}{s_{11}\tilde{s}_{11} - s_{12}^2}, \quad \tilde{\sigma}_3 = \frac{\tilde{s}_{11}U_3 - s_{12}U_2}{s_{11}\tilde{s}_{11} - s_{12}^2},$$

$$\tilde{\sigma}_4 = \frac{Q_{44}\left(\tilde{P}_2^S \tilde{P}_3^S - \tilde{P}_2\tilde{P}_3\right)}{s_{44}}, \quad (6)$$

where functions U_3 and U_2 are

$$U_3 = Q_{11}\left(\left(\tilde{P}_3^S\right)^2 - P_3^2\right) + Q_{12}\left(\left(\tilde{P}_2^S\right)^2 + \left(\tilde{P}_1^S\right)^2 - \left(\tilde{P}_2^2 + \tilde{P}_1^2\right)\right)$$
$$+ F_{12}\frac{\partial \tilde{P}_1}{\partial \tilde{x}_1}, \tag{7a}$$

$$U_2 = \begin{pmatrix} \tilde{Q}_{11}\left(\left(\tilde{P}_2^S\right)^2 - \tilde{P}_2^2\right) + \tilde{Q}_{26}\left(\tilde{P}_1^S\tilde{P}_2^S - \tilde{P}_1\tilde{P}_2\right) \\[2mm] + \tilde{Q}_{12}\left(\left(\tilde{P}_1^S\right)^2 - \tilde{P}_1^2\right) + Q_{12}\left(\left(\tilde{P}_3^S\right)^2 - \tilde{P}_3^2\right) \\[2mm] + \tilde{F}_{12}\frac{\partial \tilde{P}_1}{\partial \tilde{x}_1} + \tilde{F}_{26}\frac{\partial \tilde{P}_2}{\partial \tilde{x}_1} \end{pmatrix}. \tag{7b}$$

Hereinafter, tilda "$\sim$" defines polarization tensor components in the coordinate frame $\{\tilde{x}_1, \tilde{x}_2, \tilde{x}_3\}$. Tensors $\tilde{Q}_{11} = Q_{11} + \sin^2(2\alpha)$ $\left(\frac{Q_{44}}{4} - \frac{Q_{11}-Q_{12}}{2}\right)$, $\tilde{Q}_{12} = Q_{12} - \sin^2(2\alpha)\left(\frac{Q_{44}}{4} - \frac{Q_{11}-Q_{12}}{2}\right)$ and $\tilde{Q}_{26} = \tilde{Q}_{62} = \frac{\sin(4\alpha)}{4}\left(Q_{44} - 2\left(Q_{11} - Q_{12}\right)\right)$ are defined in rotated frame.

Elastic stresses cause band bending via several physical mechanisms, namely via the electric potential changes, deformation potential and Vegrad strains (see details in Ref. [48]) as well as rotostriction coupling.[37] Below we will concentrate our attention on the primary contribution to the band bending originated from electric potential changes. The potential $\varphi(\tilde{x}_1)$ modulates the densities of free electrons $n(\tilde{x}_1)$ accumulated in DW region and the latter can be estimated in the Boltzmann approximation as[48]:

$$n(\tilde{x}_1) \approx n_0 \exp\left(\frac{e\varphi(\tilde{x}_1)}{k_B T}\right). \tag{8}$$

Equilibrium density of electrons is $n_0 = \int_0^\infty d\varepsilon \cdot g_n(\varepsilon) \exp(-(\varepsilon + E_C - E_F)/k_B T)$; $g_n(\varepsilon)$ is density of state, $k_B = 1.3807 \times 10^{-23}$ J/K, T is the absolute temperature. E_F is the Fermi level, E_C is the bottom of the conductive band. The concentration of ionized donors in the Boltzmann approximation is

$$N_d^+(\tilde{x}_1) \approx N_d^0 \exp\left(\frac{E_d - E_F}{k_B T} - \frac{e\varphi}{k_B T}\right). \tag{9}$$

Table 1. Material parameters for bulk ferroelectric $BaTiO_3$ in tetragonal phase.

coefficient	$BaTiO_3$ (collected and recalculated mainly from Ref. [49])
ε_b	7 Ref. [14]
$a_i(C^{-2} \cdot mJ)$	$a_1 = 3.34(T - 381) \times 10^5$ (at $293°$ K -2.94×10^7)
$a_{ij}(C^{-4} \cdot m^5 J)$	$a_{11} = 4.69(T - 393) \times 10^6 - 2.02 \times 10^8,$
	$\quad a_{12} = 3.230 \times 10^8,$
	(at $293°$ K $a_{11} = -6.71 \times 10^8\, a_{12} = 3.23 \times 10^8$)
$a_{ijk}(C^{-6} \cdot m^9 J)$	(at $293°$ K $a_{111} = 82.8 \times 10^8$, $a_{112} = 44.7 \times 10^8$,
	$\quad a_{123} = 49.1 \times 10^8$)
	$a_{111} = -5.52(T - 393) \times 10^7 + 2.76 \times 10^9$
	$a_{112} = 4.47 \times 10^9$
	$a_{123} = 4.91 \times 10^9$
$Q_{ij}(C^{-2} \cdot m^4)$	$Q_{11} = 0.11, Q_{12} = -0.043, Q_{44} = 0.059$
$s_{ij}(\times 10^{-12}\, Pa^{-1})$	$s_{11} = 8.3, s_{12} = -2.7, s_{44} = 9.24$
$g_{ij}(\times 10^{-10} C^{-2} m^3 J)$	$g_{11} = 5.1, g_{12} = -0.2, g_{44} = 0.2^{17}$
$F_{ij}(\times 10^{-11} C^{-1} m^3)$	~ 100 (estimated from measurements of Ref. [50])
	$F_{11} = +2.46, F_{12} = 0.48, F_{44} = 0.05$
Bulk concentration of donors N_d^0	10^{24} m^{-3}

For mobile species Vegard strains should be included in expressions for N_d^+.[48] Equilibrium Fermi level position E_F is determined self-consistently from the electro-neutrality condition $\rho = 0$ valid in the single-domain region of ferroelectric where $\delta\varphi = 0$.

Coupled equations of state (3) and the Poisson equation (2) for the determination of electrostatic potential φ are self-consistent. The boundary conditions corresponding to the DW plane at $\tilde{x}_1 = 0$ are $\tilde{P}_3(\tilde{x}_1 = 0) = 0$, $\tilde{P}_3(\tilde{x}_1 \to +\infty) = +P_S$ and $\tilde{P}_{1,2}(\tilde{x}_1 \to \pm\infty) \to 0$.

Below we discuss the numerical results obtained for tetragonal $BaTiO_3$. Corresponding parameters used listed in the Table 1.

Results of numerical modeling for the polarization structure, strains, electric field, charge carriers redistribution in DW and DW energy are discussed in the next sections.

3. Neel Component Impact on the DW Structure and Energy in $BaTiO_3$

Results of numerical modeling for the polarization, strains, electric field, and charge carriers redistribution across the DW in tetragonal

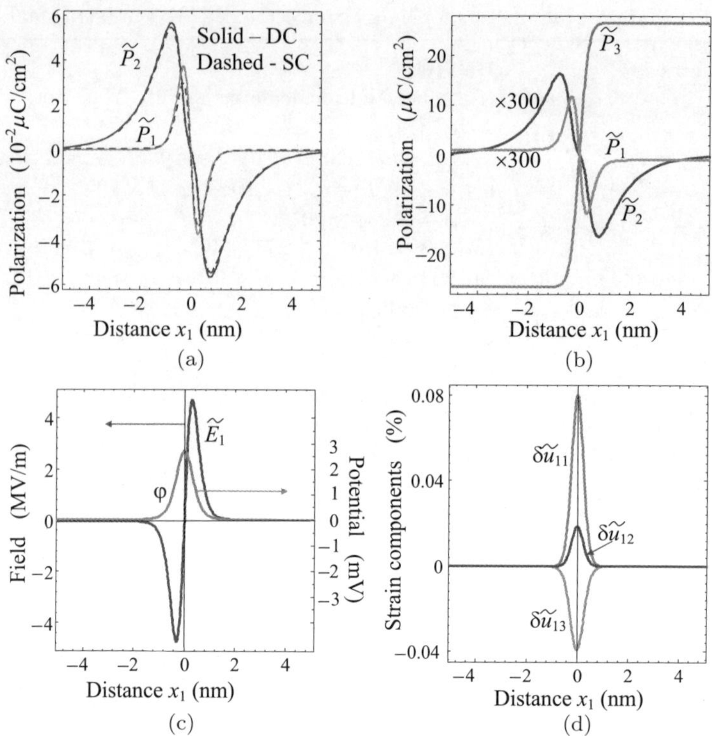

Figure 3.   Dependencies of the Neel-type $\tilde{P}_1$, Bloch-type $\tilde{P}_2$, and Ising-type $\tilde{P}_3$ polarization components (a, b), depolarization field $\tilde{E}_1$ and potential φ (c), strain variations $\delta\tilde{u}_{1i}$ (d) on the distance $\tilde{x}_1$ from the DW plane $\tilde{x}_1 = 0$ in tetragonal phase of BaTiO$_3$ calculated for DW rotation angle $\alpha = \pi/16$, room temperature, flexoelectric constants $F_{11} = 2.46$, $F_{12} = 0.48$, $F_{44} = 0.05$ in $10^{-11}\mathrm{C}^{-1}\mathrm{m}^3$, $N_d^0 = 10^{24}\mathrm{m}^{-3}$, $E_{d0} = 0.1\,\mathrm{eV}$. Other parameters are listed in Table 1.

phase at room temperature are shown in Fig. 3. Polarization vector is zero in the wall plane $\tilde{x}_1 = 0$ [Figs. 3(a) and (b)]. Neel-type and Bloch-type polarization components, $\tilde{P}_1(\tilde{x}_1)$ and $\tilde{P}_2(\tilde{x}_1)$ correspondingly, vanish with the increase of $\tilde{x}_1$. The decay scale of $\tilde{P}_2(\tilde{x}_1)$ is determined by the correlation length (about 1 nm). The decay scale of $\tilde{P}_1(\tilde{x}_1)$ is smaller then the correlation length due to the depolarization effect; it is about 0.5 nm that borders with the spatial limit of LGD continuum theory applicability.

Let us underline, that the flexoelectric coupling induces polarization components with a structure qualitatively different from the classical Bloch- and Neel-walls.[34,35,39] Polarization $\tilde{P}_2$ rotates in opposite directions on the two sides of the wall and passes through zero at the wall plane $\tilde{x}_1 = 0$, in contrast to classical Bloch-type walls, where the polarization vector draws a helix on passing from one domain to the other.[39] Rigorously speaking due to the flexoelectric coupling we deal with non-classical Bloch–Neel–Ising mixed type walls.

It is worth to underline that the value of Neel-type polarization component $\tilde{P}_1(\tilde{x}_1)$ is comparable with the Bloch-type one, $\tilde{P}_2(\tilde{x}_1)$. Maximal values of the Bloch- and Neel-type components are about 300 times smaller than the Ising-type component $\tilde{P}_3(\tilde{x}_1)$. To understand the role of free carriers we compare the results for "semiconducting" case (i.e. the charge density $\rho = e(N_d^+ + p - n)$ is non-zero) and for "dielectric case" (i.e. $\rho = 0$). It appeared that in case of Barium titanate with realistic concentration of free carriers (n-type semiconductor), Neel-type component calculated for the "semiconducting" case is still comparable with the Neel component in the "dielectric case" so the screening is weak. In the dielectric limit the depolarization field $\tilde{E}_1(\tilde{x}_1)$ is maximal and proportional to $-\tilde{P}_1(\tilde{x}_1)/\varepsilon_0\varepsilon_b$. To assure that the visual difference between these two cases are small, one can compare solid curves calculated in the dielectric case (DC), $N_d^0 = n_0 = 0$, with dashed ones calculated for semiconductor case (SC) $N_d^0 = 10^{24}\text{m}^{-3}$ in Fig. 3(a). Note that $\tilde{P}_1(\tilde{x}_1)$ is rather weakly dependent on the electron density at realistic concentration of donors $N_d^0 \leq 10^{25}\text{m}^{-3}$. Some visible increase of $\tilde{P}_1(\tilde{x}_1)$ appears for much higher density of free carriers (half-metallic or metallic), but such situation is unrealistic for all typical ferroelectrics like $BaTiO_3$. Numerical simulations assure us that electron accumulation and donor ionization in DW region rather weakly affect all the components of polarization vector, elastic strains and DW energy for donor concentration $N_d^0 \leq 10^{25}\text{m}^{-3}$.

By this we state that the flexoelectric effect provokes not only Bloch polarization component,[39] but also Neel component that value is comparable with the Bloch component Fig. 3(b).

Depolarization electric field originated from $\tilde{P}_1(\tilde{x}_1)$ is zero at the wall plane $\tilde{x}_1 = 0$, antisymmetric with respect to the wall plane and reaches sharp maximums in the wall [Fig. 3(c)]. Electric potential is maximal at the wall plane and rapidly decays with the distance from the wall. The maximal value is proportional to the flexoelectric coupling strength. The term $e\varphi \sim 2\,\mathrm{meV}$ is essentially smaller than thermal activation energy $k_B T$ at room for the values of flexoelectric coefficients $F_{11} = +2.46$, $F_{12} = 0.48$, $F_{44} = 0.05$ in $10^{-11}\mathrm{C}^{-1}\mathrm{m}^3$,[51] making Boltzmann approximation used in equations (8) and (9) self-consistent. Elastic strains variations $\delta u_{1i}(\tilde{x}_1)$ are rather small (not higher than 0.1%, see Fig. 3(d). Such strains unlikely can lead to the significant accumulation of carriers in the wall via the deformation potential mechanism. So the deformation potential mechanism (not considered here) cannot concur with the changes of carriers density caused by of electric potential variation.

Since the components $\tilde{P}_2(\tilde{x}_1)$ and $\tilde{P}_1(\tilde{x}_1)$ amplitudes are determined by the flexoelectric coupling and electrostriction coefficients, which strongly depend on the wall rotation angle α, one may expect some angular anisotropy behavior of the wall structural and electric properties. Therefore, below we discuss angular dependencies of the maximal polarization, electric potential and elastic strains.

Correlation between angular dependences of the Bloch and Neel maximal polarization values $\tilde{P}_2^{\max}(\alpha)$, $\tilde{P}_1^{\max}(\alpha)$ and maximal electric potential $\varphi(\tilde{x}_1 = 0, \alpha)$ in DW can be imagined from the Fig. 4.

$\tilde{P}_2$ can affect the potential φ only via the coupling between $\tilde{P}_1$ and $\tilde{P}_2$, but the influence of the coupling appeared small. In contrast, we would like to underline evident correlation between angular dependences of $\tilde{P}_1^{\max}(\alpha)$ and $\varphi(0, \alpha)$ (compare solid curves in Figs. 4(a) and (b) with the solid curves in Figs. 4(c) and (d). Namely, their shape, period $\pi/2$, maxima ($\alpha = 0, \pi/2, \pi, 3\pi/2$) and minima ($\alpha = \pi/4, 3\pi/4, 5\pi/4, 7\pi/4$) positions coincide (compare dashed curves in Figs. 4(a) and (b) with the solid curves in Figs. 4(c) and (d). Polar diagrams of $\tilde{P}_1^{\max}(\alpha)$ and $\varphi(0, \alpha)$ have 4 lobes located at angles $\alpha = 0, \pi/2, \pi, 3\pi/2$, while the polar diagrams of $\tilde{P}_2^{\max}(\alpha)$ has 8 lobes located at angles $\alpha = \pi/8 + n\pi/4$ [see Figs. 4(b) and (d)]. The amplitude and modulation depth of the angular dependences of $\tilde{P}_1^{\max}(\alpha)$, $\tilde{P}_2^{\max}(\alpha)$ and $\varphi(0, \alpha)$ increase with the flexoelectric coupling strength increase, but the positions of maxima and minima are

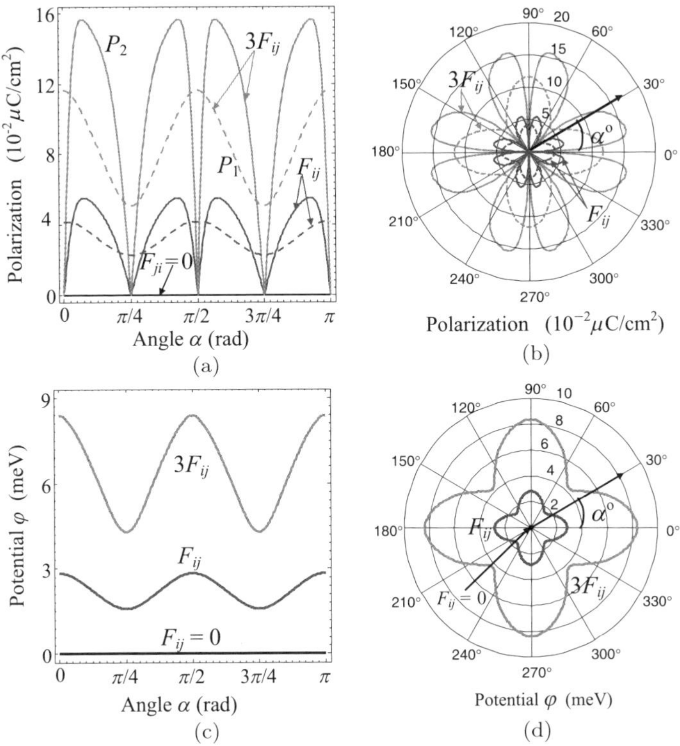

Figure 4. Angular dependence (a) and polar plot (b) of the Bloch-type (solid curves, $\tilde{P}_2$) and Neel-type (dashed curves, $\tilde{P}_1$) maximal polarization values in 180° DW located at $\tilde{x}_1 = 0$ in tetragonal BaTiO$_3$. Angular dependence (c) and polar plot (d) of the maximal potential $\varphi(\tilde{x}_1 = 0)$. Curves are calculated for different flexoelectric coefficients $F_{ij} = 0$ (zero level), F_{ij} (curves labelled as F_{ij}), $3F_{ij}$ (curves labelled as $3F_{ij}$), at room temperature. Other parameters are the same as in Fig. 3.

almost independent on the coupling (compare curves for F_{ij} and $3F_{ij}$). Note that polarization components $\tilde{P}_1^{\max}(\alpha)$, $\tilde{P}_2^{\max}(\alpha)$ and potential $\varphi(0, \alpha)$ are identically zero for $F_{ij} = 0$.

Let us introduce the strain variations as the strain u_{ij} deviation from the spontaneous strain u_{ij}^S:

$$\delta\tilde{u}_{ij} = \tilde{u}_{ij} - \tilde{u}_{ij}^S. \tag{10}$$

Angular dependences and polar plots of the strain variations $\delta u_{1i}(0)$ induced by the 180° DW located at $\tilde{x}_1 = 0$ in tetragonal phase of

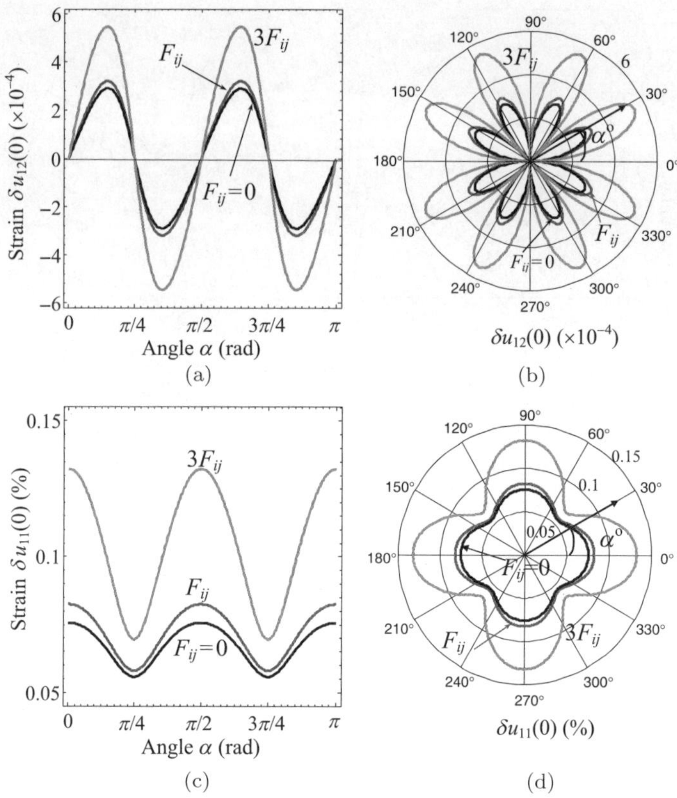

Figure 5.   Angular dependences (a,c) and polar plots (b,d) of the strain variations $\delta u_{1i}(\tilde{x}_1 = 0)$ in the 180° DW in tetragonal BaTiO$_3$. Curves were calculated at room temperature for different flexoelectric coefficients $F_{ij} = 0$ (curves labelled as $F_{ij} = 0$), F_{ij} (curves labelled as F_{ij}), $3F_{ij}$ (curves labelled as $3F_{ij}$). The wall plane is located at $\tilde{x}_1 = 0$. Other parameters are the same as in Fig. 3.

BaTiO$_3$ are shown in Fig. 5. The angular dependence of the strain variation $\delta\tilde{u}_{12}(0)$ is sinusoidal with a period $\pi/2$, i.e. $\delta\tilde{u}_{12}(0) \sim \sin(4\alpha)$. Corresponding polar diagram has 8 lobes located at angles $\alpha = \pi 8 + n\pi/4$ [Fig. 5(b)]. Angular dependence of maximal variation $\delta\tilde{u}_{11}(0)$ is periodic with a period $\pi/2$, positive and slightly anharmonic, i.e. $\delta\tilde{u}_{11}(0) \sim C_0 + C_1\cos^2(2\alpha)$ [Fig. 5(c)]. Corresponding polar diagram has 4 lobes located at angles $\alpha = 0, \pi/2, \pi, 3\pi/2$ [Fig. 5(d)]. The strain components amplitude and modulation depth increases with the flexoelectric coupling strength increase, but positions of maxima and minima are almost independent on the coupling (compare curves for $F_{ij} = 0$, F_{ij} and $3F_{ij}$). Note that the strain variation $\delta u_{13}(0) \approx -0.04\%$ is α–independent and thus not shown in Fig. 5.

To calculate the DW energy we perform the Legendre transformation of the Gibbs potential (1) to Helmholtz free energy[52] as

$$F = \int\limits_{V} \left(G + u_{ij}\sigma_{ij} + P_i E_i + \frac{\varepsilon_b \varepsilon_0 E^2}{2} \right) d\mathbf{x}. \tag{11}$$

The flexoelectric coupling contribution decreases the DW energy (11) and increases its angular anisotropy and modulation depth (see Fig. 6 and compare the curves calculated for $F_{ij} = 0$ with the ones computed for for F_{ij} and $3F_{ij}$). Such results prove that the flexoelectric coupling should be mandatory accounted in thermodynamical calculations of DW energies in ferroelectrics.[39] Without flexoelectric coupling the DW energy anisotropy is caused by elastic strains and electrostriction tensors rotation anisotropy. It is worth to underline that the flexoelectric coupling slightly shifts the wall energy minimum position from the angles $\alpha = 0, \pi/2, \pi, 3\pi/2$. Absolute values of the

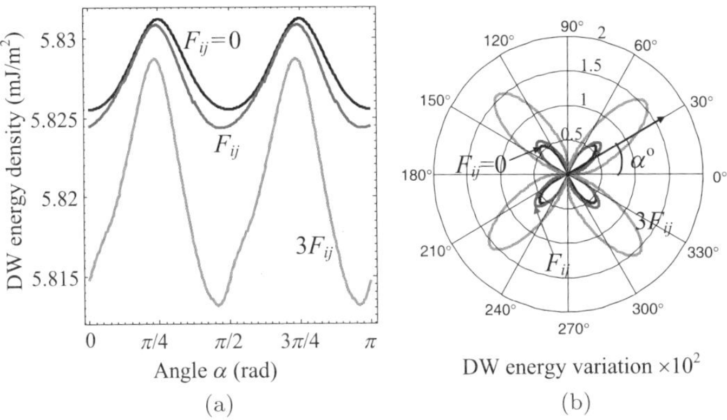

Figure 6.   (a) Angular dependence of the 180° DW energy (1b) density in tetragonal BaTiO₃. Polar plot (b) of its variation δF multiplied by the factor 100. Curves were calculated for different flexoelectric coefficients $F_{ij} = 0$ (curves labeled as $F_{ij} = 0$), F_{ij} (curves labeled as F_{ij}), $3F_{ij}$ (curves labeled as $3F_{ij}$). Other parameters are the same as in Fig. 3.

DW energy rotation anisotropy is rather small; its maximal variation δF is about 0.01–0.02%, but comparable with the electrostriction-related anisotropy [see Fig. 6(b)].

4. DW Static Conductivity in BaTiO$_3$

Analyses of the polarization, strain, and electric potential redistribution and energy variation induced by the flexoelectric coupling in DWs shows that despite the origin of Neel-type and Bloch-type polarization components presents a fundamental interest, their values are relatively small and so the question about their experimental observation or manifestation naturally arises. One possible indirect manifestation of the effect is the change of the wall electronic properties (see equation (8)). Earlier LGD-studies[28,34,35] demonstrate that Neel-type polarization component via electric potential $\varphi(\tilde{x}_1)$ can enhance the static conductivity of the nominally uncharged DWs in $Pb(Zr,Ti)O_3$ and $BiFeO_3$ by 10–100 times. Since the term $|e\varphi(\tilde{x}_1)| < k_B T$ for $BaTiO_3$ at room temperature (see e.g. Figs. 3(c) and 4(c) and (d), the potential variation induced by $\tilde{P}_1$ cannot lead to significant carriers accumulation by the DW. Also we checked numerically that the strains $\delta u_{ij}(\tilde{x}_1) \sim 0.1\%$ does not contribute essentially to the electron accumulation by the wall at realistic effective deformation potential $\Xi_{ij}^n \sim 2\text{–}5\,\text{eV}$ estimated for $BaTiO_3$, since $|\Xi_{ij}^n \delta u_{ij}(\tilde{x}_1)| << k_B T$. So that the inclusion of the term $\Xi_{ij}^n \delta u_{ij}(\tilde{x}_1)$ in equation (8) does not affect essentially on the numerical results shown in Figs. 3–6. Since the impact of stresses on the electron accumulation by the wall appears rather small (at least for the chosen flexoelectric coupling coefficients and estimated deformation potential the term) the accumulation can be caused by the potential variation related term included in equation (8).

Actually, Figs. 7(a) and (b) demonstrate at least 2–4 times electron density increase in the vicinity of 180° DW in dependence on rotation angle α with respect to the crystallographic axis x_3. Thus, in accordance with our calculations, electron density in the wall becomes at least 2–4 times higher than in the single-domain region for known values of flexoelectric coupling coefficients F_{ij}.[51]

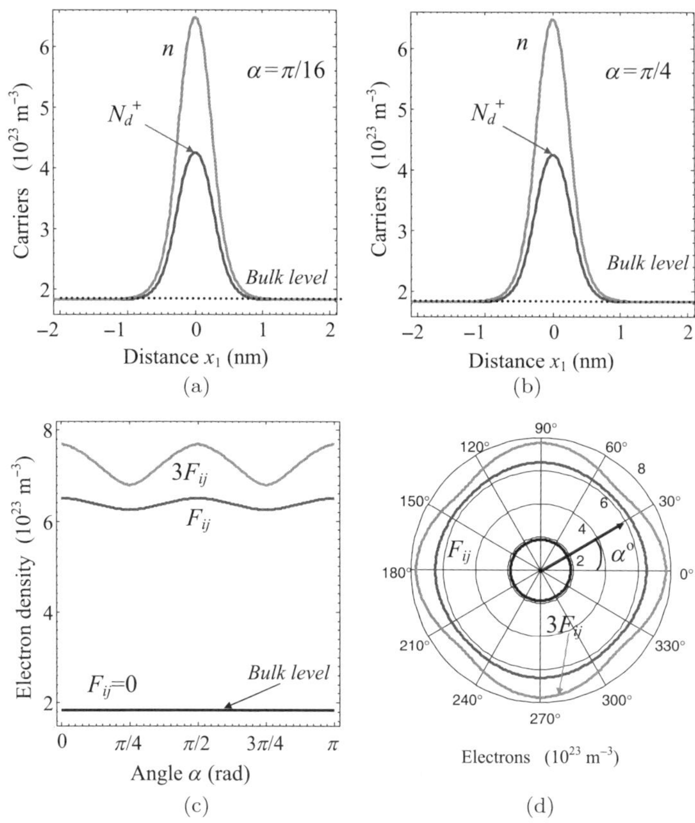

Figure 7. Dependencies of the free electron density $n(\tilde{x}_1)$ and ionized donors concentration $N_d^+(\tilde{x}_1)$ on the distance $\tilde{x}_1$ from DW plane $\tilde{x}_1 = 0$ in tetragonal ferroelectric $BaTiO_3$ calculated for the wall rotation angle $\alpha = \pi/16$ (a) and $\alpha = \pi/4$ (b), room temperature, flexoelectric coefficients $F_{11} = 2.46$, $F_{12} = 0.48$, $F_{44} = 0.05$ (in units of $10^{-11} C^{-1} m^3$). Angular dependence (c) and polar plot (d) of the maximal free electron density $n_{\max}$ in the DW calculated for different flexoelectric coefficients $F_{ij} = 0$ (curves labelled as $F_{ij} = 0$), F_{ij} (curves labelled as F_{ij}), $3F_{ij}$ (curves labelled as $3F_{ij}$). Other parameters are the same as in Fig. 3.

Such variation can be readily detected by the c-AFM.[20,23,28] For non-zero flexoelectric coupling ($F_{ij} \neq 0$) angular dependence of maximal free electron density, $\max_{\tilde{x}_1}[n(\tilde{x}_1)] \approx n(0)$, accumulated in $180°$ DW in tetragonal $BaTiO_3$, is periodic with a period $\pi/2$ with maxima positions at $\alpha = 0, \pi/2, \pi, 3\pi/2$ and minima positions at $\alpha = \pi/4, 3\pi/4, 5\pi/4, 7\pi/4$ (see Figs. 7(c) and (d).

Electron density, its angular anisotropy and modulation depth increases with the flexoelectric coupling strength increase. Without flexoelectric coupling electron density in the DW is the same as in the single-domain region and independent of the wall rotation angle α.

Notice, that the carrier accumulation in the wall does not affect the energy (11) for realistic parameter range in tetragonal $BaTiO_3$. However, virtual independence of the DW energy on the carrier density potentially allows creating arbitrary network of nanosized channels with enhanced static conductivity, which is important for applications.

5. Flexoelectric Coupling Reveals the New Symmetry of DW

As it follows from all aforementioned results, flexoelectric coupling leads to the appearance of the small mixed parity polarization component P_1 perpendicular to the wall in tetragonal ferroelectrics, since flexoelectric terms act as an external field in the right-hand side of the Euler–Lagrange equations.[34,40]

It appears that in rhombohedral (3 m) ferroelectrics the component P_1 weakly charges the wall and transforms it from the head-to-head to tail-to-tail configuration under the wall plane rotation on 180° (see Ref. [40]; wall geometry is shown in Fig. 2(b). Also the wall rotation from $\pi/6$ to $\pi/2$ (that is the symmetry element of 3 m point group) leads to the changes of flexoelectric tensor components $F_{1322}, F_{1311}, F_{1312}$, and F_{2322} signs (see Supplement in Ref. [40]). Since F_{1322} couples the gradient of the Ising component P_3 with quadratic function of all the polarization components, it gives contribution to DW energy (1b) of different signs for these two orientations. The "ground states" corresponding to rotation angles $\alpha = m\pi/3$ are energetically equivalent with or without flexoelectric coupling, but the energy maxima corresponding to $\alpha = \pi/6 + m\pi/3$ for odd and even integer m become non-equivalent (Fig. 8). This is seen from the different height of the energy maximum when the flexoelectric coupling is included (see solid curves in Fig. 8).

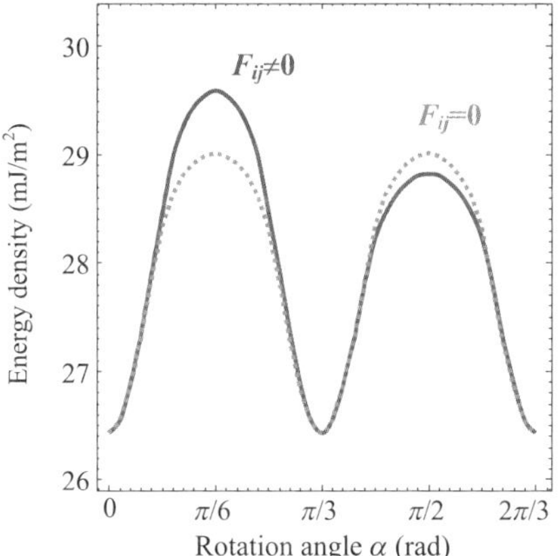

Figure 8. Angular dependence of the DW energy density in the 180° DW calculated for rhombohedral BaTiO$_3$ calculated for $F_{ij} = 0$ (dotted curves) and for non-zero flexoelectric coefficients F_{ij} (solid curves). (Adapted from Ref. [40].)

The difference in the energy between positive and negative DW is about $0.7\,\mathrm{mJ/m^2}$ for rotation angle $\pi/6$ and is absent for the angles multiple to $\pi/3$. Note that the equivalence of the minima follows from the symmetry of the problem, which contains axis of third-order along [111] direction and mirror plane {110}. For the maxima the situation is different, since there is no mirror plane at {211} and the only symmetry operation is the axis of the third order. Hence, the flexoelectric effect lifts the degeneracy of the maxima and reveals the true symmetry of the problem, which was not reflected in the approximation without the coupling (see dotted curves in Fig. 8). Note that the flexoelectric contribution in the DW energy (1b) is comparable with energy anisotropy originated from electrostriction.

Despite the amplitude of Néel component also changes under the rotations, the linear contribution of P_1 gradient into the DW energy (1b) has rather weak orientation dependence. As the consequence of the P_1 gradient contribution into the DW energy (1b), one has to distinguish between DW with different signs of the P_1-gradient in rhombohedral ferroelectrics, i.e. "negative" DW with $dP_3/dx_1 < 0$ and "positive" one with $dP_3/dx_1 > 0$. Without flexoelectric coupling energies of "negative" and "positive" DWs are the same, and so there are no reasons to distinguish them. So the presence of flexoelectric

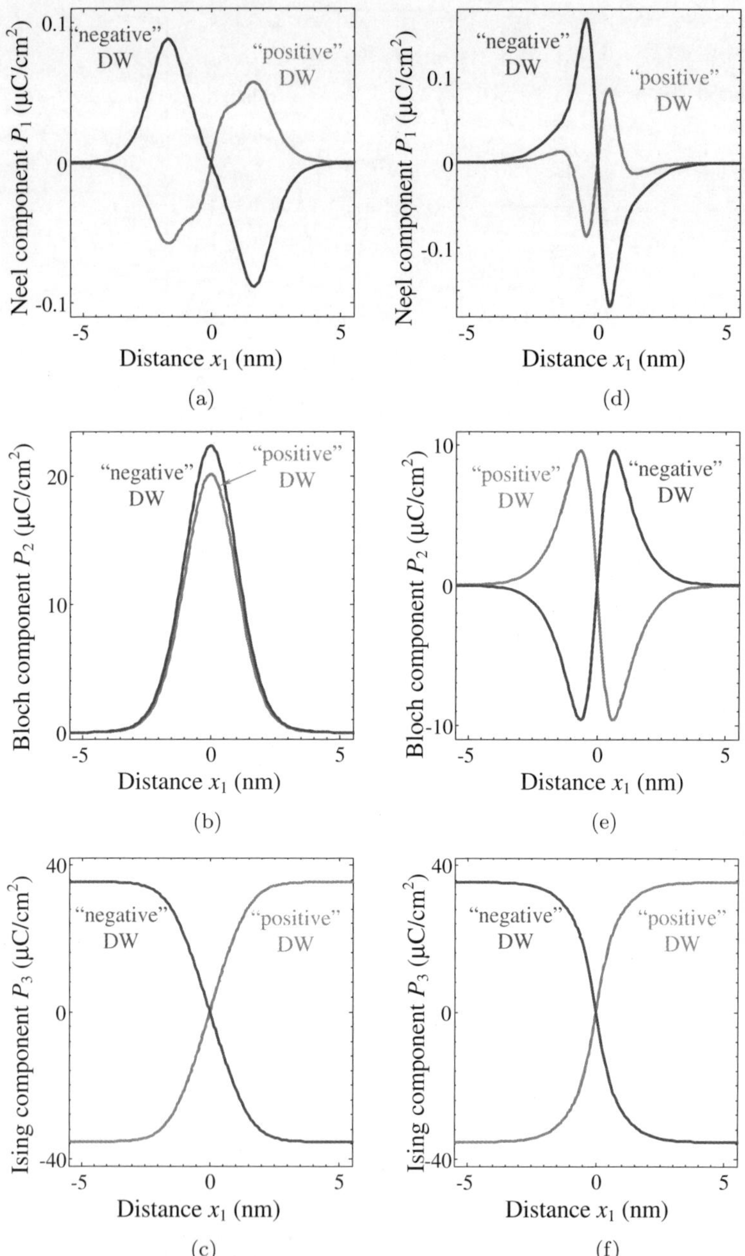

Figure 9.   Profiles of polarization components in positive and negative DWs in rhombohedral $BaTiO_3$, temperature $200\,K$. Angles: $\pi/6$ (a, b, c), $\pi/24$ (d, e, f), Flexoelectric coupling is included.

coupling is essential here, and it seems to be the only way to take into account the terms which are linear in dP_3/dx_1 in the free energy (1b). Typical profiles of polarization components in positive and negative DWs are shown in Fig. 9.

6. Summary

The influence of the flexoelectric coupling on the DWs structure, intrinsic energy, and conductivity can be studied self-consistently within the framework of LGD theory.

In the tetragonal phase, the flexoelectric coupling induces Neel component of the polarization that is perpendicular to the DW plane and leads to the depolarization field and electric potential variation across the wall. DWs with different signs of the Neel component have different energies in rhombohedral ferroelectrics.

Polarization potential leads to the increase of the wall static conductivity. Despite the origin of Neel-type polarization component presents a fundamental interest, its value is very small and the question about the experimental observation or manifestation naturally arises. In accordance with calculations, electron density at the wall becomes at least several times higher than in the single-domain region for known values of flexoelectric coupling coefficients. Such variation can be readily detected by the c-AFM.

References

1. A. K. Tagantsev, E. Courtens, and L. Arzel. *Phys. Rev. B.* **64**, 224107 (2001).
2. M. J. Haun, E. Furman, S. J. Jang, and L. E. Cross. *Ferroelectrics.* **99**, 13–25 (1989).
3. M. J. Haun, E. Furman, T. R. Halemane, and L. E. Cross. *Ferroelectrics.* **99**, 55–62 (1989).
4. B. Houchmandzadeh, J. Lajzerowicz, and E. K. H. Salje. *J. Phys. Condens. Mat.* **4**, 9779 (1992).
5. M. Daraktchiev, G. Catalan, and J. F. Scott. *Ferroelectrics.* **375**, 122 (2008).
6. V. G. Bar'yakhtar, V. A. L'vov, and D. A. Yablonskii. *JETP Lett.* **37**, 673 (1983).

7. A. P. Pyatakov and A. K. Zvezdin. *Eur. Phys. J. B.* **71**, 419–427 (2009).

8. D. Lee, R. K. Behera, P. Wu, H. Xu, S. B. Sinnott, S. R. Phillpot, L. Q. Chen, and V. Gopalan. *Phys. Rev. B.* **80**, 060102(R) (2009).

9. R. K. Behera, C.-W. Lee, D. Lee, A. N. Morozovska, S. B. Sinnott, A. Asthagiri, V. Gopalan, and S. R. Phillpot. *J. Phys. Condens. Matt.* **23**, 175902 (2011).

10. V. A. Zhirnov. *Sov. Phys. JETP.* **35**, 822 (1959).

11. W. Cao and L. E. Cross. *Phys. Rev. B.* **44**, 5 (1991).

12. B. M. Darinskii and V. N. Fedosov. *Sov. Phys. Sol. State.* **13**, 17 (1971).

13. B. Meyer and D. Vanderbilt, *Phys. Rev. B.* **65**, 104111 (2002).

14. J. Hlinka and P. Márton. *Phys. Rev. B.* **74**, 104104 (2006).

15. M. T. D. Vanderbilt, P. Marton, V. Stepkova, and J. Hlinka. *Phys. Rev. B.* **86**, 155138 (2012).

16. L. G.-Ferreira, S. A. T. Redfern, E. Artacho, and E. K. H. Salje. *Phys. Rev. Lett.* **101**, 097602 (2008).

17. P. Marton, I. Rychetsky, and J. Hlinka. *Phys. Rev. B.* **81**, 144125 (2010).

18. B. M. Vul, G. M. Guro, and I. I. Ivanchik. *Ferroelectrics.* **6**, 29 (1973).

19. T. Sluka, A. K. Tagantsev, D. Damjanovic, M. Gureev, and N. Setter. *Nat. Commun.* **3**, 748 (2012).

20. J. Seidel, L. W. Martin, Q. He, Q. Zhan, Y.-H. Chu, A. Rother, M. E. Hawkridge, P. Maksymovych, P. Yu, M. Gajek, N. Balke, S. V. Kalinin, S. Gemming, F. Wang, G. Catalan, J. F. Scott, N. A. Spaldin, J. Orenstein, and R. Ramesh. *Nat. Mater.* **8**, 229–234 (2009).

21. J. Seidel, P. Maksymovych, Y. Batra, A. Katan, S.-Y. Yang, Q. He, A. P. Baddorf, S. V. Kalinin, C.-H. Yang, J.-C. Yang, Y.-H. Chu, E. K. H. Salje, H. Wormeester, M. Salmeron, and R. Ramesh. *Phys. Rev. Lett.* **105**, 197603 (2010).

22. S. Farokhipoor and B. Noheda. *Phys. Rev. Lett.* **107**, 127601 (2011).

23. N. Balke, B. Winchester, W. Ren, Y. H. Chu, A. N. Morozovska, E. A. Eliseev, M. Huijben, R. K. Vasudevan, P. Maksymovych, J. Britson, S. Jesse, I. Kornev, R. Ramesh, L. Bellaiche, L. Q. Chen, and S. V. Kalinin. *Nat. Phys.* **8**, 81–88 (2012).

24. D. Meier, J. Seidel, A. Cano, K. Delaney, Y. Kumagai, M. Mostovoy, N. A. Spaldin, R. Ramesh, and M. Fiebig. *Nat. Mater.* **11**, 284 (2012).

25. V. Ya. Shur, A. V. Ievlev, E. V. Nikolaeva, E. I. Shishkin, and M. M. Neradovskiy. *J. Appl. Phys.* **110**, 052017 (2011).

26. M. Schröder, A. Haußmann, A. Thiessen, E. Soergel, T. Woike, and M. Lukas. *Eng. Adv. Funct. Mater.* **22**, 3936 (2012).

27. J. Guyonnet, I. Gaponenko, S. Gariglio, and P. Paruch. *Adv. Mater.* **23**, 5377–5382 (2011).

28. P. Maksymovych, A. N. Morozovska, P. Yu, E. A. Eliseev, Y. H. Chu, R. Ramesh, A. P. Baddorf, and S. V. Kalinin. *Nano Lett.* **12**, 209–213 (2012).

29. Y.-H. Chu, Q. Zhan, L. W. Martin, M. P. Cruz, P.-L. Yang, G. W. Pabst, F. Zavaliche, S.-Y. Yang, J.-X. Zhang, L.-Q. Chen, D. G. Schlom, I.-N. Lin, T.-B. Wu, and R. Ramesh. *Adv. Mater.* **18**, 2307 (2006).

30. R. K. Vasudevan, A. N. Morozovska, E. A. Eliseev, J. Britson, J.-C. Yang, Y.-H. Chu, P. Maksymovych, L. Q. Chen, V. Nagarajan, and S. V. Kalinin. *Nano Lett.* **12**(14), 5524–5531 (2012).

31. G. I. Guro, I. I. Ivanchik, and N. F. Kovtoniuk. *Sov. Sol. St. Phys.* **11**, 1956–1964 (1969).

32. M. Y. Gureev, A. K. Tagantsev, and N. Setter. *Phys. Rev. B.* **83**, 184104 (2011).

33. E. A. Eliseev, A. N. Morozovska, G. S. Svechnikov, V. Gopalan, and V. Ya. Shur. *Phys. Rev. B.* **83**, 235313 (2011).

34. E. A. Eliseev, A. N. Morozovska, G. S. Svechnikov, P. Maksymovych, and S. V. Kalinin. *Phys. Rev. B.* **85**, 045312 (2012).

35. A. N. Morozovska, R. K. Vasudevan, P. Maksymovych, S. V. Kalinin, and E. A. Eliseev. *Phys. Rev. B.* **86**, 085315 (2012)

36. M. Mostovoy. *Phys. Rev. Lett.* **106**, 047204 (2011).

37. E. A. Eliseev, A. N. Morozovska, Y. Gu, A. Y. Borisevich, L.-Q. Chen, V. Gopalan, and S. V. Kalinin. *Phys. Rev. B.* **86**, 085416 (2012).

38. R. K. Vasudevan, W. Wu, J. R. Guest, A. P. Baddorf, A. N. Morozovska, E. A. Eliseev, N. Balke, V. Nagarajan, and P. Maksymovych. *Adv. Funct. Mater.* **23**, 2592–2616 (2013).

39. P. V. Yudin, A. K. Tagantsev, E. A. Eliseev, A. N. Morozovska, and N. Setter. *Phys. Rev. B.* **86**, 134102 (2012).

40. E. A. Eliseev, P. V. Yudin, S. V. Kalinin, N. Setter, A. K. Tagantsev, and A. N. Morozovska. *Phys. Rev. B.* **87**, 054111 (2013).

41. Y. Gu, M. Li, A. N. Morozovska, Y. Wang, E. A. Eliseev, V. Gopalan, and L.-Q. Chen. *Phys. Rev. B.* **89**, 174111 (2014).

42. A. N. Morozovska, E. A. Eliseev, M. D. Glinchuk, L.-Q. Chen, and V. Gopalan. *Phys. Rev. B.* **85**, 094107 (2012).

43. A. N. Morozovska, E. A. Eliseev, M. D. Glinchuk, S. V. Kalinin, L.-Q. Chen, and V. Gopalan. *Ferroelectrics.* **438**, 32–44 (2012).

44. A. N. Morozovska, E. A. Eliseev, S. V. Kalinin, L.-Q. Chen, and V. Gopalan. *Appl. Phys. Lett.* **100**, 142902 (2012).

45. E. A. Eliseev, S. V. Kalinin, Y. Gu, M. D. Glinchuk, V. Khist, A. Borisevich, V. Gopalan, L.-Q. Chen, and A. N. Morozovska. *Phys. Rev. B.* **88**, 224105 (2013).

46. A. Y. Borisevich, E. A. Eliseev, A. N. Morozovska, C.-J. Cheng, J.-Y. Lin, Y. H. Chu, D. Kan, I. Takeuchi, V. Nagarajan, S. V. Kalinin. *Nat. Commun.* **3**, 775 (2012).
47. A. K. Tagantsev and G. Gerra. *J. Appl. Phys.* **100**, 051607 (2006).
48. A. N. Morozovska, E. A. Eliseev, A. K. Tagantsev, S. L. Bravina, L.-Q. Chen, and S. V. Kalinin. *Phys. Rev. B.* **83**, 195313 (2011).
49. A. J. Bell. *J. Appl. Phys.* **89**, 3907 (2001).
50. W. Ma and L. E. Cross. *Appl. Phys. Lett.* **88**, 232902 (2006).
51. I. Ponomareva, A. K. Tagantsev, L. Bellaiche. *Phys. Rev. B.* **85**, 104101 (2012).
52. N. A. Pertsev, A. G. Zembilgotov, and A. K. Tagantsev. *Phys. Rev. Lett.* **80**, 1988 (1998).

Chapter 10

Bending-Induced Giant Polarization in Ferroelectric MEMS Diaphragm

Zhihong Wang

Advanced Nanofabrication Core Lab,
King Abdullah University of Science and Technology,
Thuwal 23955-6900, Kingdom of Saudi Arabia

Weiguang Zhu

School of Electrical and Electronic Engineering,
Nanyang Technological University, Singapore 639798

The polarization induced by the strain gradient, i.e. the flexoelectric effect, has been observed in a micromachined $Pb(Zr_{0.52}Ti_{0.48})O_3$ (PZT) diaphragms. Applying air pressure to bend a flat diaphragm which initially does not exhibit any electromechanical coupling can induce a resonance peak in its impedance spectrum. This result supposes that bending, thus the strain gradient in the diaphragm causes polarization in PZT film. We also investigated the switching behaviors of the polarization in response to an external electric field in a bent diaphragm and further quantified the polarization induced by the strain gradient. The effective flexoelectric coefficient of the PZT film has been calculated as large as 2.0×10^{-4} C/m. A giant flexoelectric polarization of the order of $1\ \mu C/cm^2$ was characterized which is of the same order of magnitude as the normal remnant ferroelectric polarization of PZT film. The suggested explanation for the giant polarization is the large strain gradient in the diaphragm and the strain gradient induced reorientation of the polar nanodomains.

1. Introduction

Approximately half a century ago, Bursian and Zaikovskii[1] demonstrated that in a cantilever composed of a layer of $BaTiO_3$ thin film sandwiched between two electrodes, polarization induced by an electric field may cause a distortion of the lattice which alters the film curvature. Whereas the converse effect, i.e. the bending induced polarization, was not investigated in detail in their paper due to experimental difficulties. Over the last decade, L. E. Cross has shown with simple beam-bending experiments with bulk ferroelectrics that strain gradients alone can induce large changes in electrical polarization,[2-8] wherein the high dielectric constant accounts for the significant flexoelectric polarization. His work suggested that the flexoelectric effect could be exploited in real devices, stimulating successive experimental and theoretical work. Although the debate on the intrinsic value of the flexoelectric coefficients[9-11] is still ongoing, recent reports reveal that giant polarization indeed exists on both the micro- and nanoscales[12-16] due to the huge strain gradient of the order of 10^4 to $10^6 \, m^{-1}$ at film/substrate interface[14,17] or in a nanoribbon.[18] However, the direct coupling between the strain gradient and the polarization in ferroelectric thin film systems, in other words, the flexoelectric coefficient of thin films, has never been quantitatively measured and reported. Here, we report that macroscopic strain gradient can induce substantial flexoelectric polarization in a bent $Pb(Zr_{0.52}Ti_{0.48})O_3$ (PZT) diaphragm fabricated by micromachining and that this giant flexoelectric polarization can reverse the remnant ferroelectric polarization in PZT thin film. Our results clearly differentiate the contribution of flexoelectric polarization from that of ferroelectric polarization in ferroelectric thin films.

A micromachined diaphragm has two obvious advantages in an investigation of flexoelectric polarization. First, the strain gradient can be quantified by observing the curvature of the diaphragm. Second, the state of the static polarization of the film can be evaluated by using the impedance-frequency spectrum in the vicinity of the resonant frequency. If electrical polarization exists in the film, the electromechanical coupling of the diaphragm at its resonant frequency can clearly show the polarization state. As a consequence, these two advantages enable us to quantatively investigate the flexoelectric effect by correlating strain gadient and polarization. In this chapter, we will first report the microfabrication of the diaphragm, then investigate the relationship between initial polarization and strain gradient of the diaphragm, and finally the polarization switching behaviors in response to the external excitations, including external pressure and electric field which alter the strain gradient and ferroelectric polarization, respectively. The strain gradient induced by bending the diaphragm has been observed of the order of $10^2 \, \mathrm{m}^{-1}$, much larger than the strain gradient obtained in the bulk material. The observed giant flexoelectric polarization turns out to be of the same order of magnitude as the remnant polarization of the PZT film.

2. Diaphragm Fabrication and Characterization

To investigate flexoelectric polarization, we will use diaphragms with different curvature as the testing sample. Fortunately, stress usually accompanies the microfabrication process. While the tensile stress results in a flat diaphragm, the compressive stress in the diaphragm always causes bending of the diaphragm. Figure 1 shows the schematic structure and dimensions of the PZT/Si diaphragm. After fabrication, we obtained both well-bent diaphragm with compressive stress and very flat diaphragm with high tensile stress due to an unforeseen non-uniform stress distribution in the silicon on insulator (SOI) wafers. Hereafter, we will characterize these two diaphragms

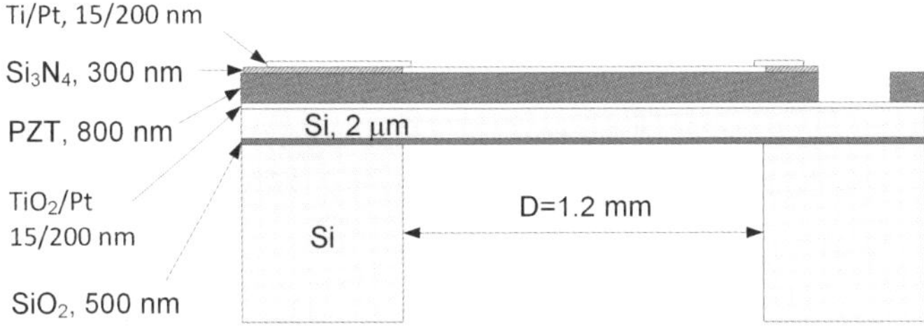

Figure 1.   Schematic multilayer structures of the micromachined circular PZT/Si diaphragm. The diameter of the PZT/Si diaphragm is 0.8 or 1.2 mm.

and refer to them as bent diaphragm and flat diaphragm. Sol-gel PZT film was used to investigate the flexoelectric effect because it has relatively high dielectric susceptibility among the known film families. The high susceptibility is closely associated with the high flexoelectric coefficient.[8]

The static profile of the released diaphragm was measured by using a 3D-view digital holographic interferometer[19] or an optical surface profiler (Zygo NewView 7300). An Agilent 4294A impedance analyzer was used to test the impedance spectrum and the C–V curves of the diaphragm. To maintain consistency in the results, during the testing, the high voltage port of the Agilent 4294A impedance analyzer was always connected to the top electrode of the diaphragm. The DC bias and AC driving signal were applied across the top and bottom electrodes of the diaphragm through the impedance analyzer. The oscillation AC level of the activating voltage was 5 mV. The impedance frequency spectra at different DC biases were measured in both the "bias on" and "bias off" states. At each applied bias, we recorded the first spectrum while the bias was applied, followed by the measurement of the diaphragm profile. After the bias was removed, we recorded the second surface profile and the spectrum, which exhibited the remnant polarization state. Instead of impedance magnitudes, phase shifts were used to qualitatively measure the polarization within the film. The C–V curve was recorded separately after we finished the impedance tests at all the biases. P–E hysteresis was measured using Precision Pro ferroelectric analyzer (Radiant Technology Inc., Albuquerque, NM).

The fabrication process of the PZT/Si diaphragms is depicted in Fig. 1. A $\langle 100 \rangle$ SOI wafer with a $2\,\mu$m device layer was used as the substrate. Firstly, thin TiO_2/Pt (15 nm/200 nm) films were deposited as the bottom electrode by sputtering. Secondly, a thin PZT film ($\sim 0.8\,\mu$m) was deposited by the sol-gel deposition technique. To open access to the bottom electrode pad, in the third step, the PZT film was wet etched in diluted HCl:HF solution. Afterwards, a Si_3N_4 layer (300 nm) was deposited by plasma-enhanced chemical vapor deposition (PECVD) and patterned by reactive ion etching (RIE) to serve as an insulation layer to minimize parasitic capacitance induced by the patterned electrode wiring. The top Ti/Pt (15 nm/200 nm) electrode was sputtered and patterned by using the lift-off technique. Finally, the diaphragm was released by deep reactive ion etching (DRIE). The diameters of the well-bent and the flat diaphragms used in measurement were 0.8 mm and 1.2 mm, respectively.

3. Microstructure of PZT Film and Multilayer Structure of the PZT Diaphragm

Figure 2 shows the SEM images of the cross-section of a fabricated bent diaphragm. The columnar structure of the PZT grain on the diaphragm can be clearly seen in the images. The composition of

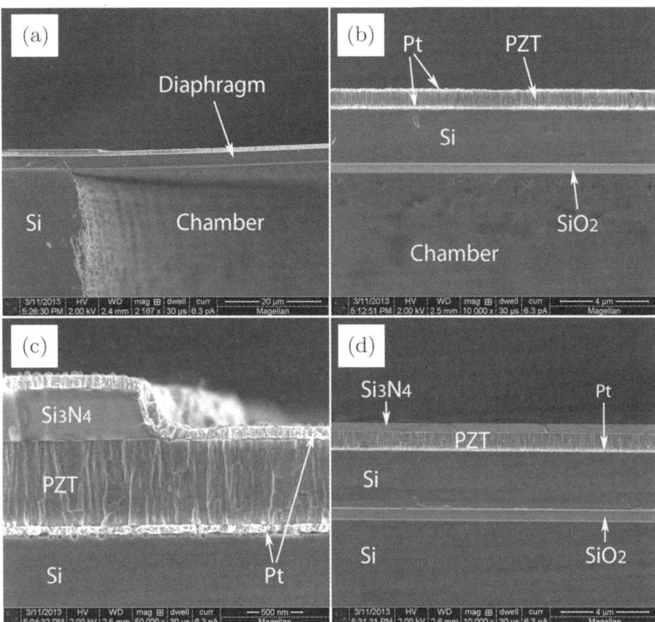

Figure 2. Cross-sectional SEM images of a fabricated circular PZT/Si diaphragm. (a) Part of the released bent diaphragm together with the supporting Si substrate. (b) Pt/PZT/Pt/Si/SiO$_2$ multilayer structure of the released PZT/Si diaphragm. (c) Detailed cross-sectional view clearly shows the insulation Si$_3$N$_4$ layer, top and bottom electrodes and the columnar grain structure of the PZT layer. (d) Multilayer structure far away from the diaphragm area. There is no Pt electrode on top of Si$_3$N$_4$ layer and the SOI substrate is clearly seen.

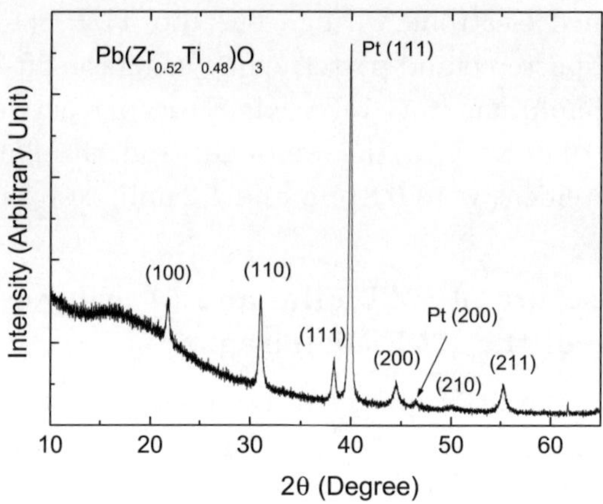

Figure 3.   XRD pattern of the multilayer structure tested on top surface of the fabricated diaphragm. The information includes the crystalline structure of PZT, top and bottom Pt electrodes and the PECVD Si_3N_4 layer. We can clearly see that the crystalline microstructure of the PZT layer is randomly oriented polycrystalline tetragonal phase (lattice constant: $a = 4.038$ Å, $c = 4.140$ Å).

the PZT film is PZT. It is located in the tetragonal phase side near the morphotropic phase boundary (MPB). We also performed an X-ray diffraction measurement to understand the crystalline structures of the multilayer top surface of the fabricated diaphragm. The tested area covers diaphragm (Fig. 2(b)) and the large, rigid supporting area (see Fig. 2(d)), and it can be seen from Fig. 3 that the PZT layer consists of a randomly oriented polycrystalline tetragonal phase.

4. Curvature of the Diaphragm and Polarization

The curvature of the diaphragm clearly affects the shape of the P–V hysteresis loop of the PZT film. Figure 4 shows the P–V hysteresis loops tested on diaphragms with different as-released curvature. Although PZT films on both the flat and the bent diaphragm exhibit normal P–V hysteresis loops, there are some features worthy of notice. First, in comparison with the bent diaphragm, the flat diaphragm with high tensile stress exhibits lower saturated and remnant polarization. Particularly, from the wide gap between onset and

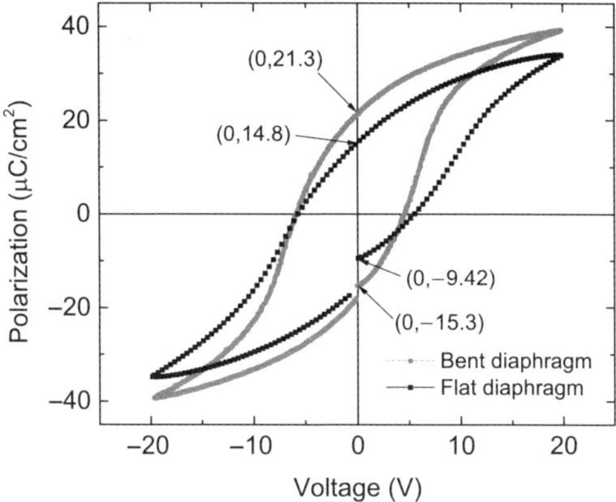

Figure 4.   P–V hysteresis loops tested on diaphragms with different as-released curvature. Flat diaphragm with high tensile stress exhibits lower remnant polarization, while bent diaphragm shows more shifted P–V loop.

ending point of the loops we can see that in flat diaphragm, remnant polarization was reduced a lot after relaxation $(-9.42\,\mu\mathrm{C/cm}^2)$, indicating that high in-plane tensile stress in the film depresses the out-of-plane polarization. Second, P–V loop of the bent diaphragm shows more negative shift, implying that a positive self-polarization exists in the PZT film. As such, we also tested the self-polarization of the as-released bent diaphragm.

If a thin film exhibits a net polarization normal to its surface, a small AC voltage applied to the film will excite an electromechanical coupling through the variation of the polarization in the film. Thus, the electromechanical coupling can be used to investigate the polarization state of the film, no matter what is the origin of the polarization. This coupling, as a measure of the polarization state of the film, can be observed as a resonant peak through the impedance-frequency spectrum of the diaphragm. We measured the impedance spectra of the diaphragms that had not yet been subject to an external field. The results show that the flat diaphragm subjected to high tensile stress does not show resonant peak in impedance spectrum, thus does not possess self-polarization, whereas, accompanying the bending of the as-released diaphragm, the PZT layers in PZT/Si diaphragms exhibit self-polarization. This is confirmed by

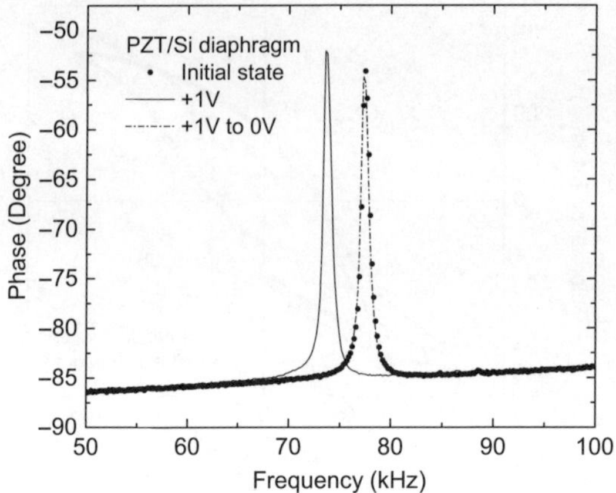

Figure 5.   Phase-frequency spectra of the bent circular PZT/Si diaphragm with diameter of 0.8 mm. The resonant peak indicates that self-polarization exists in the PZT film on the as-fabricated bent diaphragm.

occurrence of the resonant peaks in the impedance-frequency spectra (Fig. 5). In Fig. 5, a positive 1 V bias enhances the electromechanical response of the PZT/Si diaphragm, as indicated by the shift of the resonant peak and increased peak height of the phase angle at the resonant frequency. After removing the bias, the polarization state of the initial diaphragm is completely recovered. This result suggests that: (1) Self-polarization is closely related to the curvature of the as-released diaphragm; and (2) The polarization can be altered by applying DC bias. Why does the self-polarization tends to exist in the bent diaphragm? Next, we will answer this question by investigating the electromechanical coupling of the diaphragms in response to an external pressure or voltage.

5. Evolution of Polarization

5.1. *Bending Induced Electromechanical Coupling in a Flat Diaphragm Subjected to High Tensile Stress*

To understand the relationship between curvature and polarization, we designed a special test setup to control the curvature of the PZT/Si diaphragm through a small pressure chamber connected to

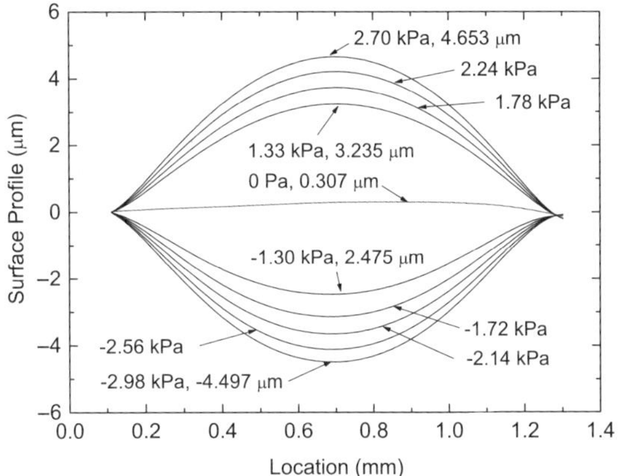

Figure 6. Cross-sectional profiles along the PZT/Si diaphragm's diameter at different chamber pressures. The positive values mean that the air pressure inside the chamber was higher than that of the ambient pressure outside the chamber.

the back of the diaphragm. Figure 6 shows the effect of the chamber pressure on the surface profile of a PZT/Si diaphragm with very high tensile stress.

We can see that the original deflection height of the diaphragm is approximately $0.307\,\mu$m when no air pressure is applied from the back chamber. Considering that the diameter of the diaphragm is $1.2\,$mm, the original diaphragm profile is actually very flat. As expected, the convexity, and thus the curvature of the diaphragm can be deformed by adjusting the air pressure from the back chamber of the diaphragm. The deflection heights are $4.65\,\mu$m at $2.70\,$kPa and $-4.50\,\mu$m at $-2.98\,$kPa. A positive pressure value means that the air pressure inside the chamber is higher than that of the outside ambient environment, and vice versa.

Figure 7 shows the impedance spectra of the PZT/Si diaphragm at different air pressures. Shown in Fig. 7(a) are the effects of the positive pressure on the impedance and phase spectra of the piezoelectric diaphragm. In its original flat state, i.e. when no pressure is applied, the diaphragm did not show a resonance peak in either the impedance spectrum or the phase-frequency spectrum, indicating that there is no electromechanical coupling in the diaphragm; in other words, the diaphragm is not piezoelectric. However, upon the application of a

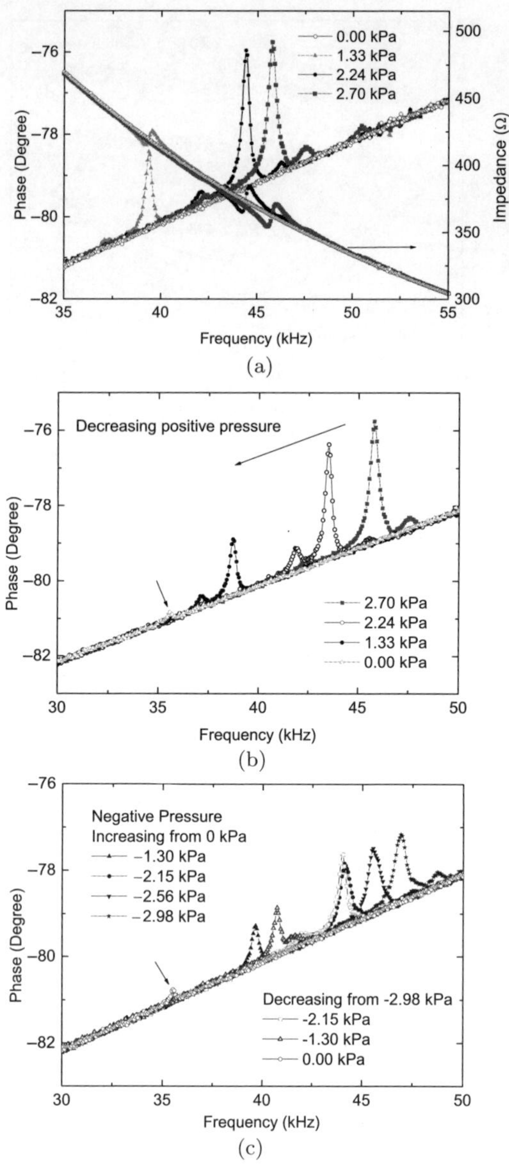

Figure 7.   Static pressure-induced electromechanical coupling in a highly stressed PZT/Si diaphragm. (a) and (b) Dependence of the impedance/phase-frequency spectra on positive pressures. (c) Dependence of the phase-frequency spectra on negative pressures.

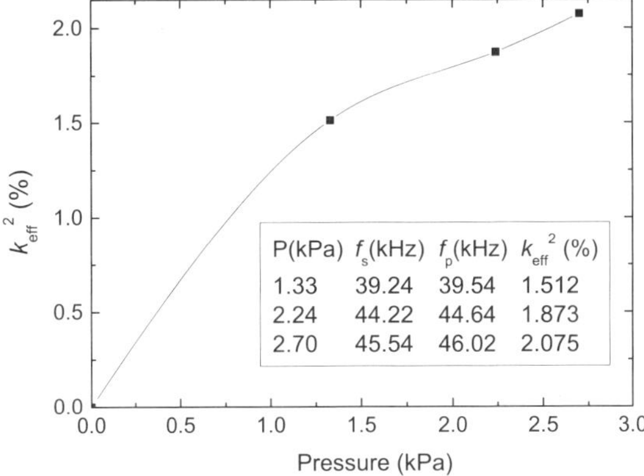

Figure 8.   Dependence of effective coupling coefficient on positive pressure. The coupling coefficient increase with the increase of the pressure. Inset table shows the resonance and antiresonance frequencies obtained from impedance spectra as shown in Fig. 7(a).

positive pressure of 1.33 kPa, a resonance peak emerged. Both the resonance peak height and the resonance frequency increased further with a higher positive air pressure of up to 2.70 kPa. Figure 8 shows the effective electromechanical coupling coefficients calculated with the tested resonance and antiresonance frequencies in the impedance spectra. It is clear that the effective coupling coefficient increases as greater pressure is applied. In comparison with the variation of resonance and antiresonance peaks in the impedance spectra, the variation of the peaks in the phase spectra due to changes in pressure are more pronounced (see Fig. 7(a)). Therefore, it is more straightforward to qualitatively indicate the electromechanical coupling in the phase spectra. Figure 7(b) shows that the induced resonance peak in the phase spectra was reversible when the positive air pressure was lowered. However, a very weak peak, as indicated by the arrow in Fig. 7(b), remained when the positive pressure was fully removed. Similar phenomena were observed when negative pressures were applied (Fig. 7(c)). That is, the induced resonance peak height and the resonance frequency increased with increasing negative pressure. The resonance peak was almost removed when the negative pressure was gradually reduced to zero from $-2.98\,\text{kPa}$ with only a very weak peak remaining.

The observed phenomena clearly indicate that the flat PZT/Si diaphragm initially does not exhibit electromechanical coupling behavior. There is no self-polarization in this highly stressed PZT film. This is probably because the high in-plane stress inhibits the formation of the out-of-plane polarization (c-domain). Upon the application of static air pressure to the diaphragm, the deflection of the diaphragm directly induced electromechanical coupling in the non-prepolarized PZT/Si diaphragm, and the electromechanical coupling was enhanced as the air pressure was increased. Therefore, strain gradient may play an important role in this phenomenon. Considering that the diaphragm consists of a PZT layer, it is reasonable to assume that strain gradient can induce electrical polarization in PZT films, thus endow the film with piezoelectric properties. However, it is currently unclear what the mechanism of the induced polarization is. From a material microstructure point of view, reorientation of grains and/or domains and phase transformation at MPB due to external poling voltage are the main causes of electrical polarization and may thus be the possible causes of the bending induced polarization. However, bending can only induce large tensile stress in diaphragm plane. Even if the large in-plane tensile stress could induce the reorientation of domains/grains or a new phase, the polarization of the induced new domain/grain/phase will direct along the plane of diaphragm so that the induced domain reorientation cannot enhance the out-of-plane polarization. Thus, the in-plane tensile stress induced reorientation of grains/domains or phase transformation are unlikely to be the main causes of the bending induced out-of-plane polarization. On the contrary, the most possible origin of the bending induced out-of-plane polarization can be attributed to strain gradient, in other word, flexoelectric effect. In addition, it is not easy for the c-axis textured columnar grains of PZT (see Fig. 2(c)) to rotate towards the in-plane direction.

The flexoelectric polarization should disappear when the strain gradient is removed, thus the very weak resonance peaks (Figs. 7(b) and 7(c)) that are observed after the removal of applied pressure are caused by the residual strain in the diaphragm. Cross[8] observed a similar residual strain in a PZT-5H specimen bar after a high stress four-point bend flexoelectric test. In comparison with the large remnant polarization after poling by an electric field in normal PZT films, the remnant component of the polarization from the residual strain gradient in highly stressed PZT films is very small.

5.2. *Reversed Remnant Polarization in a Bent Diaphragm Subjected to High Compressive Stress*

In this section, we investigated the bent PZT/Si diaphragm with self-polarization as shown in Fig. 5. Electrically biasing the diaphragm may change the polarization state of the PZT film and generate stress (or strain) in the film plane, which will alter the curvature, the electromechanical coupling and the resonant frequency of the diaphragm. Therefore, by applying DC bias, we can study the development of the polarization in the diaphragm. Hereafter, we will study the DC bias dependence of the electromechanical coupling (resonance peak of the phase-frequency spectra), central deflection and capacitance of the PZT/Si diaphragm. The experimental results will be used to estimate flexoelectric coefficient of the PZT film in Section 6.4.

After removing the positive 1 V bias, a series of negative biases were applied to the diaphragm in a step-by-step fashion. A -1 V bias was first applied and then removed after recording the impedance spectrum and static surface profile. Thereafter, the cases of stronger negative biases were investigated successively. Figure 9 shows the phase angle spectra at different negative biases. Because the negative voltage was applied to the top electrode during the testing, the applied electric field was directed from the bottom electrode to the top electrode. Figure 9(a) shows that the resonant peak decreases with increasing negative bias, indicating that the ferroelectric polarization induced by the negative bias counteracts the positive self-polarization. The self-polarization is directed from the top to the bottom electrode.

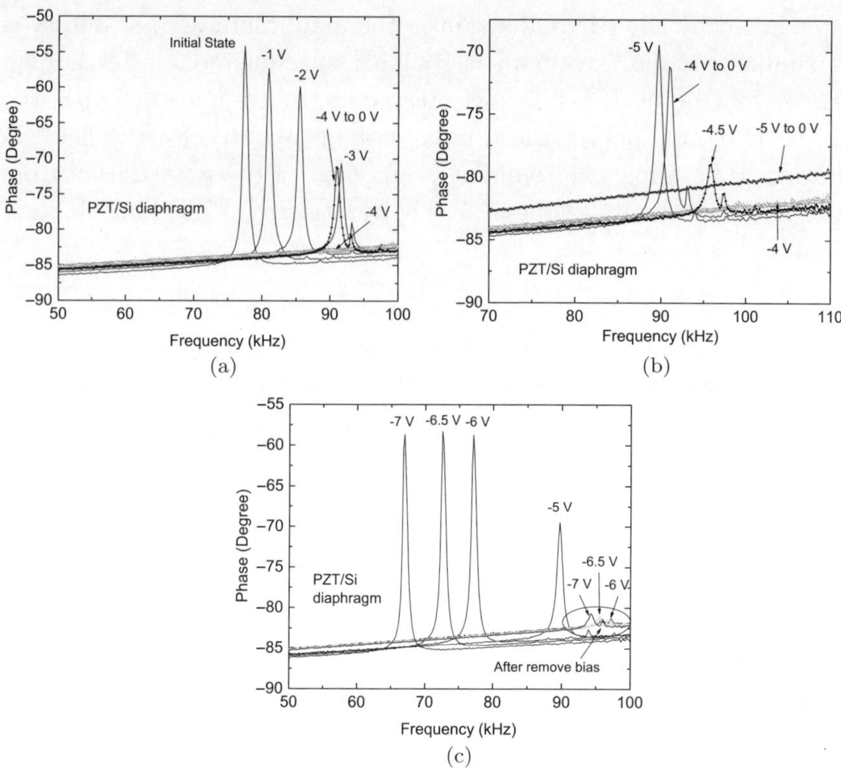

Figure 9.   The effect of negative biases on phase-frequency spectra of the bent PZT/Si diaphragm. (a) The negative bias increases gradually from $-1\,$V to $-4\,$V. The ferroelectric polarization induced by the negative bias counteracts the positive self-polarization; negative coercieve voltage is -4V; the initial state is also shown for reference. (b) The bias is increased to $-5\,$V. Negative polarization is generated in the diaphragm by the negative bias large than $-4\,$V; after removing the $-5\,$V bias, the remnant polarization becomes zero as indicated by the disappearance of the resonant peak. Thus, we can expect that the remnant polarization after removing $-4.5\,$V bias has switched back to the original positive direction. (c) The bias is further increased to $-7\,$V. The negative polarization increased dramatically when the bias was increased to $-7\,$V; after removing this bias, the negative remnant polarization became unexpectedly small. Reprinted with permission from [16]. Reprinted with permission from Wiley-VCH Verlag GmbH & Co. KgaA, Weinheim. Copyright 2012.

We can also see in Fig. 9(a) that at -4 V bias, the resonant peak disappears. It is reasonable to suppose that, at this bias, the polarization in the film becomes zero and the -4 V becomes a negative coercive voltage. After removing the -4 V bias, part of the polarization was recovered as shown by the reappearance of resonant peak.

When the negative bias was increased continuously to -4.5 V and then -5 V, the resonant peak appeared again (Fig. 9(b)) and the peak height increased with increasing bias. The resonant peak indicates that new polarization is generated in the diaphragm by the negative bias. At -5 V bias, the resulting negative polarization is larger than the remnant polarization after the -4 V bias is removed. Moreover, after removing the -5 V bias, the remnant polarization becomes zero as indicated by the disappearance of the resonant peak. Thus, we can expect that the remnant polarization after removing -4.5 V bias has switched back to the original positive direction. In other words, the negative polarization has a positive remnant polarization. We also observed this reversed remnant polarization in other bent diaphragms, indicating the possibility that this phenomenon is intrinsic in bent diaphragms.

The observed phenomenon suggests that, in addition to the bias, there must be another physical parameter that also affects the remnant polarization in this diaphragm system. In last section, we already demonstrated that bending can induce flexoelectric polarization in PZT film, thus it is reasonable to suppose that flexoelectric polarization in the diaphragm plays important role in the reversal of the remnant polarization. The curvature of the diaphragm can be deformed by applying bias, but after removing the bias, the original curvature of the diaphragm is recovered due to the internal stress. The most reasonable explanation for the reversed remnant polarization is that the negative remnant ferroelectric polarization induced by the negative bias is smaller than the positive flexoelectric polarization induced by curvature of the diaphragm, thus causing the sum of the two components of the polarization to become positive. It is the curvature or the strain gradient of the diaphragm that forces the negative remnant polarization to switch to the positive direction.

We believe that this reversed remnant polarization is direct evidence demonstrating that flexoelectric polarization exists in a curved PZT diaphragm. We therefore posit that the initial self-polarization existing in the as-released bent diaphragm also originates from the flexoelectric effect. In fact, our extensive measurements showed that the self-polarization in PZT films is always associated with the bending of the diaphragm.

Figure 9(c) shows that the reversed polarization increased dramatically when the bias was increased to $-7\,\mathrm{V}$. However, after removing this bias, the remnant polarization became unexpectedly small, which is evidenced in the very low resonant peak heights. Considering the remnant polarization was zero after removing the $-5\,\mathrm{V}$ bias, the remnant polarization is maintained in the reverse direction.

After removing the $-7\,\mathrm{V}$ bias, a series of positive biases were applied to the diaphragm. Figure 10 shows the effect of the positive bias on the phase angle spectra. In Fig. 10(a), we observe that, the very small remnant polarization caused by the $-7\,\mathrm{V}$ bias can be totally counterbalanced by applying a $1\,\mathrm{V}$ positive bias, much smaller than the negative coercive voltage of $-4\,\mathrm{V}$. The polarization direction is also clearly reversed at $1.5\,\mathrm{V}$ and $2\,\mathrm{V}$ bias. Figure 10(b) shows that the positive polarization increased as the positive bias was increased. However, the height of resonant peaks decreased as the bias was increased when the bias was higher than $4\,\mathrm{V}$. This is because the gradually saturated polarizations and increased stress above these positive voltages reduced the electromechanical coupling, just as the d_{33} coefficient decreases above a certain voltage in d_{33}-bias curves of normal ferroelectric films. When the bias increased from $6\,\mathrm{V}$ to $7\,\mathrm{V}$, the resonant frequency increased because the ferroelectric polarization induced by the bias flattens the dome and thus generates higher stress in the diaphragm. Figure 10(c) clearly shows that after removing the positive bias, the positive remnant polarization is much larger than the negative one. The remnant polarization is therefore asymmetric.

Gruverman *et al.*[15] reported that bending the substrate strongly shifts the P–E hysteresis loops of PZT thin film capacitors. As a

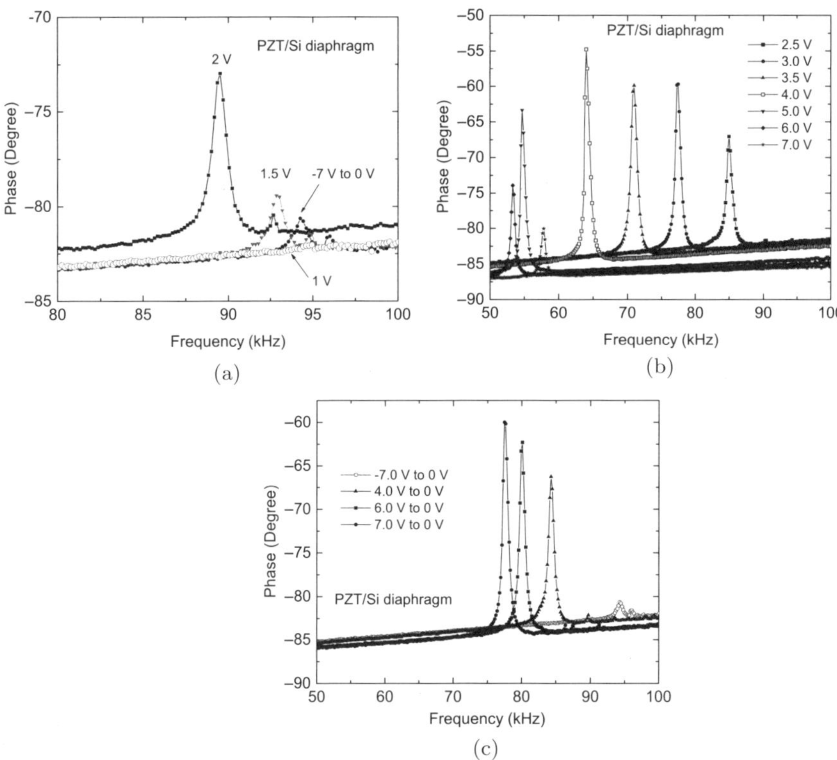

Figure 10.   The effect of positive biases on phase-frequency spectra of the PZT/Si diaphragm. (a) The negative remnant polarization is totally counterbalanced at positive coercieve voltage of 1 V, much smaller than the negative coercive voltage of −4 V. 1.5 V and 2 V bias clearly reversed the negative polarization. (b) The positive polarization increases with additional increases of the positive bias. (c) The positive remnant polarization is much larger than the negative one. Reprinted with permission from [16]. Reprinted with permission from Wiley-VCH Verlag GmbH & Co. KgaA, Weinheim. Copyright 2012.

result, the PZT film is in a single domain polarization state. The polarization direction can be easily reversed by changing the direction of the bending of the substrate. However, even after application of an opposite poling voltage, the polarization switches back to its original direction. It is therefore not the poling voltage but the substrate bending that dominates the polarization state. Lee *et al.*[14] demonstrated that the strain gradient in the thin film results in a shift in the hysteresis loop. Our observations mentioned above imply that the shift in the hysteresis loops must be strongly related to the flexoelectric effect. In our PZT/Si diaphragm, the curvature of the diaphragm partially recovers to its original state when the bias is removed due to internal stress, and flexoelectric polarization is thus generated. It is obvious that the flexoelectric polarization due to the shape of the diaphragm counterbalances the negative remnant ferroelectric polarization and enhances the positive polarizations. This is why the P–E hysteresis is asymmetric.

6. Estimation of Flexoelectric Coefficient

It should be pointed out that the origin of the self-polarization and build-in bias is complicated. In order to simplify the analysis, in the following theoretical model, we consider the flexoelectricity as the sole source of the self-polarization or the equivalent internal electric field. Since the other factors that may cause internal field have been excluded, our model estimates an effective flexoelectric coefficient that contains the contributions of all the other possible factors. Therefore, the result gives the upper limit of the intrinsic flexoelectric coefficient.

Considering only general electric polarization and flexoelectric polarization, in a non-zero field, E, the constitutive equation for direct flexoelectricity in two-dimensional diaphragm takes the form

$$P = \varepsilon_0 \chi E + \mu_{12} \frac{\partial S_{11}}{\partial x_3} \approx \varepsilon_0 \varepsilon_r E + \mu_{12} \frac{1}{R} = \frac{CV}{A} + \mu_{12} \frac{1}{R}, \quad (1)$$

where P is the polarization normal to film surface, χ the dielectric susceptibility, μ_{12} the effective flexoelectric coefficient, x_3 the position coordinate, and S_{11} the in-plane elastic strain, R the radius of curvature of the diaphragm, ε_r the relative dielectric constant, C the capacitance and A the area of the diaphragm. The approximate equality in (1) holds for films with high dielectric susceptibility where $\varepsilon_r = 1 + \chi \approx \chi$. We also assume diaphragm to have same curvature everywhere and strain gradient to be normal to the film plane and thus equal to $1/R$.

We can see from equation (1) that, to quantitatively estimate the magnitude of the flexoelectric polarization, we need to measure the radius of curvature of the diaphragm and capacitance of film at applied voltage.

6.1. *Central Deflection and Radius of the Curvature in Relation to the External Bias*

During the measurement of the impedance frequency spectra, the static surface profiles of the diaphragm were also measured at each applied bias. Correlations between the diaphragm curvature and the polarization enable us to investigate the flexoelectric effect. Figure 11 shows the measured relations between the static central deflection of the diaphragm and the bias. The solid circles show the central deflection at particular biases. The hollow circles and squares indicate the deflection after removing the bias. We can clearly see that starting from the original state, the central deflection increases as the negative bias is increased and reaches its maximum at $-4\,\mathrm{V}$. Any further increase in the bias reduces the dome height. However, when the

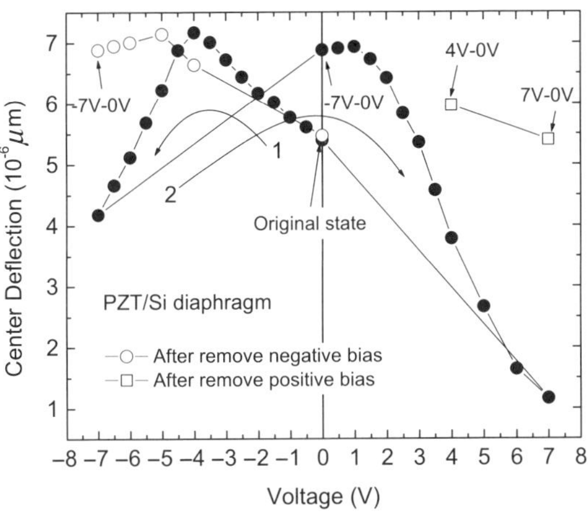

Figure 11. The bias dependence of the central deflection of the PZT/Si diaphragm. The solid circles show the central deflection at particular biases. The hollow circles and squares indicate the deflection after removing the bias. The measured data is used to calculate the radius of curvature of the diaphragm. The two maximum central deflections correspond to the coercive voltages of $-4\,\mathrm{V}$ and $+1\,\mathrm{V}$. Reprinted with permission from [16]. Reprinted with permission from Wiley-VCH Verlag GmbH & Co. KgaA, Weinheim. Copyright 2012.

$-7\,$V bias is removed, the central deflection is higher than the original deflection. While increasing the positive bias, the deflection height increases a little and reaches its maximum at $1\,$V. When the positive bias is continuously increased, the central deflection decreases. It is evident that the central deflection has a close correlation with the impedance results shown in Figs. 9 and 10. The maximum central deflection corresponds to the zero electromechanical coupling and thus to the coercive voltage.

The increased deflection height implies an increase in the flexoelectric polarization. Considering that flexoelectric polarization is opposite to ferroelectric polarization induced by the negative bias, the negative bias first needs to counterbalance the gradually increased flexoelectric polarization. Higher voltage is thus needed to fully counterbalance the positive polarization, including both the ferroelectric and flexoelectric components. Beyond the $-4\,$V coercive voltage, the reoccurrence of the negative polarization generates tensile stress within the PZT film and thus reduces the deflection height. Upon removing the $-7\,$V bias, the diaphragm rebounds to a higher dome height, such that the relatively larger flexoelectric polarization counterbalances part of the ferroelectric remnant polarization, resulting in a very small negative remnant polarization. Further application of the positive bias together with the positive flexoelectric polarization can reverse the negative remnant polarization very easily. This is why the deflection-voltage hysteresis loop shifts to the negative voltage side. Beyond a positive coercive voltage of $1\,$V, the gradually increased positive polarization flattens the diaphragm again and thus reduces the flexoelectric polarization. However, because the flexoelectric polarization is parallel to the positive ferroelectric polarization, the sum of the two components still gives a relatively large positive polarization, and therefore a large positive remnant polarization after removing $+7\,$V bias.

For a dome-shaped diaphragm with radius of r, the radius of curvature, R, and the central deflection height, h, has the relation

$$R = \frac{r^2}{2h},\tag{2}$$

Table 1.   Estimated flexoelectric polarization from the tested curvature radius of the PZT/Si diaphragm at $\mu_{12} = 2.13 \times 10^{-4}\,\mathrm{C\,m^{-1}}$.

Voltage [V]	Initial 0	−4	−7	−7 to 0	1	7	7 to 0
$h\,[\mu\mathrm{m}]$	5.473	7.170	4.184	6.886	6.933	1.168	5.403
R [m]	0.0146	0.0112	0.0191	0.0116	0.0115	0.0685	0.0148
$\mathrm{CV}/A\,[\mu\mathrm{C\,cm^{-2}}]$	0	−4.96	−6.27	0	1.21	5.29	0
$P^{flexo}\,[\mu\mathrm{C\,cm^{-2}}]$	1.46	1.90	1.11	1.83	1.85	0.311	1.44

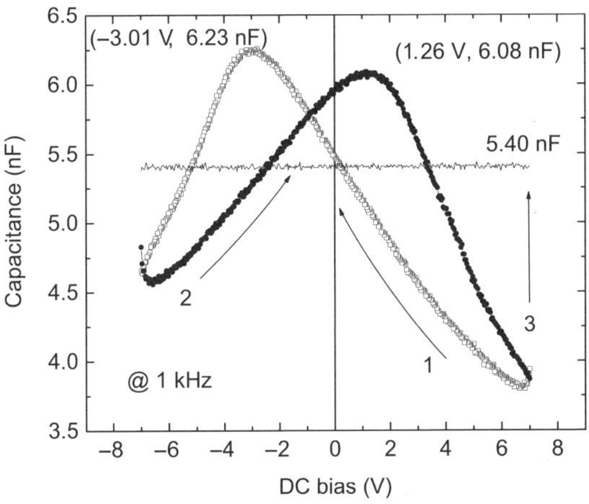

Figure 12.   C–V curves of the PZT/Si diaphragm. The maximum capacitances of 6.23 nF and 6.08 nF occurs at coercive voltages of −3.01 V and 1.26 V respectively, are used for calculating the flexoelectric coefficient of the PZT film. Reprinted with permission from [16]. Reprinted with permission from Wiley-VCH Verlag GmbH & Co. KgaA, Weinheim. Copyright 2012.

assuming that $h \ll r$. The calculated radii R for the voltages of interest are summarized in Table 1.

6.2. *Asymmetric C–V Curves*

After testing the impedance and surface profile of the diaphragm, the C–V curve was tested at 1 kHz. Figure 12 shows that when the bias varies from +7 V to −7 V, the capacitance first increases to a maximum of 6.23 nF at −3.01 V and then decreases. When the voltage is increased from −7 V to 7 V, the maximum capacitance

6.08 nF occurs at 1.26 V. When the $+7$ V bias is removed, the capacitance of the diaphragm goes back to around 5.40 nF. The flat C–V curve was measured at 0 V bias for a period of time by switching off the bias while still setting to sweep it from $+7$ V to -7 V. No bias was really applied to the diaphragm during test. It is clear that the C–V curve has similar asymmetric features as the deflection-bias loop. It also shifts to negative voltage side. The measured capacitance for the voltages of interest are also summarized in Table 1.

6.3. *Negative Shift of P–E Hysteresis at Low Maximum Voltage*

As mentioned above, the normal dynamic polarization measurement methods alone are not capable of testing the static polarization due to lack of a reference point. Fortunately, the electromechanical coupling of a diaphragm obtained from impedance spectrum provides us with a right reference. We can tell from Fig. 9(b) that after removing the -5 V bias the polarization in the film is absolute zero. Starting from this point, without applying a preset voltage cycle we can obtain an absolute P–E hysteresis loop start from absolute zero polarization, which indicates the real polarization at every test points. The result is shown in Fig. 13 as Curve 1.

It is evident that within the small testing voltage range the P–E hysteresis is absolutely not symmetric. The positive Pr $(5.10\,\mu\mathrm{C\,cm}^{-2})$ is larger than negative Pr $(-4.14\,\mu\mathrm{C\,cm}^{-2})$ and the negative Vc is -2.70 V. After that the test was repeated with a preset voltage circle to ensure the first leg of the loop starts from negative polarization state. The results were recorded as a non-centered Curve 2. Since we did not change the maximum voltage $(\pm 7\,\mathrm{V})$, the P_r and $P_{\max}$ of the P–E hysteresis would not change. Shifting the curve by aligning the $+P_r$ with that of Curve 1 we get Curve 3 from which we can obtain $+$Vc equals 1.49 V, and the relaxed negative remnant polarization is $-2.90\,\mu\mathrm{C\,cm}^{-2}$. Curve 3 gives the absolute polarization for every test points. We can see from Curve 3 that both coercive field and remnant polarization of the film are not symmetric. In comparison with the P–V loop at high voltage (Fig. 4),

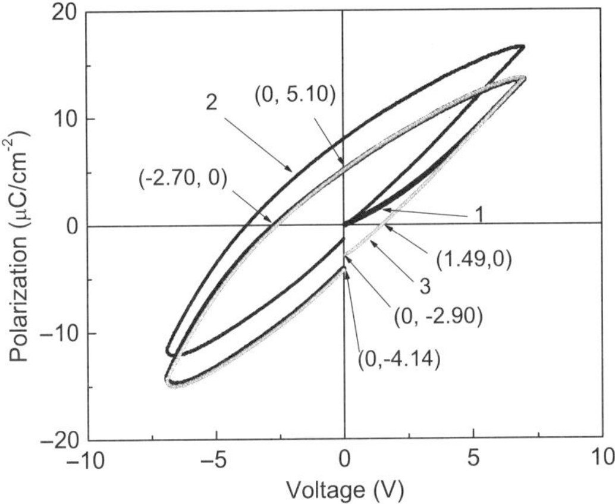

Figure 13. P–E hysteresis of PZT/Si diaphragm measured by normal virtual ground method. The film was depolarized at $-5\,$V from positive remnant polarization state. The real zero polarization state of Curve 1 was verified using impedance spectrum.

the "banana shape" of the $P\!-\!E$ hysteresis at $\pm7\,$V is due to lack of saturation.

6.4. *Calculation of Flexoelectric Coefficient*

Since PZT is also a ferroelectric material, in a small field range within the coercive field, the remnant ferroelectric polarization is initially pronounced and should also be taken into account. Therefore,

$$P = P_r^{\text{ferro}} + P^{\text{piezo}} + \varepsilon_0\chi E + P^{\text{flexo}}$$

$$\approx P_r^{\text{ferro}} + P^{\text{piezo}} + \frac{CV}{A} + \mu_{12}\frac{1}{R}, \tag{3}$$

where P_r^{ferro} is the remnant ferroelectric polarization and P^{piezo} is the contribution of conventional piezoelectricity to the polarization. If $P \neq 0$, the strain in the film will induce P^{piezo} through piezoelectric stress coefficient e_{31}. The remnant P_r^{ferro} can be considered as a constant which is determined by the highest external field, E_{max}, while testing the $P\!-\!E$ loop. The first three terms on the right side of equation (3) contribute to the hysteresis and nonlinearity of the $P\!-\!E$ loop, while the last term contributes to the asymmetric feature of the P–E loop. The above equation holds only in a small voltage range. The effect of P_r^{ferro} may vanish beyond the coercive voltage.

With the measured radius of curvature of the diaphragm and capacitance of the PZT film at coercive voltages, we can calculate the flexoelectric coefficient of PZT film by analyzing the polarization components. From Figs. 9(b), 10(a) and 11, and the data summarized in Table 1, we can see that at $V = -4\,\text{V}$ and $V = 1\,\text{V}$, the polarization, P, is 0. Thus, the contribution from piezoelectricity at these two coercive voltages also becomes zero, i.e. $P^{\text{piezo}} = 0$. We can therefore obtain

$$P_r^{\text{ferro}} - \frac{4C_{-4}}{A} + \mu_{12}\frac{1}{R_{-4}} = 0, \tag{4}$$

$$-P_r^{\text{ferro}} + \frac{C_1}{A} + \mu_{12}\frac{1}{R_1} = 0. \tag{5}$$

The capacitances at the two coercive voltages ($C_{-4} = 6.23$ nF and $C_1 = 6.08$ nF) are known from the C–V curve shown in Fig. 12. The radius of curvature is already obtained from the dome-shaped diaphragm, as listed in Table 1. The diameter of the diaphragm is $0.8\,\text{mm}$. Solving the above equations, we obtain

$$\mu_{12} = 2.13 \times 10^{-4}\,\text{Cm}^{-1}, \tag{6}$$

$$P_r^{\text{ferro}} = 3.06 \times 10^{-2}\,\text{Cm}^{-2} = 3.06\,\mu\text{C}\,\text{cm}^{-2}. \tag{7}$$

It should be pointed out that the coercive voltage obtained from the C–V curve (Fig. 12) is different from that obtained from the impedance measurement (Figs. 9(b), 10(a), and 11). The possible reason is that the C–V test is completed within seconds thus has relatively high frequency of bias loop, while the impedance spectrum is obtained at a certain DC bias. It takes minutes to obtain the impedance spectrum at each DC bias. Therefore, the difference in coercive voltage between those obtained from impedance measurment and C–V test can be attributed to the relaxation of strain and polarization in the diaphragm during the impedance measurement. Nevertheless, using the capacitance data at coercive voltage (the two peak values in Fig. 12) instead of the data at $-4\,\text{V}$

and $+1\,V$ in $C\!-\!V$ curve is reasonable and should be reasonably accurate.

Counting in the contribution of flexoelectric polarization at positive remnant state $(1.44\,\mu C\,cm^{-2})$ and negative polarization state $(1.83\,\mu C\,cm^{-2})$ in Table 1, the estimated nominal ferroelectric remnant polarization $3.06\,\mu C\,cm^{-2}$ in equation (7) gives the $+Pr$ of $4.50\,\mu C\,cm^{-2}$ and $-Pr$ of $2.23\,\mu C\,cm^{-2}$, respectively. These results agree well with the measured remnant polarization (see Fig. 13) in the order of magnitude.

It is worth mentioning that the contribution of flexoelectric polarization in equation (3) shifts the $P\!-\!E$ hysteresis loop along the P-axis. P is no longer symmetric about the voltage axis. Because a reference polarization state is lacking, it is impossible to accurately locate the $P\!-\!E$ hysteresis loop along the P-axis and further to determine the static polarization value by using only conventional dynamic polarization measurement methods. The conventional Sawyer–Tower technique and the virtual ground method can only measure the shape of a $P\!-\!E$ loop but these methods are not capable of indicating the exact location of the loop if the loop is not symmetric. Normally, the given values of the remnant polarization and the coercive field in the asymmetric loop are not accurate. Unfortunately, this point is neglected in most of the published work when dealing with asymmetric $P\!-\!E$ loops. On the contrary, the electromechanical coupling method accurately determines the coercive field, which enables us to calculate the contribution of the flexoelectric polarization accurately.

The calculated value of the nominal remnant polarization is smaller than the normal value for PZT films $(21.3\,\mu C/cm^2$ in Fig. 4). However, this result is reasonable considering that the tests were conducted in a small field range $(\pm 7\,V)$ in which the polarization was not fully saturated.

It can be seen from Table 1 that the initial self-polarization induced by the strain gradient at $0\,V$ is $1.46\,\mu C\,cm^{-2}$, and the flexoelectric polarization at $-4\,V$ is $1.90\,\mu C\,cm^{-2}$. These giant flexoelectric polarization values confirm our prediction that the flexoelectric polarization in thin film systems can be dramatically large; indeed, they are on the same order of magnitude as the normal remnant ferroelectric polarization of PZT film.

To crosscheck the result under equations (6) and (7), in addition to the analysis based on polarization, we also estimate the flexo-electric coefficient from the internal electric field induced by the flexoelecric effect and the observed coercive electric field. A pure ferroelectric P–E loop is symmetric. We can thus define a coercive electric field for ferroelectricity, E_c^{ferro}, which is a constant for a given maximum external electric field. Flexoelectric polarization generates an equivalent internal electric field, E^{flexo}, which shifts the P–E loop along the E-axis. The real coercive field, E_c, in the resulting P–E loop will take the form

$$E_c = E_c^{\text{ferro}} + E^{\text{flexo}}, \tag{8}$$

$$E^{\text{flexo}} = \frac{\mu_{12}}{\varepsilon_0 \chi} \cdot \frac{1}{R} \approx \frac{\mu_{12} A}{Ct} \cdot \frac{1}{R}, \tag{9}$$

where A is the area of the diaphragm/capacitor and $t = 0.8 \times 10^{-6}$ m is the thickness of the PZT film. Again, using the tested radius of curvature R of the diaphragm and capacitance C in Figs. 11 and 12 and Table 1 at coercive voltages of $-4\,$V and $1\,$V, we obtain

$$\frac{-4}{0.8 \times 10^{-6}} = -E_c^{\text{ferro}} - \frac{\mu_{12} A}{0.8 \times 10^{-6} C_{-4}} \cdot \frac{1}{R_{-4}}, \tag{10}$$

$$\frac{1}{0.8 \times 10^{-6}} = E_c^{\text{ferro}} - \frac{\mu_{12} A}{0.8 \times 10^{-6} C_1} \cdot \frac{1}{R_1}. \tag{11}$$

Solving equations (10) and (11), we obtain

$$\mu_{12} = 2.08 \times 10^{-4}\,\text{Cm}^{-1}, \tag{12}$$

$$E_c^{\text{ferro}} = 3.12\,\text{MVm}^{-1}. \tag{13}$$

The equivalent internal electric field induced by flexoelectricity at $-4\,$V is

$$E_{-4}^{\text{flexo}} = 1.88\,\text{MVm}^{-1}. \tag{14}$$

By comparing equations (6) and (12), we can see that the two methods give the same flexoelectric coefficient value, suggesting that the proposed model explains the observed reversed remnant polarization phenomenon in piezoelectric diaphragms with flexoelectric effects. The flexoelectric effect shifts the P–E hysteresis loop not only along the E-axis but also along the P-axis. This result agrees well with the measured P–E hysteresis loop as shown in Fig. 13. The phenomenon of self-polarization[20–23] has been observed in various ferroelectric thin films for decades. Our work clearly demonstrated that strain gradient can induce self-polarization and thus is one of the sources of the built-in bias or imprint of P–E hysteresis loops. Very recently, Jeon *et al.*[24] reported that in epitaxial $BiFeO_3$ (BFO) thin films, flexoelectric effect also plays an important role in determining the orientation of the self-polarization.

The obtained flexoelectric coefficient μ_{12} in PZT thin films is of the order of $10^{-4}\,\mathrm{C\,m^{-1}}$. This value exceeds theoretical expectations by several orders of magnitude.[8] However, it is on the same order of magnitude as that measured in $Ba_{0.67}Sr_{0.33}TiO_3$ ceramics,[4] which has the highest flexoelectric coefficient, μ_{12}, among the reported high-susceptibility ferroelectric ceramics. It is interesting to note that, among the reported results about flexoelectric coefficient in perovskites, only the experimental result of $SrTiO_3$ single crystal agrees with the intrinsic theoretical prediction[9–11] on the order of magnitude. The difference between $SrTiO_3$ and other perovskites is that it is a centrosymmetric paraelectric material at moderate temperature. Neither of the phases of $SrTiO_3$ is polar or ferroelectric thus there is no polar nanodomain in the crystal. As such, we believe that the measured giant bending-induced polarization in most ferroelectrics may not solely result from the intrinsic (lattice distortion induced) flexoelectricity. It may be due to bending induced reorientation of polar nanodomains in polarizable ferroelectrics.[25] On the one hand, strain gradient generates flexoelectric polarization inside polar nanoregions, it is also capable of switching the existing nanodomains in ferroelectrics. On the other hand, the flexoelectric polarization in polar nanoregions persists once it is generated by strain gradient, which in turn enhance the bending-induced polarization in ferroelectrics and relaxors. The strain gradient induced reorientation of nanodomains may generate much larger bending-induced polarization than the intrinsic flexoelectric polarization induced by lattice distortion.

7. Conclusions

The polarization induced by the strain gradient, i.e. the effective flexoelectric effect, has been directly observed in bent PZT diaphragms. Observation of the electromechanical coupling of a ferroelectric diaphragm has revealed a new way to test the static electrical polarization state. This method makes it possible to determine accurately the coercive electric field at which the polarization is truly zero. The strain gradient induced in a dome-shaped micromachined PZT diaphragm is measured to be of the order of $10^2 \, \mathrm{m}^{-1}$, much larger than that obtained in the bulk material. From the measured coercive electric field and the strain gradient, the effective flexoelectric coefficient of PZT thin films is estimated to be $2.0 \times 10^{-4} \, \mathrm{C\,m}^{-1}$, which is much larger than the theoretical upper limit estimated by intrinsic flexoelectricity. A possible explanation for the giant flexoelectric polarization of the order of $1 \, \mu\mathrm{C\,cm}^{-2}$ which is comparable to the remnant ferroelectric polarization of PZT thin films may be the large strain gradient together with the reorientation of polar nanodomains.

References

1. J. M. Bursian and O. I. Zaikovskii. *Soviet Physics-Solid State* **10**, 1121–1124 (1968).
2. W. H. Ma and L. E. Cross. *Appl. Phys. Lett.* **78**, 2920–2921 (2001).
3. W. H. Ma and L. E. Cross. *Appl. Phys. Lett.* **79**, 4420–4422 (2001).
4. W. H. Ma and L. E. Cross. *Appl. Phys. Lett.* **81**, 3440–3442 (2002).
5. W. H. Ma and L. E. Cross. *Appl. Phys. Lett.* **82**, 3293–3295 (2003).
6. W. H. Ma and L. E. Cross. *Appl. Phys. Lett.* **86**, 072905 (2005).
7. W. H. Ma and L. E. Cross. *Appl. Phys. Lett.* **88**, 232902 (2006).
8. L. E. Cross. *J. Mater. Sci.* **41**, 53–63 (2006).

9. P. Zubko, G. Catalan, A. Buckley, P. R. L. Welche, and J. F. Scott. *Phys. Rev. Lett.* **99**, 167601 (2007).

10. R. Maranganti and P. Sharma. *Phys. Rev. B.* **80**, 054109 (2009).

11. J. W. Hong, G. Catalan, J. F. Scott, and E. Artacho. *J. Phys.-Condes. Matter* **22**, 112201 (2010).

12. H. Lu, C. W. Bark, D. E. de los Ojos, J. Alcala, C. B. Eom, G. Catalan, and A. Gruverman. *Science* **336**, 59–61 (2012).

13. G. Catalan, A. Lubk, A. H. G. Vlooswijk, E. Snoeck, C. Magen, A. Janssens, G. Rispens, G. Rijnders, D. H. A. Blank, and B. Noheda. *Nature Materials* **10**, 963–967 (2011).

14. D. Lee, A. Yoon, S. Y. Jang, J. G. Yoon, J. S. Chung, M. Kim, J. F. Scott, and T. W. Noh. *Phys. Rev. Lett.* **107**, 057602 (2011).

15. A. Gruverman, B. J. Rodriguez, A. I. Kingon, R. J. Nemanich, A. K. Tagantsev, J. S. Cross, and M. Tsukada. *Appl. Phys. Lett.* **83**, 728–730 (2003).

16. Z. H. Wang, X. X. Zhang, X. B. Wang, W. S. Yue, J. Q. Li, J. Miao, and W. Zhu. *Adv. Funct. Mater.* **23**, 124–132 (2013).

17. G. Catalan, B. Noheda, J. McAneney, L. J. Sinnamon, and J. M. Gregg. *Phys. Rev. B.* **72**, 020102 (2005).

18. Y. Qi, J. Kim, T. D. Nguyen, B. Lisko, P. K. Purohit, and M. C. McAlpine. *Nano Lett.* **11**, 1331–1336 (2011).

19. V. R. Singh, J. M. Miao, Z. H. Wang, G. Hegde, and A. Asundi. *Opt. Commun.* **280**, 285 (2007).

20. E. Sviridov, I. Sem, V. Alyoshin, S. Biryukov, and V. Dudkevich. *Res. Soc. Symp. Proc.* **361**, 141–146 (1995).

21. A. L. Kholkin, K. G. Brooks, D. V. Taylor, S. Hiboux, and N. Setter. *Integr. Ferroelectr.* **22**, 1045–1053 (1998).

22. S. Sun, Y. M. Wang, P. A. Fuierer, and B. A. Tuttle. *Integr. Ferroelectr.* **23**, 25–43 (1999).

23. V. P. Afanasjev, A. A. Petrov, I. P. Pronin, E. A. Tarakanov, E. J. Kaptelov, and J. Graul. *J. Phys.-Condes. Matter* **13**, 8755–8763 (2001).

24. B. C. Jeon, D. Lee, M. H. Lee, S. M. Yang, S. C. Chae, T. K. Song, S. D. Bu, J. Chung, J. Yoon, and T. W. Noh. *Adv. Mater.* **25**, 5643–5649 (2013).

25. M. Wallace, R. L. Johnson-Wilke, G. Esteves, C. M. Fancher, R. H. T. Wilke, J. L. Jones, and S. Trolier-McKinstry. *J. Appl. Phys.* **117**, 054103 (2015).

Chapter 11

Quasi-Amorphous Materials

David Ehre, Ellen Wachtel, and Igor Lubomirsky

Department of Materials & Interfaces,
Weizmann Institute of Science, Israel

Quasi-amorphous films are the only known inorganic, non-crystalline, polar materials. The conditions under which they are formed and the origin of their polarity set these materials apart from other classes of inorganic materials. The unique feature of the quasi-amorphous phase is that its polarity is the result of flexoelectric-induced orientational ordering of local bonding units without any detectable spatial periodicity. However, unlike classical flexoelectricity — the reversible coupling of polarity and strain gradients — the strain gradient imposed on quasi-amorphous films leads to permanent polarization, i.e. the polarization is retained even after the strain gradient has been eliminated. This mechanism permits compounds that do not have polar crystalline polymorphs, such as $SrTiO_3$ and $BaZrO_3$, to form polar, non-crystalline solids. In this chapter, we describe the essential features of quasi-amorphous materials including preparation, structure, and chemical composition.

1. Introduction

Pyro- and piezo-electricity in inorganic materials require a non-centrosymmetric arrangement of the constituent ions.[1] Therefore, only materials with crystal structures that belong to the 20 non-centrosymmetric crystallographic classes[2] or centrosymmetric crystals with polar layers at the crystal surface[3-5] are expected to exhibit such activity. However, restricting pyro- or piezo-electricity to crystalline, inorganic solids has had to be reconsidered following reports that non-crystalline, pyro- and piezo-electric thin films of $BaTiO_3$, $BaZrO_3$, and $SrTiO_3$ can be prepared on a variety of substrates.[6–13] The term *quasi-amorphous* has been applied to compounds that form polar non-crystalline phases; included are even those compounds that do not have polar crystalline polymorphs. X-ray absorption fine structure spectroscopy (XAFS)[8,11] and X-ray photoelectron spectroscopy (XPS)[12,14] revealed that the structure of the non-crystalline phases of quasi-amorphous materials is significantly different from that of all known amorphous, inorganic materials. The XAFS and XPS data, together with other experimental observations such as the unusually large (>10%) volume expansion that occurs prior to nucleation and crystallization,[15] led to the development of a phenomenological theory that describes the formation and microscopic origin of polarity in quasi-amorphous films. Further investigation showed that the source of the permanent electric polarization is a *flexoelectric* effect arising from the strain gradient that develops in the film during its passage through a temperature gradient as part of the preparation procedure.[13]

This chapter summarizes what is currently known about quasi-amorphous materials. While avoiding excessive detail, the most important features of the preparation of quasi-amorphous materials, their thermodynamics, and the microscopic origin of their polarity will be described.

2. Preparation of Quasi-Amorphous Thin Films

The preparation of quasi-amorphous films requires two steps. First, a 50–200 nm thick amorphous, non-polar film of $BaTiO_3$, $SrTiO_3$, or $BaZrO_3$ is deposited by Radio Frequency (RF) sputtering directly onto a substrate (Si, SiO_2, or sapphire) or onto a substrate covered with an electrically conductive layer ($SrRuO_3$, Cr, or W).[6,13] The film is then pulled at a rate of 1–5 mm/hr through a narrow (1–3 mm) hot zone with a peak temperature of 550–650°C (Fig. 1).[16] The exact combination of pulling rate and peak temperature that is required to convert the amorphous film into a non-crystalline, polar film depends on the compound chemistry ($BaTiO_3$, $SrTiO_3$, or $BaZrO_3$), the substrate and the film thickness. Absence of crystallization can be verified by transmission electron microscopy (TEM), X-ray diffraction (XRD), or scanning electron microscopy (SEM). Of the three methods, XRD and SEM are the more reliable since the sample area examined in these methods is much larger than in the case of TEM techniques. It should also be noted that if the as-deposited, amorphous film is subjected to homogeneous isotropic heating, it always crystallizes. In addition, if the thickness of the $BaTiO_3$ or $SrTiO_3$ films exceeds 250 nm or if they are deposited on a substrate that readily produces nucleation, e.g. crystalline MgO, or if the films have been removed from the substrate (flakes or self-supported films), crystallization occurs irrespective of the mode of thermal treatment.

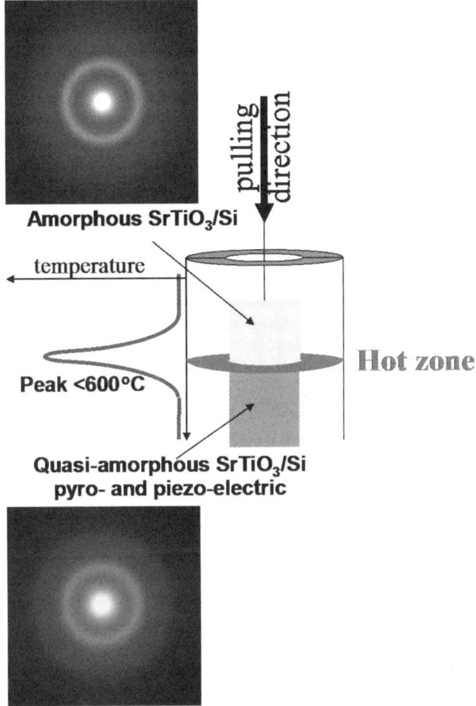

Figure 1. Preparation of a quasi-amorphous film and electron diffraction patterns of the amorphous and quasi-amorphous films. Reproduced from Ref. [16] with permission.

3. Physical Properties of Quasi-Amorphous Thin Films

Following heating in a thermal gradient, as described above, the non-crystalline films are found to exhibit both pyroelectric and piezoelectric effects i.e. they are "quasi-amorphous".[9] The magnitude of the pyroelectric coefficient may reach $\alpha = 10^{-9} \mathrm{C\,cm^{-2}K^{-1}}$ (for $SrTiO_3$), which is comparable with commonly used crystalline, pyroelectric compounds such as bulk $BaTiO_3$ or $LiTaO_3$.[17,18] The magnitude of the piezoelectric coefficient may reach $d_{33} = 0.1\,\text{Å/V}$ (for $SrTiO_3$), which is comparable with commonly used crystalline piezoelectric compounds such as bulk $LiTaO_3$.[19,20] Generally, α, $d_{33}(SrTiO_3) > \alpha$, $d_{33}(BaTiO_3) \gg \alpha$, $d_{33}(BaZrO_3)$.

Transformation of films from the as-deposited, amorphous to the quasi-amorphous state is accompanied by changes in the elastic constant C_{44}, the dielectric constant, refractive index, and in-plane stress. The elastic constants (C_{44}) of quasi-amorphous films of $SrTiO_3$ and $BaTiO_3$ have been measured using nanoindentation and acoustic resonance techniques.[21] It was found that C_{44} of quasi-amorphous (*qa*) films is significantly smaller than that of the corresponding crystalline (*cry*) phases: $C_{44(SrTiO_3)}^{qa} = 24\,\mathrm{GPa}$, $C_{44(SrTiO_3)}^{cry} = 68\,\mathrm{GPa}$; $C_{44(BaTiO_3)}^{qa} = 41\,\mathrm{GPa}$, $C_{44(BaTiO_3)}^{cry} = 117\,\mathrm{GPa}$. The dielectric constant increases by 15–40% (to $\sim$22 and $\sim$33 for quasi-amorphous $SrTiO_3$ and $BaTiO_3$, respectively[6,9]), but remains much smaller than the dielectric constant of the corresponding crystalline phases. The refractive indices of as-deposited, amorphous films (n_{am}) of $SrTiO_3$, $BaTiO_3$, and $BaZrO_3$ are isotropic, indicating that the films are stress free. Using the Clausius–Mossotti equation, one can estimate that the density of the amorphous phase of all three materials is $\sim$82–86% of the density of the corresponding crystalline phases.[10] In contrast to the dielectric constant, following transformation to the quasi-amorphous state, the refractive index, n_{qa}, decreases and is no longer isotropic (Fig. 2[10]): $n_{qa\|\perp} > n_{qa\|\,\|} > n_{qa\perp}$, where $n_{qa\perp}$ is the out-of-plane refractive index; $n_{qa\|\perp}$ and $n_{qa\|\,\|}$ are the in-plane refractive indices perpendicular and parallel, respectively, to the pulling direction. For $BaTiO_3$,

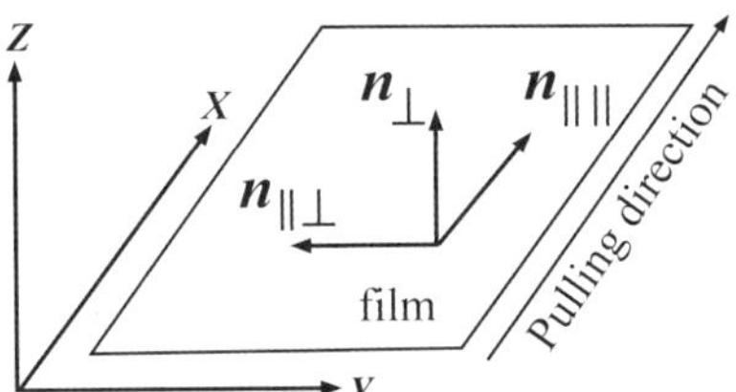

Figure 2.   Scheme of the notation used for the refractive indices and axes. Reproduced from Ref. [10] with permission.

$n_{am} = 1.91$; $n_{aq\perp} = 1.76(-12\%)$; $n_{qa\|\,\|} = 1.82(-7.2\%)$; and $n_{qa\|\perp} = 1.88(-2.3\%)$.[10] These changes in refractive index clearly indicate that the transformation into the quasi-amorphous phase is accompanied by a strongly anisotropic expansion. They are also consistent with the fact that the quasi-amorphous films are observed to develop a highly anisotropic, in-plane compressive stress of 1–4 GPa with $\sigma_{yy} < \sigma_{xx}$, where x is the direction in which the film is pulled through the temperature gradient (Fig. 2).[10,13] The important role of this anisotropic compressive stress in producing the stability and polarity characteristic of quasi-amorphous films is discussed in the next section.

4. The Role of Stress and Strain in the Formation of Quasi-amorphous Films

4.1. *Pre-nucleation Expansion*

The development of in-plane, compressive stress, and the reduction of the refractive index of as-deposited titanate films, following thermal treatment, both indicate that these films expand upon heating. This process was studied in detail for $BaTiO_3$ using self-supported films,[15] i.e. films for which the substrate had been locally removed, leaving a $200\,\mu m \times 200\,\mu m$ section of the film tethered at the edges (Fig. 3[15,16]). Upon being uniformly heated above $200°C$, the self-supported films buckled, a clear indication of expansion. The maximum expansion (10–12%), calculated from both the extent of buckling and from the change in film thickness inferred from the internal interference pattern measured by transmission spectroscopy, is achieved upon

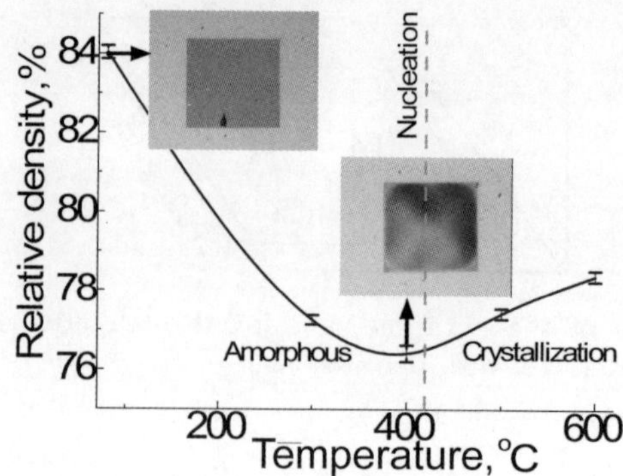

Figure 3.   Pre-nucleation expansion of a self-supported film of $BaTiO_3$. The insets show optical micrographs of an amorphous self-supported $BaTiO_3$ film (left) and the same film after heating to 400°C (right). The widths of the films are 200 μm. After heating, the film buckles as a result of pre-nucleation expansion. Reproduced from Ref. [15, 16] with permission.

heating to 400°C (Fig. 3). Heating above this temperature produces nucleation and crystallization, at which point the films begin to contract.

Transformation from a high density amorphous (HDA) state to a low density amorphous (LDA) state prior to nucleation of the crystalline phase, resembles a phenomenon called *polyamorphism*.[22,23] Polyamorphism was first observed in amorphous ice deposited at T < 30 K. Upon being heated to 120 K, the ice undergoes expansion of ~24%[24] before crystallization begins (~130 K).[25] The microscopic origin of pre-nucleation expansion in $BaTiO_3$, $SrTiO_3$, and $BaZrO_3$, and its importance for the development of a permanent dipole moment, are discussed below.

4.2. *Suppression of Crystallization in Amorphous Titanate Films*

The fact that nucleation can only take place in the LDA phase is a potential explanation for the suppression of crystallization in substrate-supported, amorphous, titanate films. In the absence of external mechanical constraints, as is the case for self-supported amorphous films[15] or fine amorphous flakes,[26] the HDA phase undergoes volume expansion to the LDA state upon heating to

400°C. Above this temperature, nucleation occurs, followed by crystallization. Substrate-supported films are not free to expand (clamped): the substrate prevents in-plane expansion and permits only out-of-plane expansion, which is limited due to the finite Poisson ratio of the expanding phase. As a result, high in-plane compressive stress accumulates. Amorphous solids do not support dislocation movement or grain boundary sliding, which are the major mechanisms of stress relaxation in polycrystalline solids.[27] If the yield stress of the amorphous phase is not reached, the compressive stress will suppress crystal nucleation; the stressed amorphous phase may persist indefinitely, i.e. it is in a "jammed" state between the HDA and LDA phases.

The maximum volume difference between the as-deposited HDA and the pre-nucleation LDA can be estimated. Nucleation requires expansion and compression of the surrounding amorphous matrix, which is associated with mechanical work, A_E, given by[7,16]:

$$A_E = \frac{B \times \alpha^2}{2 \times \rho_{HDA}} \times \frac{\delta}{(\delta + 1)^2}, \tag{1}$$

where B is the bulk modulus of the amorphous phase; α is the relative difference in density between the HDA and the LDA phases just prior to nucleation; ρ_{HDA} is the molar density of the HDA phase; and δ is related to the Poisson ratio ν as $\delta = 2(1 - 2\nu)/(1 + \nu)$. The energy necessary to perform this work comes at the expense of the crystallization enthalpy $\Delta H^{a \to c}$. For amorphous $BaTiO_3$ prepared by sputtering, the crystallization enthalpy is $\Delta H^{a \to c} = 7$–$9 \, \text{kJ} \cdot \text{mol}^{-1}$,[26] which is very low in comparison with other amorphous oxides. When $A_E = \Delta H^{a \to c}$, α reaches its maximum value and crystallization begins. This critical value (α_{cr}) can be calculated for $BaTiO_3$ (B≈115 GPa; $\nu \approx 0.325$; $\rho_{HDA} \approx 21427 \, \text{mol} \times \text{m}^{-3}$)[7,16]:

$$\alpha_{cr(BTO)} = (1 + \delta) \times \sqrt{\frac{2 \times \rho_{HDA}}{B \times \delta} \times \Delta H^{a \to c}} \approx 11.4\%, \tag{2}$$

which agrees with experiment.[15] One should note that the suppression of crystallization by compressive mechanical stress is reminiscent of the suppression of crystallization in some glass ceramics and green (raw) ceramic phases due to tensile stress.[28] In those systems, crystal nucleation is accompanied by a large reduction in volume (10–20%) and the crystal nucleus cannot grow due to the stressed regions surrounding it. In the absence of a means of stress relaxation, crystal growth may be prevented indefinitely.

In light of the above, one might conclude that amorphous, substrate-supported films will not crystallize, even with homogeneous, isotropic heat treatment. However, as noted in Section 2, this is not true, and for several reasons. Any real, as-deposited film has a number of voids. As the film is heated, two competing processes occur: (i) volume expansion that tends to eliminate the free volume of the voids and (ii) nucleation in the regions where sufficiently low density is reached. If the free volume associated with voids is eliminated due to volume expansion, then no further expansion is possible in the absence of stress relaxation; nucleation is suppressed. However, if nucleation does begin, then it will proceed, because crystallites provide a very efficient stress relaxation mechanism via grain boundary sliding and dislocation movement. There are in fact three cases in which nucleation can be expected to occur: (a) If the fraction of voids is large, then volume expansion cannot eliminate all the voids and nucleation will proceed. (b) If the film has a large surface roughness, then nucleation is unobstructed at the film surface, and (c) If the nucleation is heterogeneous, as in the case of titanate films deposited on a layer of MgO,[7] then this layer provides a nucleation (seeding) surface. In fact, supported titanate films with a large number of voids or a rough surface and films deposited on substrates that initiate heterogeneous nucleation are observed to crystallize, irrespective of the mode of thermal treatment. One should add that substrate clamping as a means to suppress crystallization imposes a limit on the maximum thickness, t, of the titanate films because they will remain stable only as long as the film-substrate adhesion is strong enough to prevent film delamination, i.e. $t < Y_B^{HDA/qa} \times \gamma/\sigma^2$,[29] where σ is the in-plane compressive stress (typically 1–4 GPa), γ is the energy of the interface, and $Y_B^{HDA/qa}$ is the biaxial modulus of the HDA or the quasi-amorphous phase. For both phases, $Y_B^{HDA/qa}$ equals 100–200 GPa.[21] For a typical value of the interface energy $\gamma = 0.1\text{–}1\,\mathrm{J}\cdot\mathrm{m}^{-2}$, $t \sim 200\,\mathrm{nm}$, which is close to the observed maximum value of $\sim\!250\,\mathrm{nm}$.

4.3. *The Essential Role of the Temperature Gradient During Heat Treatment: Flexoelectricity*

Slow heating is a necessary but not a sufficient condition for the amorphous to quasi-amorphous transformation of substrate-supported films. Although homogeneous heating can in some restricted cases suppress nucleation, it certainly cannot transform isotropic films into anisotropic polar films. In fact, as described in Section 2, the amorphous to quasi-amorphous transformation occurs only when samples are pulled through a furnace hot zone at the controlled rate of 1–5 mm/hr.[6,7,9] Therefore, the temperature gradient plays an essential role in the transformation of the amorphous into the quasi-amorphous phase. The resulting in-plane strain gradient produces: (i) suppression of crystallization, even in the presence of voids in the as-deposited films; and (ii) formation of a permanent electric polarization due to flexoelectric coupling.

4.3.1. *Suppressing crystallization in the presence of as-deposited voids*

As noted in Section 4.2, all as-deposited voids must be eliminated in order to suppress crystallization. Expansion of a substrate-clamped film creates "planar stress" with axial symmetry. Such an isotropic stress field cannot lead to the elimination of voids,[30–33] and even an anisotropic stress field that does not have a shear component would not cause the elimination of voids unless the stress is very high ($\sigma > Y/3$,[30,34] where Y is the elastic modulus). Void closure is associated with a change of shape, which occurs only due to shear stress.[30–33] That shear stress develops as the sample moves through the temperature gradient can be demonstrated as follows. The region that has been converted into the quasi-amorphous phase has density $\rho_{ga}(\rho_{HDA} > \rho_{qa} > \rho_{LDA})$ whereas the density of the amorphous part of the wafer remains ρ_{HDA}. These regions are separated by a distance equal to the width of the hot zone. Therefore, the gradient of the in-plane stress in the direction of pulling (see Fig. 2) is non-zero ($\partial\sigma_{xx}/\partial x \neq 0$). Since in clamped thin films σ_{xx}, $\sigma_{yy} \gg \sigma_{zz}$ and only the out-of-plane displacement, U_z, is not equal to zero ($U_x = U_y = 0$; $U_z \neq 0$), and since all the in-plane stress gradients in the direction perpendicular to the direction of pulling are equal to zero ($\partial\sigma_{ik}/\partial y = 0$; $i, k = x, y, z$), the condition for mechanical equilibrium[32] is[29]:

$$\frac{\partial\sigma_{xx}}{\partial x} + \frac{\partial\sigma_{xz}}{\partial z} = 0, \tag{3}$$

or equivalently:

$$-\frac{\partial\sigma_{xx}}{\partial x} = \frac{\partial\sigma_{xz}}{\partial z} \neq 0. \tag{4}$$

Thus, pulling through the temperature gradient breaks the axial symmetry of the stress field and creates stress with a shear component, which leads to the elimination of the voids. The larger the temperature gradient, the larger the shear stress and the more probable the formation of the quasi-amorphous phase. The magnitude of the shear stress can be estimated as[7]:

$$\sigma_{xz} \approx -t\frac{\partial\sigma_{xx}}{\partial x}, \tag{5}$$

where t is the film thickness. The in-plane stress σ_{xx} in the HDA phase is zero while in the quasi-amorphous phase it is of the order of $1\,\mathrm{GPa}$. The transition region, L where the in-plane stress rises from zero to its maximum value is much smaller than the width of the hot zone ($\sim 10^{-1}\,\mathrm{cm}$), where the expansion occurs.[7] We approximate L as $10^{-2}\,\mathrm{cm}$. Thus, if $t \sim 200\,\mathrm{nm}$, then from equation (5), $\sigma_{xz}, \sim (t/L)\sigma_{xx} \approx 2\,\mathrm{MPa}$. This value is sufficiently large to cause collapse of the voids.[30]

4.3.2. *Formation of a permanent electric polarization*

The polarity of the quasi-amorphous phase is directly related to the strain gradient that develops in the amorphous film during heat treatment in a temperature gradient.[13] Not only the magnitude but also the direction of the permanent electric polarization, as monitored

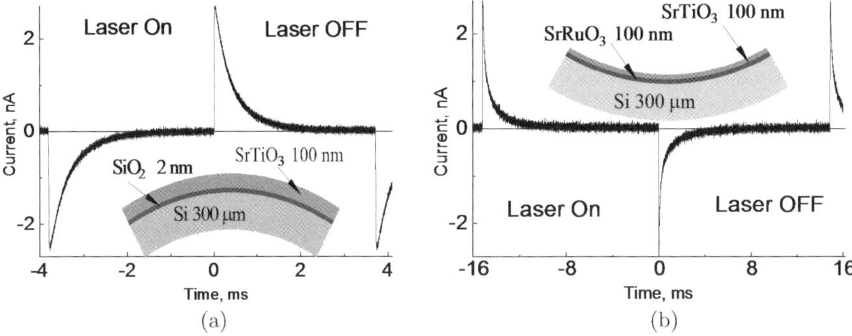

Figure 4. Pyroelectric current generated in response to a periodic temperature change induced by a modulated 1310 nm laser in (a) Si/SrTiO$_3$ and (b) Si/SrRuO$_3$/SrTiO$_3$ structures pulled through the temperature gradient. To measure the pyroelectric current top 1.5 × 1.5 mm Ag contacts were deposited. In the Si/SrTiO$_3$ structure, the stress developing in the structure as a result of the passage through the temperature gradient is compressive (a). The pyroelectric response is negative (top surface charges negatively upon heating). In the Si/SrRuO$_3$/SrTiO$_3$ structure, the net stress developing in the structure as a result of passage through the temperature gradient is tensile. In (b) the pyroelectric response is positive (top surface charges positively upon heating). Reproduced from Ref. [13] with permission.

by pyroelectric measurements, depends on the sample structure as shown in Fig. 4.[13]

The onset of polarization may be described as follows.[13] As already noted, when the supported film is pulled through the temperature gradient, large stress/strain gradients develop at the boundary between the HDA and the quasi-amorphous phase, the density of which is trapped between that of the HDA and the LDA. These strain gradients are able to align microscopic electric dipoles. The poling and resulting polarization are retained even after the strain gradient is removed. The influence of the sign of the in-plane strain gradient can be demonstrated by comparing Si/SrTiO$_3$ and Si/SrRuO$_3$/SrTiO$_3$ substrate-supported films. During the transformation from the HDA into the quasi-amorphous phase in the Si/SrTiO$_3$ structure, the in-plane strain at the interface with Si is equal to zero, $u_{xx}^{\mathrm{SrTiO_3 bottom}} = 0$ since the film, thickness t, is clamped to the substrate. However, at the free surface, the in-plane strain $u_{xx}^{\mathrm{SrTiO_3 top}} = \Delta l/l$ (Fig. 5a[13]) is given by simple geometric

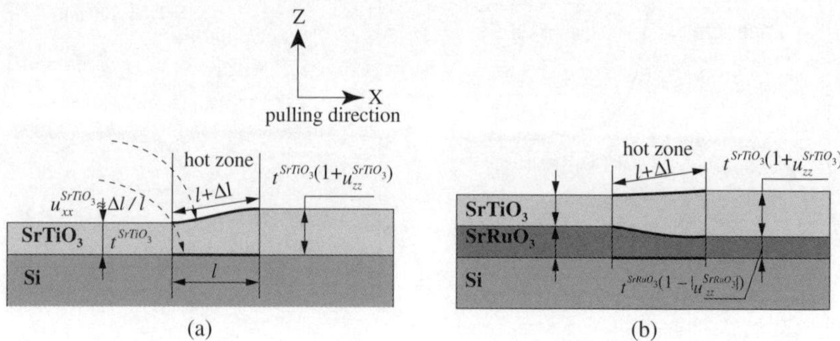

Figure 5.   Schematic representation of the strain developing in the Si/SrTiO₃ and Si/SrRuO₃/SrTiO₃ structures during pulling through the temperature gradient. The interfaces at which the in-plane strain is considered are highlighted. Reproduced from Ref. [13] with permission.

considerations[13]:

$$u_{xx}^{\mathrm{SrTiO_3}\mathrm{top}} = \frac{\Delta l}{l} \approx \frac{\left(\frac{t^{\mathrm{SrTiO_3}} \times u_{zz}^{\mathrm{SrTiO_3}}}{l}\right)^2}{2}, \tag{6}$$

$$\left.\frac{\partial u_{xx}^{\mathrm{SrTiO_3}}}{\partial z}\right|_{\mathrm{Si/SrTiO_3}} \approx \frac{u_{xx}^{\mathrm{SrTiO_3}\mathrm{top}} - u_{xx}^{\mathrm{SrTiO_3}\mathrm{bottom}}}{t^{\mathrm{SrTiO_3}}}$$

$$= \frac{t^{\mathrm{SrTiO_3}} \times \left(\frac{u_{zz}^{\mathrm{SrTiO_3}}}{l}\right)^2}{2}. \tag{7}$$

However, in the case of Si\SrRuO₃\SrTiO₃, the presence of a SrRuO₃ layer modifies the distribution of the strain gradient, because SrRuO₃ contracts in the hot zone. Assuming that the in-plane strain in the SrTiO₃ layer is imposed by the SrRuO₃ layer (Fig. 5b[13]), one obtains in analogy to equation (6)[13]:

$$u_{xx}^{\mathrm{SrTiO_3}\mathrm{bottom}} = u_{xx}^{\mathrm{SrRuO_3}\mathrm{top}} \approx \frac{\left(\frac{t^{\mathrm{SrRuO_3}} \times u_{zz}^{\mathrm{SrRuO_3}}}{l}\right)^2}{2}. \tag{8}$$

By combining equations (6) and (8) one can approximate $u_{xx}^{\mathrm{SrTiO_3}\mathrm{top}}$ for a Si\SrRuO₃\SrTiO₃ multilayer[13]:

$$u_{xx}^{\mathrm{SrTiO_3}\mathrm{top}} = \frac{1}{2} \times \left(\frac{t^{\mathrm{SrTiO_3}} \times u_{zz}^{\mathrm{SrTiO_3}} - t^{\mathrm{SrRuO_3}} \times |u_{zz}^{\mathrm{SrRuO_3}}|}{l}\right)^2. \tag{9}$$

Therefore, in this case the strain gradient in the $SrTiO_3$ quasi-amorphous layer will be[13]:

$$\frac{\partial u_{xx}^{SrTiO_3}}{\partial z}\bigg|_{Si/SrRuO_3/SrTiO_3}$$

$$= \frac{1}{2} \times \frac{\left(t^{SrTiO_3} \times u_{zz}^{SrTiO_3} - t^{SrRuO_3} \times \left|u_{zz}^{SrRuO_3}\right|\right)^2 - \left(t^{SrRuO_3} \times \left|u_{zz}^{SrRuO_3}\right|\right)^2}{t^{SrTiO_3} \times l^2}$$

$$= \frac{t^{SrTiO_3}}{2} \times \left(\frac{u_{zz}^{SrTiO_3}}{l}\right)^2 \times \left(1 - \frac{2 \times t^{SrRuO_3} \times \left|u_{zz}^{SrRuO_3}\right|}{t^{SrTiO_3} \times u_{zz}^{SrTiO_3}}\right). \tag{10}$$

Combining equations (7) and (10) gives[13]:

$$\frac{\partial u_{xx}^{SrTiO_3}}{\partial z}\bigg|_{Si/SrRuO_3/SrTiO_3}$$

$$= \frac{\partial u_{xx}^{SrTiO_3}}{\partial z}\bigg|_{Si/SrTiO_3} \times \left(1 - \frac{2 \times t^{SrRuO_3} \times \left|u_{zz}^{SrRuO_3}\right|}{t^{SrTiO_3} \times u_{zz}^{SrTiO_3}}\right). \tag{11}$$

In the case of $t^{SrTiO_3} \approx t^{SrRuO_3}$:

$$\frac{\partial u_{xx}^{SrTiO_3}}{\partial z}\bigg|_{Si/SrRuO_3/SrTiO_3}$$

$$= \frac{\partial u_{xx}^{SrTiO_3}}{\partial z}\bigg|_{Si/SrTiO_3} \times \left(1 - 2 \times \frac{\left|u_{zz}^{SrRuO_3}\right|}{u_{zz}^{SrTiO_3}}\right). \tag{12}$$

It is apparent from equation (12) that if $u_{zz}^{SrTiO_3} < 2u_{zz}^{SrRuO_3}$, the sign of the out-of-plane gradient of the in-plane strain in the titanate layer will be opposite to that in the case of the $Si/SrTiO_3$ structure.

Measurements of pyroelectric current and film curvature are consistent with the conclusion that the vertical gradient of the in-plane strain defines the direction of the out-of-plane component of the thermally induced polarization in the quasi-amorphous films[13] (Fig. 4). Because such polarization is obviously coupled to a strain gradient, the coupling is therefore classified as "flexoelectric". However, in contrast to classical flexoelectricity, which is reversible, the strain gradient in the quasi-amorphous films leads to permanent polarization, i.e. the polarization is retained even after the strain gradient generated by pulling through the temperature gradient is removed. The fact that a flexoelectric field in supported titanate films has a role in aligning microscopic dipoles during the expansion from the HDA to the LDA state suggests that the latter can be identified as ferroelectric. An experimental verification has been presented by Wang *et al.*[35] who showed that an ultra thin layer of amorphous $BaTiO_3$ (3–4 nm), following 1 hr annealing in O_2 at 700°C, exhibits ferroelectricity with coercive field of 2000 kV/cm. Although the electric field induced by the flexoelectric effect during the HDA to quasi-amorphous transformation is probably much smaller than the coercive field,[36–42] the total effect of the strain gradient should be sufficient to align a few percent of the microscopic dipoles[8] and create a permanent macroscopic dipole. There are two reasons for this claim: (1) the electric field develops during the transformation when the local bonding units (LBUs) are already mobile; (2) the electric field is accompanied by a driving force to decrease the strain gradient and the elastic energy. Quasi-amorphous films represent the first known example of a "permanent" flexoelectric effect. The closest analogy is the flexoelectric-induced symmetry breaking of centrosymmetric perovskites[43] the strain-gradient-induced, irreversible domain wall motion in ferroelectric $Pb(Zr,Ti)O_3$ capacitors, which results in permanent polarization (self-poling).[41]

5. The Structure of Quasi-Amorphous Materials: Theory of the Random Network of Local Bonding Units

5.1. *Local Bonding Units in High Density Amorphous (HDA) and Quasi-Amorphous Films*

The fact that $SrTiO_3$ and $BaZrO_3$ form non-crystalline, polar, quasi-amorphous films but do not have any polar, crystallographic polymorphs demonstrates that the origin of polarity in these quasi-amorphous films is different from that in pyroelectric, inorganic crystals. Most standard characterization techniques are not applicable to non-crystalline thin films since they either require periodicity/ordered structures or more material than is present in a thin film. Two techniques used to characterize the structure of non-crystalline thin films are XPS and XAFS. These techniques were used to characterize the structure of both HDA and quasi-amorphous films of $SrTiO_3$ and $BaTiO_3$.[8,11,12,14] Measurements reveal that the structures of HDA and quasi-amorphous titanates cannot be characterized as random, close-packed, hard spheres — the typical arrangement for ionic and metallic glasses. Rather, these two phases have a statistically distributed, random network (RN) structure, despite the fact that both $SrTiO_3$ and $BaTiO_3$ are primarily ionic. Yet the RN structure of the non-crystalline titanates also differs markedly from the RN described for common (covalent) glasses, e.g. SiO_2. It does not satisfy Zachariasen's rules[27] for glass formation because: (1) the network consists of TiO_6 octahedra, which have six apices, instead of a maximum of four; and (2) the octahedra in HDA and quasi-amorphous films, which are similar to the LBUs in crystalline $SrTiO_3$ and $BaTiO_3$, are connected not only apex to apex but also share edges and, probably, faces (Fig. 6[16]). Therefore, each LBU in the RN is distorted. As a result of these distortions, the Ti ions are displaced from the center of the oxygen octahedra. The Ti shift can be calculated from the XAFS data.[8,11] It reaches 0.3 Å

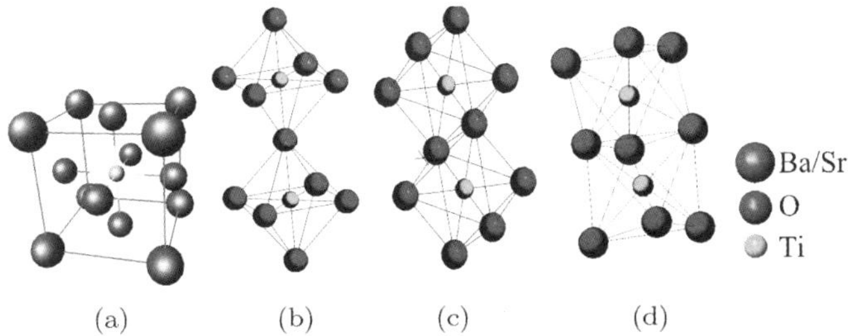

Figure 6. (a) Unit cell of crystalline perovskite (cubic) $BaTiO_3$ or $SrTiO_3$; (b) scheme of the TiO_6 octahedra connected apex–apex as in the crystalline phase; (c) and (d) TiO_6 octahedra sharing edges and faces, respectively. Reproduced from Ref. [16] with permission.

in $SrTiO_3{}^{11}$ and $0.44\,\text{Å}^8$ in $BaTiO_3$. The off-center displacement of the Ti-ions in amorphous and quasi-amorphous $BaTiO_3$, $0.44\,\text{Å}$, is much larger than in crystalline ferroelectric $BaTiO_3$, $0.23\,\text{Å}$.[8] Using these data, one can calculate that the dipole moment of the TiO_6 groups in non-crystalline $BaTiO_3$ is a relatively large 1.5–2 Debye. In contrast to Ti, the local environments of the Ba and Sr cations are not phase invariant. The volume available for Sr and Ba ions in the non-crystalline phases is much larger than that in the crystal.[11,12,14] This allows the Sr and Ba cations increased motion, leading to variation in the local environment from ion to ion. Their motional freedom, together with other characteristics discussed below, make the Sr and Ba cations essential for RN stability.

5.2. *The Role of Ba and Sr as Random Network Stabilizers in the HDA*

Since the non-crystalline titanates do not satisfy Zachariasen's rules[27] for glass formation, a question is raised concerning the source of stability of the TiO_6 RN. In the crystalline perovskite phase, the TiO_6 octahedra share apices and each oxygen ion has two Ti neighbors and four Sr neighbors. In the HDA and quasi-amorphous phases, the stoichiometry is the same as in the crystalline phases.[10] Therefore, the observation that TiO_6 LBUs share not only apices but also edges and faces implies that some apices remain unshared. The oxygen ions in the HDA have different numbers of Ti^{4+} neighbors[11] as compared to the crystalline phase (Fig. 7^{16}). Those oxygen ions that participate in apex sharing have two Ti^{4+} neighbors as in a crystalline perovskite; those ions that participate in edge or face sharing of TiO_6 octahedra have more than two Ti^{4+} neighbors, while the oxygen ions located at the unshared apices have only one Ti^{4+} neighbor.

Despite the difference in the coordination numbers, XPS data[12,14] clearly show (Fig. 8^{12}) that the valence states of all oxygen ions are the same, which implies that in the HDA there is a mechanism that equilibrates the charges on all oxygen ions and, thereby, stabilizes the network of TiO_6 LBUs. Although the details of this mechanism have yet to be clarified, relevant experimental observations have been

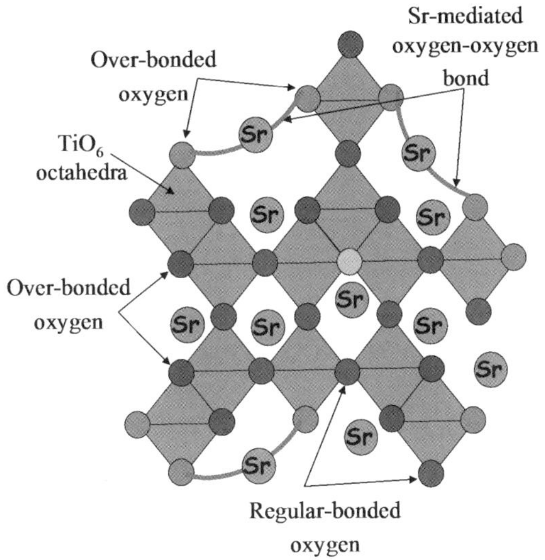

Figure 7. Cartoon of the amorphous phase of SrTiO$_3$ (BaTiO$_3$ is similar). Reproduced from Ref. [16] with permission.

made. The oxygen XPS spectra[12] of the HDA phases of BaTiO$_3$ and SrTiO$_3$ each contain a strong satellite (shake-up) peak with almost the same energy, i.e. 13.0 eV and 13.9 eV above the main peak for SrTiO$_3$ and BaTiO$_3$ respectively (Fig. 8). This peak was interpreted as belonging to a Ba- or Sr-mediated oxygen–oxygen complex which is not present in either the LDA or quasi-amorphous phases.

The connection between the satellite peak and mediation by cations is inferred from the fact that, according to extended X-ray absorption fine structure spectroscopy (EXAFS) data[11] (Fig. 9[11]), the local environment of Sr changes considerably during the transformation from the HDA to the quasi-amorphous phase, whereas the local environment of Ti does not change at all. Therefore, the RN of the TiO$_6$ LBUs in the HDA phase appears to be stabilized by a redistribution of Sr and Ba that equilibrates the effective charge on all oxygen ions. We note that both Ba and Sr easily form peroxides: they belong to a small group of metals (Na, K, Rb, Cs, Sr, and Ba), the oxides of which absorb molecular oxygen, thereby forming peroxides or superoxides.[44–47] While the presence of BaO and SrO stabilizes

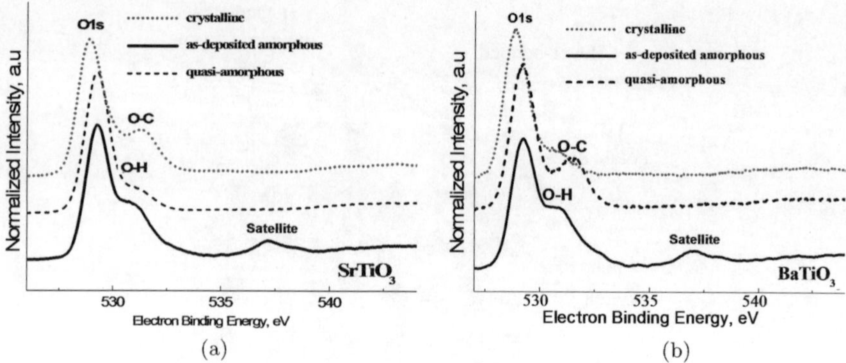

Figure 8.   XPS spectra of O1s: (a) $SrTiO_3$; (b) $BaTiO_3$. There are two important features: (1) all oxygen ions have the same ($\pm 0.5\,eV$) binding energy; (b) the peak marked "Satellite", is present in the spectra of the amorphous phase but is absent in the quasi-amorphous phase. Reproduced from Refs. [12, 16] with permission.

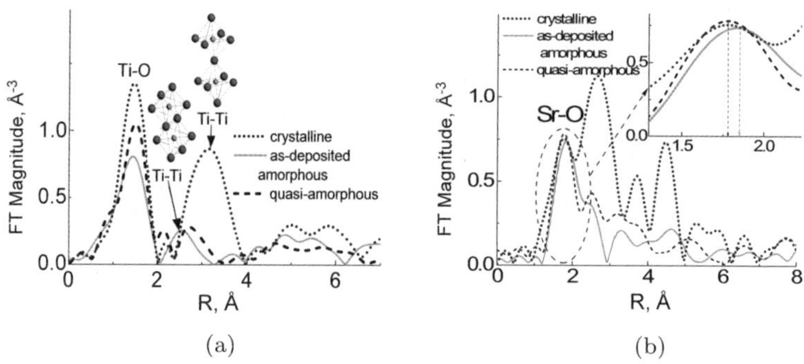

Figure 9.   (a) Radial distribution function deduced from EXAFS measurements of the Ti L_{III}-edge indicating that the first coordination shell, the TiO_6 octahedra, remain intact in all three phases. The amorphous phase clearly contains a large fraction of edge-sharing TiO_6 octahedra. The quasi-amorphous phase still contains edge-sharing TiO_6 octahedra but their amount is obviously reduced because the peak has shifted towards that of the apex-shearing octahedra. (b) Radial distribution function deduced from EXAFS measurements of the Sr L_{III}-edge. The first coordination shell of Sr is not preserved in the amorphous and quasi-amorphous phases and there is a considerable difference in the distance to the first (see the position of the peak maximum in the inset) and second neighbors between these two phases. Reproduced from Ref. [11] with permission.

the RN of TiO_6 LBUs, as in TeO_2-based glasses,[48] the addition of BaO or SrO to the silica-based glasses increases their tendency to crystallize.[49]

5.3. *Charge Redistribution in the HDA to LDA Transformation*

Allowing for charge redistribution on a RN of LBUs provides a phenomenological explanation for the HDA to LDA transformation prior to crystallization. As described above, the charge imbalance on the network in the HDA phase is mitigated by the Ba/Sr mediated complexes. Upon heating to 300°C, these complexes dissociate, an event which is clearly detected by the disappearance of the XPS satellite peak and by EXAFS-detected changes in the local environment of the cations.[11,12] Once the complexes have been disrupted, charge imbalance causes the TiO_6 octahedra to repel each other, thereby resulting in the low density state. Indeed, the fact that nucleation only begins after volume expansion ceases, suggests that nucleation requires full transformation to the LDA state. If so, when substrate clamping succeeds in suppressing nucleation of the quasi-amorphous films, this may be a direct result of the fact that TiO_6 octahedra are difficult to rearrange. This "difficulty" is reminiscent of the very slow transition from the hexagonal to the cubic modification of $BaTiO_3$ upon cooling.[50] In the hexagonal polymorph the octahedra share apices and faces. Although this polymorph is only stable above 1316°C, the transition is so slow that the hexagonal polymorph can exist below 1000°C, if quenched.

6. Compositional Criteria for the Quasi-Amorphous Phase

It is possible to derive a set of compositional criteria for the formation of a quasi-amorphous oxide.[14] The material must contain (i) a cation (M1) forming stable LBUs that can be interconnected in multiple ways, and (ii) a second cation (M2) that can stabilize under-bonded oxygen yet can easily release it upon heating. Candidates for M1 are Ti, Zr, Sn, Nb, and Ta, all of which form stable LBUs with oxygen (octahedral or tetrahedral coordination). The criteria for choosing M2 are more complex but all metals that form stable peroxides and/or easily change oxidation states are suitable. Therefore, one may expect that a quasi-amorphous phase can be formed by (Ca, Sr, Ba, Pb, Bi)(Zr, Ti, Sn)O_3 and (Na, K, Rb, Cs)(Nb, Ta)O_3. Indeed, Zhao *et al.*[51] discovered a piezoelectric amorphous phase in $(Na_{0.5}Bi_{0.5})_{0.94}Ba_{0.06}TiO_3$–$Bi_{12}TiO_{20}$ composite ceramics.

The stoichiometric relationship between M1 and M2 must also be controlled: however, this requirement has been investigated in detail only for $BaTiO_3$.[10] It was found that substrate-supported amorphous films with a Ti/Ba ratio 0.95–1.1 form a polar, quasi-amorphous phase after pulling through a temperature gradient (Fig. 10,[10] Fig. 11a). While pulling films with a Ti/Ba ratio of 1.1–1.5 through a temperature gradient may not result in crystallization (Fig. 10[10]), films with this composition exhibit neither pyro- nor piezo-electricity.

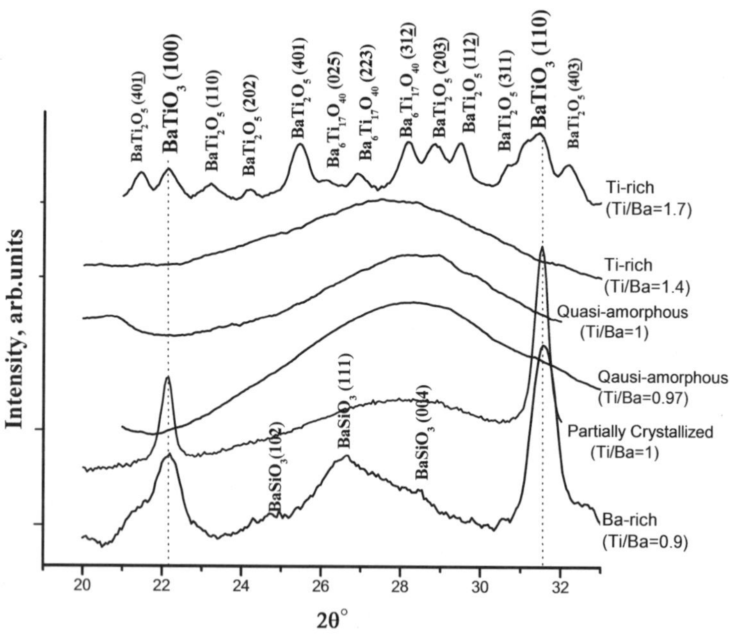

Figure 10.   10 XRD patterns (scans) of barium titanium oxide films of different stoichiometry deposited on Si substrates. Reproduced from Ref. [10] with permission.

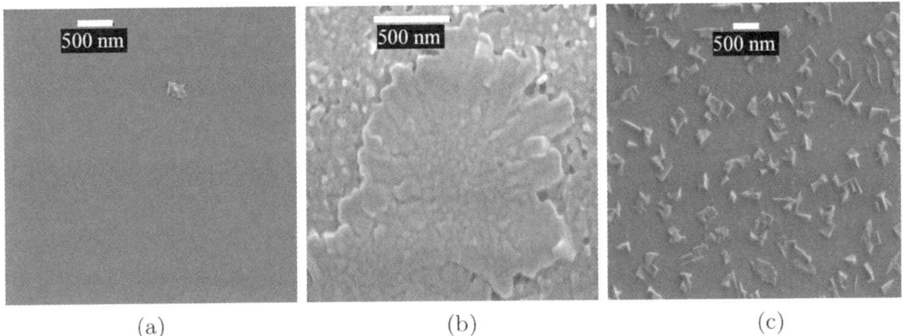

Figure 11. 10 SEM images (secondary electron mode) of barium titanium oxide films. (a) Film with Ti/Ba ratio 1.0 pulled through the temperature gradient. The area with the dust particle on the surface enabled proper focusing, which would not be otherwise possible. (b) Film with Ti/Ba ratio 0.9 pulled through the temperature gradient. (c) Film with Ti/Ba ratio 1.7 pulled through the temperature gradient. Reproduced from Ref. [10] with permission.

Pulling films with ratio Ti/Ba > 1.5 or Ti/Ba < 0.95 through a temperature gradient results in crystallization (Fig. 10, Fig. 11). The fairly broad range of Ti/Ba ratios within which quasi-amorphous, pyro- and piezo-electric thin films are formed implies that this phase is significantly more tolerant of a deviation from stoichiometry than its crystalline counterparts. At equilibrium below 800°C, bulk $BaTiO_3$ does not tolerate any deviation from the 1:1 cation ratio to the Ba-rich region and can only tolerate deviation of $\leq 1.5\%$ to the Ti-rich region, above which it separates into two phases.[52,53] On the other hand, quasi-amorphous films are significantly less tolerant than covalent glasses, which may exist within a very large concentration range: for instance, the SiO_2 content in silica-based glasses may vary from 54% to 100%.[54] The crystallization of Ba-rich films (Fig. 11b) is promoted by the $BaSiO_3$ phase that is formed at the film-substrate interface during heat treatment, reducing the development of stress and strain gradients. Traces of this phase can clearly be seen in the XRD pattern (Fig. 10) and ellipsometry data.[10] A possible explanation for the crystallization of the Ti-rich film (Fig. 11c) is as follows. A low level of Ba destabilizes the TiO_6 RN. As a result, some parts of the network organize into non-perovskite, crystal-like clusters. These clusters function as nucleation sites during heat treatment. Evidence for the suggested mechanism can be seen in the fact that, following heat treatment, the Ti-rich films are rich in non-perovskite crystalline phases (Fig. 10).

7. Conclusion

Quasi-amorphous materials demonstrate that polarity does not necessarily require a periodic structure or long-range order. In the formation of quasi-amorphous films, polyamorphism is used to create a strain gradient during the HDA to LDA transformation, which partially aligns polar bonding units, breaks the material isotropy and forms a permanent macroscopic dipole. This is a unique example of polarity and strain gradient coupling, since unlike classical flexoelectricity — the reversible coupling of polarity and strain gradients — the strain gradient which develops during the formation of the quasi-amorphous film leads to permanent polarization, i.e. the polarization is retained even after the strain gradient has been eliminated.

References

1. P. Muralt, R. G. Polcawich, and S. Trolier-McKinstry. *Mrs Bull.* **34**, 658–664 (2009).
2. S. B. Lang. *Phys. Today.* **58**, 31–36 (2005).
3. S. Piperno, E. Mirzadeh, E. Mishuk, D. Ehre, S. Cohen, M. Eisenstein, M. Lahav, and I. Lubomirsky. *Angew. Chem. Int. Ed.* **52**, 6513–6516 (2013).
4. A. K. Tagantsev. *Uspekhi Fizicheskikh Nauk.* **152**, 423–448 (1987).
5. A. K. Tagantsev. *Phase Transit.* **35**, 119–203 (1991).
6. V. Lyahovitskaya, I. Zon, Y. Feldman, S. R. Cohen, A. K. Tagantsev, and I. Lubomirsky. *Adv. Mater.* **15**, 1826–1828 (2003).
7. V. Lyahovitskaya, Y. Feldman, I. Zon, E. Wachtel, K. Gartsman, A. K. Tagantsev, and I. Lubomirsky. *Phys. Rev. B.* **71**, 94205-1-9 (2005).
8. A. I. Frenkel, Y. Feldman, V. Lyahovitskaya, E. Wachtel, and I. Lubomirsky. *Phys. Rev. B.* **71**, 024116 (1–4) (2005).
9. D. Ehre, V. Lyahovitskaya, A. Tagantsev, and I. Lubomirsky. *Adv. Mater.* **19**, 1515–1517 (2007).
10. D. Ehre, V. Lyahovitskaya, and I. Lubomirsky. *J. Mater. Res.* **22**, 2742–2746 (2007).

11. A. Frenkel, D. Ehre, V. Lyahovitskaya, L. Kanner, E. Wachtel, and I. Lubomirsky. *Phys. Rev. Lett.* **99**, 215502 (1–4) (2007).
12. D. Ehre, H. Cohen, V. Lyahovitskaya, and I. Lubomirsky. *Phys. Rev. B.* **77**, 184106 (1–6) (2008).
13. V. Shelukhin, D. Ehre, E. Lavert, E. Wachtel, Y. Feldman, A. Tagantsev, and I. Lubomirsky. *Adv. Funct. Mater.* **21**, 1403–1410 (2011).
14. D. Ehre, H. Cohen, V. Lyahovitskaya, A. Tagantsev, and I. Lubomirsky. *Adv. Funct. Mater.* **17**, 1204–1208 (2007).
15. I. Ebralidze, V. Lyahovitskaya, I. Zon, E. Wachtel, and I. Lubomirsky. *J. Mater. Chem.* **15**, 4258–4261 (2005).
16. E. Wachtel and I. Lubomirsky. *Adv. Mater.* **22**, 2485–2493 (2010).
17. A. Hossain and M. H. Rashid. *IEEE T. Ind. Appl.* **27**, 824–829 (1991).
18. S. G. Porter. *Ferroelectrics.* **33**, 193–206 (1981).
19. Y. Wang and Y. J. Jiang. *ISAF: 2009 18th IEEE International Symposium on the Applications of Ferroelectrics.* 350–353 (2009).
20. R. T. Smith and F. S. Welsh. *J. Appl. Phys.* **42**, 2219–2230 (1971).
21. S. Sobol, MSc Thesis, "Surface acoustic wave (SAW) resonator based on quasi-amorphous thin films", Weizmann Institute of Science (2008).
22. C. A. Angell. *Chem. Rev.* **102**, 2627–2649 (2002).
23. C. A. Angell. *Annu. Rev. Phys. Chem.* **55**, 559–583 (2004).
24. M. A. Floriano, Y. P. Handa, D. D. Klug, and E. Whalley. *J. Chem. Phys.* **91**, 7187–7192 (1989).
25. O. Mishima and Y. Suzuki. *J. Chem. Phys.* **115**, 4199–4202 **2001**.
26. J. P. Chu, S. F. Wang, S. J. Lee, and C. W. Chang. *J. Appl. Phys.* **88**, 6086–6088 (2000).
27. R. Zallen. *The Physics of Amorphous Solids.* Wiley, New York (1983).
28. H. Kessler, H. J. Kleebe, R. W. Cannon, and W. Pompe. *Acta Metall. Mater.* **40**, 2233–2245 (1992).
29. O. Milton. *The Materials Science of Thin Films.* Academic Press, New York (1992).
30. R. Abeyaratne and H. S. Hou, *J. Elasticity.* **26**, 23–42 (1991).
31. N. G. Savin. *Stress Distribution Around Holes (Translated from Russian).* NASA, Washington DC, (1970).
32. S. Timoshenko and J. N. Goodier. *Theory of Elasticity,* 3rd ed. McGraw-Hill, New York, (1970).
33. H. W. Westergaard. *Theory of elasticity and plasticity.* Harvard University Press, Cambridge, (1952).
34. Z. Suo and W. Wang, *J. Appl. Phys.* **76**, 3410–3421 (1994).
35. J. L. Wang, A. Pancotti, P. J'egou, G. Niu, B. Gautier, Y. Y. Mi, L. Tortech, S. Yin, B. Vilquin, and N. Barrett. *Phys. Rev. B.* **84**, 205426 (2011).

36. J. Y. Fu, W. Y. Zhu, N. Li, N. B. Smith, and L. E. Cross. *Appl. Phys. Lett.* **91**, (2007).

37. P. V. Yudin and A. K. Tagantsev. Fundamentals of flexoelectricity in solids. *Nanotechnology.* **24**, p. 432001 (2013).

38. W. H. Ma and L. E. Cross. *Appl. Phys. Lett.* **81**, 3440–3442 (2002).

39. W. H. Ma and L. E. Cross. *Appl. Phys. Lett.* **82**, 3293–3295 (2003).

40. W. H. Ma and L. E. Cross. *Appl. Phys. Lett.* **88**, (2006).

41. A. Gruverman, B. J. Rodriguez, A. I. Kingon, R. J. Nemanich, A. K. Tagantsev, J. S. Cross, and M. Tsukada. *Appl. Phys. Lett.* **83**, 728–730 (2003).

42. P. Zubko, G. Catalan, A. Buckley, P. R. L. Welche, and J. F. Scott. *Phys. Rev. Lett.* **99**, (2007).

43. A. Biancoli, C. M. Fancher, J. L. Jones, and D. Damjanovic. *Nat. Mater.* **14**, 224–229 (2015).

44. J. D. Bernal, E. Djatlowa, I. Kasarnowsky, S. Reichstein, and A. G. Ward. *Z. Kristallogr. Cryst. Mater.* **92**, 344–354 (1935).

45. P. W. Gilles and J. L. Margrave. *J. Phys. Chem.* **60**, 1333–1335 (1956).

46. Y. Sun, Z. Liu, P. Pianetta, and D. I. Lee. *J. Appl. Phys.* **102**, (2007).

47. Y. N. Zhuravlev and O. S. Obolonskaya. *J. Struct. Chem.* **51**, 1005–1013 (2010).

48. G. S. Murugan and Y. Ohishi. *J. Non-Cryst. Solids.* **351**, 364–371 (2005).

49. J. J. Shyu and J. R. Wang. *J. Amer. Cer. Soc.* **83**, 3135–3140 (2000).

50. O. Eibl, P. Pongratz, P. Skalicky, and H. Schmelz. *Philos. Mag. A.* **60**, 601–612 (1989).

51. M. Zhao, L. Wang, C. Wang, J. Zhang, Z. Gai, C. Wang, and J. Li. *Scr. Mater.* **63**, 207–210 (2010).

52. D. E. Rase and R. Roy. *J. Amer. Cer. Soc.* **38**, 102–113 (1955).

53. X. Lu and Z. Jin. *Calphad.* **24**, 319–338 (2000).

54. W. D. Callister. *Materials Science and Engineering, An Introduction,* 3rd edn. John Wiley & Sons, Inc., Salt Lake City, Utah, 1993.

Index